Advances in Intelligent Systems and Computing

Volume 910

Series editor

Janusz Kacprzyk, Systems Research Institute, Polish Academy of Sciences,
Warsaw, Poland
e-mail: kacprzyk@ibspan.waw.pl

The series "Advances in Intelligent Systems and Computing" contains publications on theory, applications, and design methods of Intelligent Systems and Intelligent Computing. Virtually all disciplines such as engineering, natural sciences, computer and information science, ICT, economics, business, e-commerce, environment, healthcare, life science are covered. The list of topics spans all the areas of modern intelligent systems and computing such as: computational intelligence, soft computing including neural networks, fuzzy systems, evolutionary computing and the fusion of these paradigms, social intelligence, ambient intelligence, computational neuroscience, artificial life, virtual worlds and society, cognitive science and systems, Perception and Vision, DNA and immune based systems, self-organizing and adaptive systems, e-Learning and teaching, human-centered and human-centric computing, recommender systems, intelligent control, robotics and mechatronics including human-machine teaming, knowledge-based paradigms, learning paradigms, machine ethics, intelligent data analysis, knowledge management, intelligent agents, intelligent decision making and support, intelligent network security, trust management, interactive entertainment, Web intelligence and multimedia.

The publications within "Advances in Intelligent Systems and Computing" are primarily proceedings of important conferences, symposia and congresses. They cover significant recent developments in the field, both of a foundational and applicable character. An important characteristic feature of the series is the short publication time and world-wide distribution. This permits a rapid and broad dissemination of research results.

More information about this series at http://www.springer.com/series/11156

Sabu M. Thampi · Ljiljana Trajkovic ·
Sushmita Mitra · P. Nagabhushan ·
Jayanta Mukhopadhyay · Juan M. Corchado ·
Stefano Berretti · Deepak Mishra
Editors

Intelligent Systems, Technologies and Applications

Proceedings of ISTA 2018

Springer

Editors
Sabu M. Thampi
School of Computer Science
and Information Technology
Indian Institute of Information Technology
and Management—Kerala (IIITM-K)
Trivandrum, Kerala, India

Sushmita Mitra
Machine Intelligence Unit
Indian Statistical Institute
Kolkata, West Bengal, India

Jayanta Mukhopadhyay
Department of Computer Science
and Engineering
Indian Institute of Technology Kharagpur
Kharagpur, West Bengal, India

Stefano Berretti
Dipartimento di Ingegneria
dell'Informazione (DINFO)
Università degli Studi di Firenze
Florence, Italy

Ljiljana Trajkovic
School of Engineering Science
Simon Fraser University
Burnaby, BC, Canada

P. Nagabhushan
Indian Institute of Information Technology,
Allahabad (IIIT-A)
Allahabad, Uttar Pradesh, India

Juan M. Corchado
Departamento de Informática y Automática
Universidad de Salamanca
Salamanca, Spain

Deepak Mishra
Indian Institute of Space Science
and Technology
Trivandrum, Kerala, India

ISSN 2194-5357 ISSN 2194-5365 (electronic)
Advances in Intelligent Systems and Computing
ISBN 978-981-13-6094-7 ISBN 978-981-13-6095-4 (eBook)
https://doi.org/10.1007/978-981-13-6095-4

Library of Congress Control Number: 2018967420

This Springer imprint is published by the registered company Springer Nature Singapore Pte Ltd.
The registered company address is: 152 Beach Road, #21-01/04 Gateway East, Singapore 189721, Singapore

Organization

Chief Patron

M. R. Doreswamy, Chancellor, PES University

Patrons

D. Jawahar, Pro-chancellor, PES University
Ajoy Kumar, COO, PES Institutions

General Chairs

Sushmita Mitra, Machine Intelligence Unit of the Indian Statistical Institute, Kolkata, India
Ljiljana Trajkovic, School of Engineering Science, Simon Fraser University, Canada
Jayanta Mukhopadhyay, Department of Computer Science and Engineering, IIT Kharagpur, India

General Executive Chair

Sabu M. Thampi, IIITM-Kerala, India

Steering Committee Chairs

J. Surya Prasad, Director, PESIT, Bangalore South Campus
T. S. B. Sudarshan, Associate Director, PESIT, Bangalore South Campus

Organizing Chair

Shikha Tripathi, Research Head, PESIT, Bangalore South Campus

Organizing Co-chairs

D. Annapurna, PESIT-BSC
Subhash Kulkarni, PESIT-BSC
B. J. Sandesh, PESIT-BSC

Technical Program Chairs

P. Nagabhushan, Indian Institute of Information Technology, Allahabad, India
Juan M. Corchado Rodriguez, University of Salamanca, Spain
Stefano Berretti, University of Florence, Italy
Deepak Mishra, Indian Institute of Space Science and Technology (IIST), Trivandrum, India

Organizing Secretaries

S. N. R. Ajay, PESIT-BSC
Pooja Agarwal, PESIT-BSC
K. S. V. Krishna Srikanth, PESIT-BSC

TPC Members/Additional Reviewers

http://www.acn-conference.org/ista2018/committee.html

Preface

This volume of proceedings provides an opportunity for readers to engage with a selection of refereed papers that were presented at the Fourth International Symposium on Intelligent Systems Technologies and Applications (ISTA'18). ISTA aims to bring together researchers in related fields to explore and discuss various aspects of intelligent systems technologies and their applications. This edition was hosted by PES Institute of Technology, Bangalore South Campus, India, during September 19–22, 2018. ISTA'18 was colocated with the International Conference on Applied Soft Computing and Communication Networks (ACN'18).

All submissions were evaluated on the basis of their significance, novelty, and technical quality. A double-blind review process was conducted to ensure that the author names and affiliations were unknown to TPC. These proceedings contain 20 papers selected for presentation at the symposium.

We are very grateful to many people who helped with the organization of the symposium. Our sincere thanks go to all the authors for their interest in the symposium and to the members of the program committee for their insightful and careful reviews, all of which were prepared on a tight schedule but still received in time. We are grateful to General Chairs for their support. We express our most sincere thanks to all keynote speakers who shared with us their expertise and knowledge.

We thank PES Institute of Technology, Bangalore, for hosting the conference. Sincere thanks to Dr. M. R. Doreswamy, Chancellor, PES University, and Dr. D. Jawahar, Pro-chancellor, PES University, for their valuable suggestions and encouragement. We would like to thank the organizing committee and many other volunteers who worked behind the scenes to ensure the success of this symposium. Dr. J. Surya Prasad, Dr. Sudarshan T. S., and Dr. Shikha Tripathi deserve special mention for their unwavering support. The EDAS conference system proved very helpful during the submission, review, and editing phases. Finally, we would like to acknowledge Springer for active cooperation and timely production of the proceedings.

Trivandrum, India	Sabu M. Thampi
Burnaby, Canada	Ljiljana Trajkovic
Kolkata, India	Sushmita Mitra
Allahabad, India	P. Nagabhushan
Kharagpur, India	Jayanta Mukhopadhyay
Salamanca, Spain	Juan M. Corchado
Florence, Italy	Stefano Berretti
Trivandrum, India	Deepak Mishra

Contents

Dynamic Mode-Based Feature with Random Mapping for Sentiment Analysis 1
S. Sachin Kumar, M. Anand Kumar, K. P. Soman and Prabaharan Poornachandran

Efficient Pre-processing and Feature Selection for Clustering of Cancer Tweets 17
P. G. Lavanya, K. Kouser and Mallappa Suresha

Intrinsic Evaluation for English–Tamil Bilingual Word Embeddings 39
J. P. Sanjanasri, Vijay Krishna Menon, S. Rajendran, K. P. Soman and M. Anand Kumar

A Novel Approach of Augmenting Training Data for Legal Text Segmentation by Leveraging Domain Knowledge 53
Rupali Sunil Wagh and Deepa Anand

Sarcasm Detection on Twitter: User Behavior Approach 65
Nitin Malave and Sudhir N. Dhage

Personalized Recommender Agent for E-Commerce Products Based on Data Mining Techniques 77
Veer Sain Dixit and Shalini Gupta

Pep—Personalized Educational Platform 91
Vivek M. Jude, A. Nayana, Reshma M. Pillai and Jisha John

Maneuvering Black-Hole Attack Using Different Traffic Generators in MANETs 101
Fahmina Taranum and Khaleel Ur Rahman Khan

Variant of Nearest Neighborhood Fingerprint Storage System by Reducing Redundancies 117
K. Anjana, K. Praveen, P. P. Amritha and M. Sethumadhavan

Evaluation of Water Body Extraction from Satellite Images Using Open-Source Tools 129
K. Rithin Paul Reddy, Suda Sai Srija, R. Karthi and P. Geetha

A Modification to the Nguyen–Widrow Weight Initialization Method 141
Apeksha Mittal, Amit Prakash Singh and Pravin Chandra

Impact Analysis of LFM Jammer Signals on Stepped Frequency PAM4 RADAR Waveforms 155
K. Keerthana and G. A. Shanmugha Sundaram

IoT-Enabled Air Monitoring System 173
Chavi Srivastava, Shyamli Singh and Amit Prakash Singh

Phase-Modulated Stepped Frequency Waveform Design for Low Probability of Detection Radar Signals 181
R. Vignesh, G. A. Shanmugha Sundaram and R. Gandhiraj

Effect of Waveform Coding on Stepped Frequency Modulated Pulsed Radar Transmit Signals 197
G. Priyanga and G. A. Shanmugha Sundaram

A Decision-Making Approach Based On L-IVHFSS Setting 219
AR. Pandipriya, J. Vimala, Xindong Peng and S. Sabeena Begam

Battery Assisted, PSO-BFOA based Single Stage PV Inverter fed Five Phase Induction Motor Drive for Green Boat Applications 227
Y. Suri Babu and K. Chandra Sekhar

Intelligent Refrigerator 241
Ishank Agarwal

Online Knowledge-Based System for CAD Modeling and Manufacturing: An Approach 259
Jayakiran Reddy Esanakula, J. Venkatesu Naik, D. Rajendra and V. Pandu Rangadu

Hepatoprotective Activity of the Biherbal Extract in Carbon Tetrachloride (CCl_4) Induced Hepatoxicity—A Study of Histopathological Image Analysis 269
K. Sujatha, V. Karthikeyan, R. S. Ponmagal, N. P. G. Bhavani, V. Srividhya, Rajeswari Hari and C. Kamatchi

Author Index 289

About the Editors

Sabu M. Thampi is a Professor at the Indian Institute of Information Technology and Management-Kerala (IIITM-K), Trivandrum, India. His main research areas are sensor networks, IoT, social networks, and video surveillance. He has authored and edited few books and published papers in academic journals and international proceedings. He has served as a guest editor for special issues in journals and as a program committee member for many conferences. He has co-chaired several workshops and conferences. He has initiated and is also involved in the organisation of several annual conferences. He is a senior member of IEEE and a member of IEEE CS and ACM.

Ljiljana Trajkovic received the Dipl. Ing. degree from the University of Pristina, Yugoslavia, in 1974, the M.Sc. degrees in Electrical Engineering and Computer Engineering from Syracuse University, Syracuse, NY, in 1979 and 1981, respectively, and the Ph.D. degree in Electrical Engineering from the University of California at Los Angeles, in 1986. She is currently a Professor in the School of Engineering Science at Simon Fraser University, Canada. She served as 2007 President of the IEEE Circuits and Systems Society. She is General Co-Chair of SMC 2016 and HPSR 2014, Technical Program Co-Chair of ISCAS 2005, and Technical Program Chair and Vice General Co-Chair of ISCAS 2004. She served as an Associate Editor of the IEEE Transactions on Circuits and Systems and the IEEE Circuits and Systems Magazine. She was a Distinguished Lecturer of the IEEE Circuits and Systems Society. She is a Professional Member of IEEE-HKN and a Life Fellow of the IEEE.

Sushmita Mitra is a full Professor at the Machine Intelligence Unit (MIU), Indian Statistical Institute, Kolkata. She is the author of many books published by reputed publishers. Dr. Mitra has guest edited special issues of several journals, is an Associate Editor of "IEEE/ACM Trans. on Computational Biology and Bioinformatics", "Information Sciences", "Neurocomputing", and is a Founding Associate Editor of "Wiley Interdisciplinary Reviews: Data Mining and Knowledge

Discovery". She has more than 75 research publications in referred international journals. Dr. Mitra is a Fellow of the IEEE, and Fellow of the Indian National Academy of Engineering and The National Academy of Sciences. She has served in the capacity of Program Chair, Tutorial Chair, Plenary Speaker, and as member of programme committees of many international conferences. Her current research interests include data mining, pattern recognition, soft computing, image processing, and Bioinformatics.

P. Nagabhushan, Director, Indian Institute of Information Technology, Allahabad is a Professor in Computer Science and Engineering, University of Mysore, Mysore. He has more than three and half decades of teaching experience. He is actively associated with various academic, statutory and professional bodies. He is the member of academic councils and Board of studies of several universities. He has successfully guided 28 numbers of Ph.D.s and has authored more than 500 numbers of Research Papers out of which more than 100 papers are DBLP indexed. The citation index of 1856 with H-index of 21 and i10 index of 53 reflects his academic accomplishments.

Jayanta Mukhopadhyay received his Ph.D. degree in Electronics and Electrical Communication Engineering from the Indian Institute of Technology (IIT), Kharagpur. He was a Humboldt Research Fellow at the Technical University of Munich in Germany for one year in 2002. He has published about 250 research papers in journals and conference proceedings in these areas. He received the Young Scientist Award from the Indian National Science Academy in 1992. Dr. Mukherjee is a Senior Member of the IEEE. He is a fellow of the Indian National Academy of Engineering (INAE).

Dr. Juan M. Corchado is a Spanish scientist, teacher and researcher. He is currently Vice President for Research and Transfer from December 2013 and a Professor at the University of Salamanca. He is Director of the Science Park of the University of Salamanca and Director of the School of Doctorate from the same university. He holds a Ph.D. in Computer Science from the University of Salamanca and also is also Doctor in Artificial Intelligence from the University of the West of Scotland. Corchado is also Editor-in-Chief of the magazine ADCAIJ and IJDCA.

Stefano Berretti is an Associate Professor in the Department of Information Engineering (DINFO) of the University of Firenze, Italy. He received the Laurea degree in Electronics Engineering and the Ph.D. in Informatics Engineering and Telecommunications from the University of Firenze, in 1997 and 2001, respectively. His current research interests are mainly focused on content modeling, retrieval, and indexing of image and 3D object databases. He is the author of more than 100 publications appeared in conference proceedings and international journals in the area of pattern recognition, computer vision and multimedia. He is in the program committee of several international conferences and serves as a frequent reviewer of many international journals. Stefano is the member of GIRPR affiliated to the IAPR, and member of the IEEE.

Dr. Deepak Mishra pursued his Ph.D. at IIT Kanpur in the Electrical Engineering Department. His Thesis title was "Novel Biologically Inspired Neural Network Models". Later he joined as a postdoc researcher at the University of Louisville, KY, USA in the field of signal processing and system neuroscience. After a brief stint 2009–2010 as a senior software engineer at CMC Limited, Hyderabad, he opted to work as an academic faculty at IIST, Trivandrum in 2010 and continued to work as Associate Professor in the Department of Avionics. He was also awarded Young Scientist Award from System Society of India for his research work in the year 2012. His research interest includes Neural networks and Machine learning, Computer vision and Graphics, Image and Video Processing. He has published research papers both in International and National Journal of repute and presented his research work in various international and national conferences.

Dynamic Mode-Based Feature with Random Mapping for Sentiment Analysis

S. Sachin Kumar, M. Anand Kumar,
K. P. Soman and Prabaharan Poornachandran

Abstract Sentiment analysis (SA) or polarity identification is a research topic which receives considerable number of attention. The work in this research attempts to explore the sentiments or opinions in text data related to any event, politics, movies, product reviews, sports, etc. The present article discusses the use of dynamic modes from dynamic mode decomposition (DMD) method with random mapping for sentiment classification. Random mapping is performed using random kitchen sink (RKS) method. The present work aims to explore the use of dynamic modes as the feature for sentiment classification task. In order to conduct the experiment and analysis, the dataset used consists of tweets from SAIL 2015 shared task (tweets in Tamil, Bengali, Hindi) and Malayalam languages. The dataset for Malayalam is prepared by us for the work. The evaluations are performed using accuracy, F1-score, recall, and precision. It is observed from the evaluations that the proposed approach provides competing result.

Keywords Sentiment analysis · Polarity identification · Dynamic mode decomposition · Random mapping · Random kitchen sink

1 Introduction

Sentiment analysis or SA is an important task in text analysis. The sentiment in the text data carries information about love, happiness, anger, sadness, etc. It carries opinions related to events such as sports, politics, products, and movies. In this way, the data in

---

S. Sachin Kumar (✉) · K. P. Soman
Center for Computational Engineering & Networking (CEN),
Amrita School of Engineering, Amrita Vishwa Vidyapeetham, Coimbatore, India
e-mail: sachinnme@gmail.com

M. Anand Kumar
Department of Information Technology,
National Institute of Technology, Surathkal, Karnataka, India

P. Poornachandran
Amrita Center for Cyber Security & Networks, Amrita Vishwa Vidyapeetham,
Amritapuri, Kerala, India

S. M. Thampi et al. (eds.), *Intelligent Systems, Technologies and Applications*,
Advances in Intelligent Systems and Computing 910,
https://doi.org/10.1007/978-981-13-6095-4_1

text form become the source for sentiment. Internet has massive volumes of data [1] in the form of audio, video, images, text, etc. Huge volume of metadata also exists. In a big picture, the data in text form exist as two categories (i) structured and (ii) unstructured. As the text data contain several opinions, vital suggestions/recommendations, it is favorite for researchers, companies, organizations, and government. The social media websites like google+, twitter, facebook, etc. provide options for sharing views or starting discussions on the web. The users of these social media sites write in their mother tongue (native language) or English. Hence, the sentiment analysis work has been carried out in both English and non-English text data.

Extensive research is ongoing in SA of text data in English, German, Chinese, Japanese, Spanish, etc. [2–4]. The research work on social media contents in Indian languages, such as Hindi, Marathi, Bengali, Punjabi, Manipuri, Malayalam, and Tamil has started. Due to the complexity of text content in social media data in Indian languages, a requirement of a good system to perform SA motivates the research. The research work in this direction needs to incorporate annotated datasets, lexicon dictionaries, parsers and taggers, and SentiWordNet in different Indian languages. Constructing such materials in each language is itself a tremendous challenge. Also, different contexts exist where same word occur, and it poses a challenge as the meaning itself will change. Hence, the text content for SA task needs to perform disambiguation as discussed in [5]. In [5], a methodology for SA of text data in Hindi using SentiWordNet is proposed. It improves the SentiWordNet by updating with missing words (sentiwords). The words missing are translated to English initially, and its polarity value is obtained from English SentiWordNet. In [6], the authors developed an annotated dataset on movie reviews in Hindi. It shares views on adding more words to Hindi SentiWordNet. The authors proposed rules to take care of negations and discourse relations. The article [7] proposes three different approaches to perform SA. (1) Using machine-learning approach on annotated dataset. The dataset consists of reviews related to movies in Hindi. (2) The text content in Hindi was translated to English first and then performs sentiment classification. (3) A strategy based on score with Hindi SentiWordNet is used to find sentiments. The paper [8] proposed different ways to create SentiWordNets in Indian languages using Wordnet, dictionary, and corpus. The paper shares the distribution of text contents in Bengali using data collected from news and blogs. The articles [9–11] give a survey on opinion mining and SA of text data in Hindi. The authors in [12] proposed SA on movie review data in Tamil. Their work considers the frequency as the feature, and data were manually tagged into positive, negative category. They used machine-learning algorithms such as multinomial and Bernolli naive Bayes, logistic regression (LR), random kitchen sink (RKS), support-vector machines (SVM) for classification.

The work on SA in Malayalam text data is proposed in [13] initially. Their proposed approach finds positive and negative moods from the given text. A wordset with negative and positive responses was developed. The authors used unsupervised method for classifying the text. In [14], the authors proposed convolutional neural nets (CNN) and long short-term memory (LSTM) based SA on tweets written in Malayalam. The tweets were collected, annotated manually. For the experiments, three different activation functions were used, namely Rectified Linear Units (ReLU),

Exponential Linear Units (ELU), and Scaled Exponential Linear Units (SELU). In the article [15], the authors proposed a feature mapping using RKS method and performed regularized least square-based classification for sentiment analysis of tweets in Indian languages such as Hindi, Bengali, and Tamil. The dataset used for the experiments was obtained from shared task [16]. The authors showed that the proposed approach was effective for the SA classification task. The paper [17] proposed deep learning-based tagger for tweets in Malayalam. Coarse-grained tagset was used for annotation. The results obtained are competing and can be used for SA by incorporating part-of-speech tags. In [18], the authors provided a comparison of SA on tweets in Malayalam based on CNN, LSTM, SVM, k-nearest neighbor (KNN), LR, and decision tree (DT). The authors showed that the deep learning approach has competing evaluation scores compared to the other methods.

The current paper proposes dynamic mode-based features with random mapping for sentiment classification. The dynamic modes are obtained from DMD method [19], a popular approach in fluid dynamics area. For the current work, the dynamic modes are used as features for SA task. This is the first work which uses dynamic mode-based features for NLP tasks. The structure of the paper is organized as follows. Section 2 discusses the methods used in the current work. The proposed approach is discussed in Sect. 3. Section 4 describes the dataset used for the experiments. The results are discussed in Sect. 5, and Sect. 6 concludes.

2 Background

2.1 Dynamic Mode Decomposition (DMD)

DMD originated from fluid dynamics is an effective, interpretable, spatio-temporal decomposition computed using basic linear algebra formulations and certain nonlinear optimization procedures. This equation-free, matrix-decomposition algorithm is a hybrid method combines the advantages of principal component analysis (PCA) in spatial domain and Fourier transforms in time domain. It is a regression kind of algorithm which captures the inherent dynamics of the underlying data by creating observation matrices or snapshot matrices. Singular value decomposition (SVD) is the core idea used in DMD for the computation of dynamic modes. DMD algorithm is related to the linear Koopman operator, which represents the nonlinear system dynamics through linear infinite-dimensional operator. The ability to reveal the underlying dynamics of the data through spatio-temporal mode identification is the key feature of DMD and is now widely being used in several domains for prediction, state estimation, diagnostics, and control applications. The major advantage of DMD like data-driven methods is its ability to identify the characteristic or the dynamics by mining the data which are hidden otherwise [19].

2.1.1 DMD Algorithm

For a given time, the observation matrices are created using the gathered p measurements with q dimensions as follows,

$X_1 = \begin{bmatrix} | & | & | & & | \\ x_1 & x_2 & x_3 & \dots & x_{p-1} \\ | & | & | & & | \end{bmatrix}$ and $X_2 = \begin{bmatrix} | & | & | & & | \\ x_2 & x_3 & x_4 & \dots & x_p \\ | & | & | & & | \end{bmatrix}$,

Generally, these two are snapshot matrices with overlapping measurements over time. The objective of DMD algorithm is to find the eigendecomposition of unknown linear operator A. The relation between A and observation matrices is derived as,

$$x_2 = Ax_1, x_3 = A^2 x_1 = Ax_2, \dots, x_p = A^{p-1} x_1 \tag{1}$$

The above relation implies that,

$$AX_1 \approx X_2 \Rightarrow A = X_2 X_1^{\dagger} \tag{2}$$

However, computation of eigendecomposition of A is heavy for many of the practical systems, and hence, a similar matrix, S, with reduced dimension is introduced. S is a rank-reduced representation for A obtained via proper orthogonal decomposition (POD). The relation between S and observation matrices X_1 and X_2 is defined as,

$$X_2 = X_1 S + r e_{m-1}^T \Rightarrow X_2 \approx X_1 S \tag{3}$$

where $r \in R^q$ represent the residual error and $e_{p-1}^T \in R^{p-1}$. The matrix S is defined as,

$$S = \begin{bmatrix} 0 & 0 & \dots & 0 & 0 & a_1 \\ 1 & 0 & \dots & 0 & 0 & a_2 \\ \vdots & \vdots & \ddots & \vdots & \vdots & \vdots \\ 0 & 0 & \dots & 1 & 0 & a_{p-2} \\ 0 & 0 & \dots & 0 & 1 & a_{p-1} \end{bmatrix} \tag{4}$$

where $\begin{bmatrix} a_1, a_2, \dots, a_{p-1} \end{bmatrix}$ are the unknown coefficients. The steps involved in the computation of dynamic modes using DMD algorithm are given below.

1. **Compute SVD of X_1 and represent X_2 in terms of X_1**
   - SVD of X_1 is computed as, $X_1 \approx U \Sigma V^H$. Where $U \in C^{q \times r}$, $\Sigma \in C^{r \times r}$, $V \in C^{p \times r}$, and r represents the rank of the reduced SVD approximation to X_1.
   - Represent matrix X_2 in terms of X_1 as $X_2 \cong X_1 S \Rightarrow X_2 \cong U \Sigma V^H S$
2. **Compute the similarity transformation matrix S and $\tilde{S}$**
   - S is formulated from X_2 as $S = V \Sigma^{\dagger} U^H X_2$
   - $\tilde{S}$ is computed using the similarity transformation of S as $\tilde{S} = U^H S U \Rightarrow U^H X_2 V \Sigma^{\dagger}$
3. **Compute the eigendecomposition of $\tilde{S}$**

$$\tilde{S}W = W\Lambda \tag{5}$$

where W is the eigenvectors matrix, and Λ is the diagonal matrix of eigenvalues.

4. **Compute the dynamic modes**

$$\Phi = X_2 V \Sigma^{\dagger} W \tag{6}$$

where each column ϕ_i of the Φ matrix represents the DMD modes with corresponding eigenvalue λ_i. In the present work, the ϕ_i is taken as the feature vectors.

2.1.2 Snapshot Matrix

Embedding matrix X is created for each tweet vector by converting to a matrix form. Originally for time series data, the column vectors of X can be considered as the time shifted representation of the data.

$$X = \begin{bmatrix} x_1 & x_2 & \cdots & x_L \\ x_2 & x_3 & \cdots & x_{L+1} \\ \vdots & \vdots & \ddots & \vdots \\ x_K & x_{K+1} & \cdots & x_{K+L-1} \end{bmatrix} \tag{7}$$

Here, $K + L - 1$ represents the length of each tweet vector; L and K represent the length of each segment and the total number of shifted segments, respectively. However, in case of text data the column has the shifted form of token index numbers. Let the tweet vector corresponding to a tweet with three token (w_1, w_2, w_3) is of the form $\begin{bmatrix} 0 \; 0 \; w_1 \; w_2 \; w_3 \end{bmatrix}$. Zeros are padded to make the vectors corresponding to each tweet of the same length. The embedding matrix X of the tweet vector is of the form given in Eq. 8.

$$X = \begin{bmatrix} 0 & 0 & w_1 & w_2 & w_3 \\ 0 & w_1 & w_2 & w_3 & 0 \\ w_1 & w_2 & w_3 & 0 & 0 \\ w_2 & w_3 & 0 & 0 & 0 \\ w_3 & 0 & 0 & 0 & 0 \end{bmatrix} \tag{8}$$

The columns of the above matrix represent a shifted form of the tweet vector. Originally, the columns of X representing the snapshot of the data at different time instance. In the current scenario, the columns represent a portion of the tweet. The shift operation rearranges the position of each token in the column. Hence, the matrix columns captures the evolution of tokens in the tweet. The dynamic modes are obtained using Eq. 6. The present work aims in using the idea of dynamic modes for sentiment classification for the first time.

2.2 Random Mapping Approach Using Random Kitchen Sink (RKS)

RKS method is an effective approach for classification of data having nonlinear nature [20]. The method has its advantage to work on larger nonlinear datasets when compared to traditional nonlinear kernel approaches. In RKS approach, an explicit mapping of features, $\psi(x)$, obtained from radial basis function (RBF) kernel is performed. The authors in [20] have shown that the data points in this feature space can be separated linearly. A regularized least square regression acts as a simple classifier in this context for real-time applications. To obtain random mapped vector corresponding to a data instance, RBF kernel, which is Gaussian in nature, is used. An important property of Gaussian function is utilized. That is, the Fourier transform (FT) of real Gaussian function is another real Gaussian function. This could be interpreted as a multivariate Gaussian PDF (probability density function). This interpretation provides an explicit method to derive feature function [20]. It can be understood mathematically as,

$$K(x_1, x_2) = \langle \psi(x_1), \psi(x_2) \rangle = e^{\frac{-1}{\sigma}\|x_1-x_2\|_2^2} = e^{\frac{-1}{2}(x_1-x_2)^T \Sigma^{-1}(x_1-x_2)} \tag{9}$$

Here, K denotes the RBF kernel and $\Sigma = \begin{bmatrix} 2\sigma & 0 & \dots & 0 \\ 0 & 2\sigma & \dots & 0 \\ \vdots & \vdots & \ddots & 0 \\ 0 & 0 & 0 & 2\sigma \end{bmatrix}$.

The kernel can be expressed as a Gaussian PDF. As the covariance matrix have only diagonal elements, the Gaussian PDF could be written as the product of n Gaussian function. The kernel function is defined as,

$$f(z) = e^{\frac{-1}{2}z^T \Sigma^{-1} z}, z = x_1 - x_2 \tag{10}$$

The Fourier transform, $F(\Omega)$, of kernel function $f(z)$ is expressed as,

$$F(\Omega) = \frac{1}{2\pi} \int_{-\infty}^{\infty} f(z) e^{-jz^T \Omega} \mathrm{dz} \tag{11}$$

Here, $F(\Omega)$ could be interpreted as a multivariate Gaussian PDF. The expected value of $e^{jz^T \Omega}$ is expressed as follows.

$$E(e^{jz^T \Omega}) = \int_{-\infty}^{\infty} F(\Omega_i) e^{jz^T \Omega} \mathrm{dz} \tag{12}$$

A generic expression for $\psi(x)$ can be obtained as follows.

$$E(e^{jz^T\Omega}) = \frac{1}{k}\sum_{i=1}^{k} e^{jz^T\Omega_i} = \frac{1}{k}\sum_{i=1}^{k} e^{j(x-y)^T\Omega_i} = \psi(x_1)^T\psi(x_2) = K(x_1, x_2) \quad (13)$$

$$K(x_1, x_2) = \frac{1}{k}\begin{bmatrix} e^{j(x_1-x_2)^T\Omega_1} \\ e^{j(x_1-x_2)^T\Omega_2} \\ . \\ e^{j(x_1-x_2)^T\Omega_k} \end{bmatrix} = \left\langle \begin{bmatrix} \sqrt{1/k}e^{jx_1^T\Omega_1} \\ \sqrt{1/k}e^{jx_1^T\Omega_2} \\ . \\ \sqrt{1/k}e^{jx_1^T\Omega_k} \end{bmatrix}, \begin{bmatrix} \sqrt{1/k}e^{jx_2^T\Omega_1} \\ \sqrt{1/k}e^{jx_2^T\Omega_2} \\ . \\ \sqrt{1/k}e^{jx_2^T\Omega_k} \end{bmatrix} \right\rangle \quad (14)$$

This means that, $\psi(x) = \begin{bmatrix} \sqrt{1/k}e^{jx^T\Omega_1} \\ \sqrt{1/k}e^{jx^T\Omega_2} \\ . \\ \sqrt{1/k}e^{jx^T\Omega_k} \end{bmatrix}$

To avoid complex number computation, we can equivalently take $\psi(x)$ as,

$$\psi(x) = \sqrt{1/k}\begin{bmatrix} \mathrm{Cos}(x^T\Omega_1) \\ \vdots \\ \mathrm{Cos}(x^T\Omega_k) \\ \mathrm{Sin}(x^T\Omega_1) \\ \vdots \\ \mathrm{Sin}(x^T\Omega_k) \end{bmatrix} \quad (15)$$

In the present work, each dynamic mode vector is mapped to 500 and 1000 dimensions.

2.3 Regularized Least Square-Based Classification (RLSC)

The regularized least square method for supervised classification is explored for sentiment analysis in the past [15]. The randomly mapped dynamic features of each tweet forms a new feature matrix upon which the regularized least square approach is applied. In general, the n-dimensional data vectors for training instances are $X = \{x_1, x_2 \ldots, x_n\} \in R^n$ and the corresponding class labels are $Y = \{y_1, y_2, \ldots y_n\} \in R^n$. The method of regularized least square finds a weight matrix W which is used during the testing stage. The objective function to find W is given as,

$$\min_{W \in R^{n\times T}} \left\{ \|Y - WX\|_F^2 + \lambda \|W\|_F^2 \right\} \quad (16)$$

Here, λ acts as a control parameter. The objective function provides a trade-off to find an appropriate weight matrix as shown in the second term and minimizes the discrepancy in the first term. It can be solved based on the concept of trace as below.

$$\text{Trace}\,\|Y - WX\|_F^2 + \lambda\text{Trace}\,\|W\|_F^2$$
$$= \text{Trace}\left(Y^TY - Y^TXW - W^TX^TY + W^TX^TXW + \lambda W^TW\right) \quad (17)$$

Now, differentiating with W,

$$\frac{\partial}{\partial W}\left(\text{Trace}\left(Y^TY - Y^TXW - W^TX^TY + W^TX^TXW + \lambda W^TW\right)\right) = 0 \quad (18)$$

Differentiating each term separately and combine the result gives

$$-2X^TY + 2X^TXW + 2\lambda W = 0 \quad (19)$$

$$\Rightarrow 2(X^TX + \lambda I)W = 2X^TY \quad (20)$$

$$\Rightarrow W = \left(X^TX + \lambda I\right)X^TY \quad (21)$$

In the testing stage, the test data is projected on to W.

3 Proposed Approach

This section discusses the approach proposed for sentiment classification of tweets using dynamic mode-based features with random mapping. For the current work, the dataset of SAIL 2015 shared task [16] is taken. It contains tweets in Hindi, Bengali, and Tamil languages. Along with this, we have prepared a dataset of tweets in Malayalam language with labeled sentiment. Figure 1 represents the flow of the proposed approach for SA task. The following section describes the steps proposed in the current work.

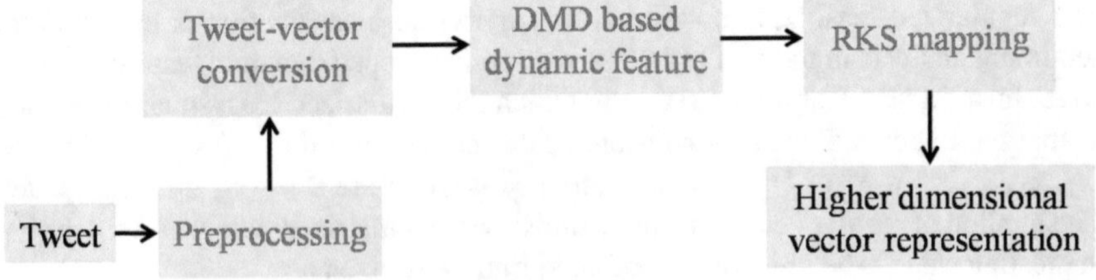

Fig. 1 Block diagram view of higher dimensional vector representation of the tweet

3.1 Step 1: Preprocessing

It consists of replacing each URL's (http/https) in the tweet to URL, each @-mentions to @, each #-mentions to #. For the current work, the smiles corresponding to happy, love, sad, wink, drool, and flirt are detected using regular expression and replaced with its name. The dates in the format, dd/mm/yy, dd-mm-yy, and dd.mm.yy, are replaced with <date>. The symbols such as ..., ——, ????, !!!!. ****, ???+++++++++, .!!, ...>>, ♡, …?,;//,??,?. are replaced with empty space. All the unnecessary white spaces are replaced with a single space.

3.2 Step 2: Tweet Vector

After preprocessing operation, a vocabulary, V, with the unique token as entries is prepared from the collected tweets. In order to prepare the initial vector representation of the tweet, tokens in the tweet are replaced by its corresponding token index number in V. As the number of tokens in each tweet will be different (variable length), zeros are padded to make all the tweet vector of the same length.

3.3 Step 3: DMD-Based Dynamic Mode Feature Extraction

The central idea of using dynamic mode-based feature starts from arranging tweet vector to form a snapshot matrix. The tweet vector in shifted form is arranged to form the snapshots (observations). Consider the rearrangement of the sequence "I like cricket" as an example. During the experiment, tokens in the sequence is changed with its corresponding index number from the vocabulary of tokens.

$$\begin{bmatrix} 0 & 0 & I & \text{like} & \text{cricket} \\ 0 & I & \text{like} & \text{cricket} & 0 \\ I & \text{like} & \text{cricket} & 0 & 0 \\ \text{like} & \text{cricket} & 0 & 0 & 0 \\ \text{cricket} & 0 & 0 & 0 & 0 \end{bmatrix} \tag{22}$$

DMD method captures the change provided in the column-wise shift of tokens. We can observe that the first column represents the entire tweet with zeros padded which is the initial tweet vector representation. The last column vector contains the last token and the rest of the values in it will be zeros. The ϕ_i obtained from Eq. 6 is taken as the feature representation for the particular tweet. For the current work, the first three ϕ_i vectors are appended together to form a single dynamic mode-based feature vector (instead of three, we can choose any number for the experiment). These selected modes are corresponding to the largest three eigenvalues.

3.4 Step 4: Random Mapping with RKS

Random mapping using RKS method performs a nonlinear feature mapping to a Fourier feature compact space. The kernel in RKS method, $k(x, y) = \langle \psi(x), \psi(y) \rangle \approx z(x)^T z(y)$ where $z : \Re^d \rightarrow \Re^D$. Here, ψ denotes implicit mapping, and Z denotes explicit mapping. A high computational cost is associated with the explicit mapping. RKS provides a solution to perform the same in a cost effective way [20]. The mapping using RKS method is a explored in [20].

3.5 Step 5: Regularized Least Square-Based Classification

As discussed in Sect. 2.3 the weight matrix, W, is calculated as per Eq. 21. The test vector is denoted as t_{test}. For two class problem, y and w will be vectors. The projection of t_{test} on to w will be a dot-product operation, $\langle w^T, t_{\text{test}} \rangle$, and obtains a single value. For multi-class problem (more than two classes), Y and W will be matrices and the projection of t_{test} results a vector. The position of the maximum element in the resultant vector denotes the class label.

4 Dataset Description

The tweets used for the experiment are in Tamil (Ta), Bengali (Bn), Hindi (Hi), and Malayalam (Ma). The tweets in Tamil, Bengali, and Hindi are obtained from the shared task [16]. The examples are shown in Fig. 2. The shared task provides separate

Bengali: দেশে এখনো আড়াই কোটি লোক নিরক্ষর: প্রাথমিক ও গণশিক্ষামন্ত্রী (*Negative*)
Transliteration: Deshe ekhono adai koti lok nirakhor: prathamik o ganasikhamontri
Translation: Till date, two and half crores of people are illiterate in the Nation : Primary and Mass Education Minister

Hindi: भारत के पूर्व राष्ट्रपति एपीजे अब्दुल कलाम को चीन की पेकिंग यूनिवर्सिटी में पढ़ाने का न्यौता मिला है। (*Positive*)
Transliteration: Bharat ke purb rastrapati APJ Abdul Kalam ko China ki Peking University main padane ka niyota mila hai |
Translation: The Former President of India APJ Abdul Kalam got an invitation for teaching at Peking University, China.

Tamil: கடவுள் என்று தனியாக ஒருவர் இல்லை...!! உன் கைகாசில் ஒரு குழந்தைக்குசாக்லேட் வாங்கிக்குடுத்துவிட்டு அந்த குழந்தையின் முகத்தைபார் !! (*Positive*)
Transliteration: kadavul endru thaniyaga oruvar illai ... !! un kaikasil oru kizhandaikku chaaklet vanghikkuduththuvittu andha kuzhandaiyin mugaththaippaar!!
Translation: There is no one like GOD ... !! Offer a chocolate to a kid by spending your own money and then look at the face of the kid !!

Fig. 2 Examples of tweets in Indian languages as shown in [16]

Table 1 Train data from SAIL 2015 shared task

	Positive	Negative	Neutral	Total
Bengali (Bn)	277	354	368	999
Hindi (Hi)	168	559	494	1221
Tamil (Ta)	387	316	400	1103

Table 2 Test data from SAIL 2015 shared task

	Positive	Negative	Neutral	Total
Bengali (Bn)	213	151	135	499
Hindi (Hi)	166	251	50	467
Tamil (Ta)	208	158	194	560

Table 3 Distribution of tweets in Malayalam language

Polarity	Positive	Negative	Neutral	Total
No. of tweets	3183	3137	6680	13,000

tweets for training and testing. The distribution of the dataset is shown in Tables 1 and 2, respectively. The tweets in Malayalam are manually annotated into positive, neutral, and negative. The examples for tweets in Malayalam in phonetic form are given in ref. [18]. Table 3 represents the number of tweets in each polarity category. The Malayalam tweets do not contain data in code-mixed form. The datasets for tweets in Malayalam are collected from twitter using available API's.

5 Results

In this section, the result obtained for the proposed approach on SAIL 2015 shared task dataset and the Malayalam sentiment dataset is discussed. For evaluation, popular metrics such as precision, recall, F1-score, and accuracy are used.

5.1 Result of Evaluation for SAIL 2015 Shared Task Dataset

The result obtained for SAIL 2015 shared task dataset using the proposed approach is shown in Table 4. The evaluation is done for tweets from Bengali, Tamil, and Hindi languages. The attribute 'Dim' denotes the dimension to which data are mapped using RKS method. For the experiment, we have mapped the dynamic mode features

Table 4 Evaluation score obtained for SAIL 2015 shared task dataset using the proposed approach

Dataset	Dim	Accuracy	Precision	Recall	F1-measure
Bengali (Bn)	500	0.9048	0.9081	0.8982	0.9015
	1000	0.9536	0.9553	0.9504	0.9524
Tamil (Ta)	500	0.8989	0.9015	0.8948	0.8954
	1000	0.9598	0.9593	0.9525	0.9532
Hindi (Hi)	500	0.8068	0.8564	0.8145	0.8328
	1000	0.8880	0.9241	0.8217	0.8464

Table 5 Performance comparison for SAIL 2015 shared task dataset based on accuracy measure

Dataset	Amrita-CENNLP	JUTeam-KS	IIT-TUDA	ISMD	AMRITA-CEN	SVM-SMO	ML
Bengali (Bn)	31.40	41.20	43.20	-	33.60	45	51.25
Tamil (Ta)	32.32	–	–	–	39.28	–	56.96
Hindi (Hi)	47.96	50.75	49.68	40.47	55.67	–	45.24

to 500 and 1000 dimensional vectors. The shared task had provided training and testing data separately for the evaluation.

Further, the performance of the proposed method for SAIL 2015 dataset is compared with the existing approaches for sentiment analysis, and the results are tabulated in Table 5. The overall accuracy obtained by Amrita-CENNLP [15], JUTeam-KS [21], IIT-TUDA [22], ISMD [23], AMRITA-CEN [24], SVM-SMO [25], and ML [26] methods is shown in Table 5. Among the referred methods, first five methods have acquired top positions in SAIL 2015 contest. It is clear from Table 5 that the proposed dynamic feature with the RKS mapping method has obtained the highest overall accuracy than existing benchmark models.

5.2 *Result of Evaluation for Malayalam Dataset*

For the experiment with Malayalam dataset, we used the same approach as discussed in the previous section. For the experiment, 10% of the dataset is chosen for testing and the rest of the dataset is chosen for training. The result is obtained by iterating the experiment for 10 times. Table 6 shows the average result obtained during the experiment.

The performance of the proposed dynamic feature-random mapping method is compared with conventional methods on Malayalam dataset. Table 7 shows the result obtained using conventional methods such as LR, KNN, random forest (RF), and DT. The feature used is the tweet-index vector. The tokens in the tweet are changed with

Table 6 Results on Malayalam dataset using the proposed approach

Dim	Accuracy	Precision	Recall	F1-measure
500	0.9883	0.9880	0.9879	0.9878
1000	0.9995	0.9994	0.9995	0.9995

Table 7 Result on Malayalam dataset using conventional methods

Algorithm	Kernel function	Accuracy (%)
RKS	RBF	86.5
	Randfeats	89.3
Classical	LR	82.9
	KNN	79.1
	RF	83.3
	DT	78.3

its index number from the prepared unique vocabulary. In [14], the authors have applied deep learning approaches such as CNN and LSTM for the same dataset and obtained a competing score above 0.95 for the evaluation measures. From the results shown in Sects. 5.1 and 5.2, it can be observed that the proposed approach provides competing result with the existing methods.

5.3 Discussion

In general, the results tabulated in Tables 4, 5, and 6 indicates the effectiveness of the proposed approach as a strategy for the sentiment classification task. Dynamic mode-based feature can be joined with any classifiers. From the experiment results obtained, it can be observed that the proposed approach is competitive in nature. Even though the proposed approach gives competing result, there is misclassification and this may have occurred due to the lack of data or variability in the train and test dataset. The current work shows that the proposed approach works for the sentiment analysis task on tweets in Hindi, Bengali, Tamil, and Malayalam languages. The number of ϕ_i vectors chosen to make dynamic mode feature is fixed as three for the current experiment. These are the dynamic modes corresponding to the first three largest eigenvalues. The length of ϕ_i vectors depends on the length of tweet vector. For the present work, the tweet vector length is fixed as 50. In RKS mapping (well known and promising method), the dynamic mode-based feature is mapped to 500 and 1000 dimensional vector space where the data points can be linearly separated. For this purpose, regularized least square regression (ridge regression) is used.

The current work does not claim dynamic mode-based feature as sentence-level representation as it can only be verified using task like sentence similarity. However, the current work proposes to use dynamic modes as features for classifiers.

6 Conclusion

In this paper, we discuss the sentiment analysis work as a data-driven model approach. A novel dynamic feature is used for the sentiment classification task. The dynamic feature is obtained from DMD method, a popular approach in the fluid dynamics area. One of the key advantages of DMD is its data-driven nature which does not rely on any prior assumption about the form of data. This is the first time it is been applied to NLP tasks. The experiments are performed using SAIL 2015 shared task dataset. Along with that we prepared a manually annotated twitter dataset in Malayalam language for sentiment analysis. The results obtained show that the proposed dynamic mode-based feature with RKS is effective and provides competing evaluation scores. As future work, we intend to explore the effectiveness of dynamic mode-based features on larger dataset by providing a comparison with the existing methods and sentence similarity tasks.

References

1. Zikopoulos, P., Eaton, C., DeRoos, D., Deutch, T., Lapis, G.: Understanding Big Data: Analytics for Enterprise Class Hadoop and Streaming Data. McGraw-Hill Osborne Media (2011)
2. Agarwal, A., Xie, B., Vovsha, I., Rambow, O., Passonneau, R.: Sentiment analysis of twitter data. In: Proceedings of ACL 2011 Workshop on Languages in Social Media, pp. 30–38 (2011)
3. Saif, H., He, Y., Alani, H.: Semantic smoothing for twitter sentiment analysis. In: Proceeding of the 10th International Semantic Web Conference (ISWC) (2011)
4. Kiritchenko, S., Zhu, X., Mohammad, S.M.: Sentiment analysis of short informal texts. J. Artif. Intell. Res. 723–762 (2014)
5. Pandey, P., Govilkar, S.: A framework for sentiment analysis in Hindi using HSWN. Int. J. Comput. Appl. **119**(19) (2015)
6. Mittal, N., Agarwal, B., Chouhan, G., Bania, N., Pareek, P.: Sentiment analysis of Hindi review based on negation and discourse relation. In: International Joint Conference on Natural Language Processing, Nagoya, Japan (2013)
7. Joshi, A., Balamurali, A.R., Bhattacharyya, P.: A fall-back strategy for sentiment analysis in Hindi: a case study. In: Proceedings of the 8th ICON (2010)
8. Das, A., Bandyopadhyay, S.: SentiWordNet for Indian Languages, Asian Federation for Natural Language Processing (COLING), China, pp. 56–63 (2010)
9. Gupta, S.K., Ansari, G.: Sentiment analysis in Hindi Language: a survey. Int. J. Mod. Trends Eng. Res. (2014)
10. Sharma, R., Nigam, S., Jain, R.: Opinion mining in Hindi Language: a survey. Int. J. Found. Comput. Sci. Technol. (IJFCST) **4**(2) (2014)
11. Pooja, P., Sharvari, G.: A survey of sentiment classification techniques used for Indian regional languages. Int. J. Comput. Sci. Appl. (IJCSA) **5**(2), 13–26 (2015)

12. Arunselvan, S.J., Anand Kumar, M., Soman, K.P.: Sentiment analysis of Tamil movie reviews via feature frequency count. In: International Conference on Innovations in Information, Embedded and Communication Systems (ICIIECS 15). IEEE (2015)
13. Neethu, M., Nair, J.P.S., Govindaru, V.: Domain specific sentence level mood extraction from Malayalam text. In: Advances in Computing and Communications (ICACC), pp. 78–81 (2012)
14. Sachin Kumar, S., Anand Kumar, M., Soman, K.P.: Sentiment Analysis on Malayalam Twitter data using LSTM and CNN. Lecture Notes in Computer Science (including subseries Lecture Notes in Artificial Intelligence and Lecture Notes in Bioinformatics), vol. 9468, pp. 320–334. Springer, Hyderabad India (2017)
15. Sachin Kumar, S., Premjith, B., Anand Kumar, M., Soman, K.P.: AMRITA_CEN-NLP@SAIL2015: Sentiment Analysis in Indian Language Using Regularized Least Square Approach with Randomized Feature Learning. Lecture Notes in Computer Science (including subseries Lecture Notes in Artificial Intelligence and Lecture Notes in Bioinformatics), vol. 9468, pp. 671–683. Springer, Hyderabad India (2015)
16. Patra, B.G., Das, D., Das, A., Prasath, R.: Shared task in sentiment analysis in Indian Languages (SAIL) tweets—an overview. In: The Proceedings of Mining Intelligence and Knowledge Exploration. Springer, ISBN: 978-3-319-26832-3
17. Sachin Kumar, S., Anand Kumar, M., Soman, K.P.: Deep learning based part-of-speech tagging for Malayalam Twitter data (Special issue: deep learning techniques for natural language processing). J. Intell, Syst (2018)
18. Sachin Kumar, S., Anand Kumar, M., Soman, K.P.: Identifying Sentiment of Malayalam Tweets using Deep learning, Digital Business, pp.391-408. Springer, Cham (2019)
19. Schmid, P.J.: Dynamic mode decomposition of numerical and experimental data. J. Fluid Mech. **656**, 5–28 (2010)
20. Rahimi, A., Recht, B.: Random features for large-scale kernel machines. In: Advances in Neural Information Processing Systems, pp. 1177–1184 (2008)
21. Sarkar, K., Chakraborty, S.: A sentiment analysis system for Indian language tweets. In: International Conference on Mining Intelligence and Knowledge Exploration, pp. 694-702. Springer, Cham (2015)
22. Kumar, A., Kohail, S., Ekbal, A., Biemann, C.: IIT-TUDA: system for sentiment analysis in indian languages using lexical acquisition. In: International Conference on Mining Intelligence and Knowledge Exploration, pp. 684–693. Springer, Cham (2015)
23. Prasad, S.S., Kumar, J., Prabhakar, D.K., Pal, S.: Sentiment classification: an approach for Indian language tweets using decision tree. In: International Conference on Mining Intelligence and Knowledge Exploration, pp. 656–663. Springer, Cham (2015)
24. Se, S., Vinayakumar, R., Anand Kumar, M., Soman, K.P.: AMRITA-CEN@ SAIL2015: sentiment analysis in Indian languages. In: International Conference on Mining Intelligence and Knowledge Exploration, pp. 703–710. Springer, Cham (2015)
25. Sarkar, K., Bhowmick, M.: Sentiment polarity detection in bengali tweets using multinomial Nave Bayes and support vector machines. In: 2017 IEEE Calcutta Conference (CALCON), pp. 31–36. IEEE (2017)
26. Phani, S., Lahiri, S., Biswas, A.: Sentiment analysis of Tweets in three Indian Languages. In: Proceedings of the 6th Workshop on South and Southeast Asian Natural Language Processing (WSSANLP2016), pp. 93–102 (2016)

Efficient Pre-processing and Feature Selection for Clustering of Cancer Tweets

P. G. Lavanya, K. Kouser and Mallappa Suresha

Abstract The impact of social media in our daily life cannot be overlooked. Harnessing this rich and varied data for information is a challenging job for the data analysts. As each type of data from social media is unstructured, these data have to be processed, represented and then analysed in different ways suitable to our requirements. Though retail industry and political people are using social media to a great extent to gather feedback and market their new ideas, its significance in other fields related to public like health care and security is not dealt with effectively. Though the information coming from social media may be informal, it contains genuine opinions and experiences which are very much necessary to improve the healthcare service. This work explores analysing the Twitter data related to the most dreaded disease 'cancer'. We have collected over one million tweets related to various types of cancer and summarized the same to a bunch of representative tweets which may give key inputs to healthcare professionals regarding symptoms, diagnosis, treatment and recovery related to cancer. This, when correlated with clinical research and inputs, may provide rich information to provide a holistic treatment to the patients. We have proposed additional pre-processing to the raw data. We have also explored a combination of feature selection methods, two feature extraction methods and a soft clustering algorithm to study the feasibility of the same for our data. The results have proved our intuition right about underlying information and also show that there is a tremendous scope for further research in the area.

Keywords Cancer · Centroid · Clustering · Feature selection · Feature extraction · Fuzzy C-means · K-means · Laplacian score · PCA · Summarization · SVD · Tweets · Twitter · Variance · Visualization

P. G. Lavanya (✉) · M. Suresha
DoS in Computer Science, University of Mysore, Mysuru, India
e-mail: lavanyarsh@compsci.uni-mysore.ac.in

K. Kouser
Government First Grade College, Gundlupet, India

S. M. Thampi et al. (eds.), *Intelligent Systems, Technologies and Applications*, Advances in Intelligent Systems and Computing 910,
https://doi.org/10.1007/978-981-13-6095-4_2

1 Introduction

Twitter, which is the most popular microblogging site on the Internet, has become the latest and hot medium for exchange of ideas, sharing of news and opinions in all fields like sports, politics, media, entertainment, etc. A tremendous amount of data is generated every day as on an average 6000 tweets are tweeted every second which are 350,000 tweets per minute and amounts to an average of 500 million tweets per day. The growth in the volume of tweets is around 30% per year [1]. Country-wise, USA is the leading user of Twitter (22.08%) followed by Japan (12.26%) and UK (5.82%) [2]. It has attracted the entire world ranging from a common man to the celebrities in every field. In fact, the popularity of a celebrity is being gauged by the number of followers he/she has. Such being its influence in the real life, the content which flows from the Twitter cannot be neglected. Hence, analysis of Twitter data which is highly diverse in nature is not a trivial task. Analysing this enormous data automatically and deriving insightful information can be termed as a herculean task. Extracting useful information from this massive data is like searching for a needle in a haystack. But, in the present era of 'Big Data', this is the challenge data analysts have to face every day. The challenges posed by Big Data are its sheer volume along with noise and veracity.

As this huge data cannot be stored forever, extracting summaries periodically on a particular topic or area of interest would be a desirable task. Clustering tweets and extracting representatives from these clusters seem a natural way to summarize Twitter data. But it is an extremely difficult task as the Twitter data is having limited number of words and need not follow grammatical rules. This poses a challenge for clustering as inter-cluster distances may be low and intra-cluster distances may be high.

Twitter data related to various topics like sports, politics and natural disasters has been collected and analysed as this data which is informal also contains genuine feelings and opinions related to a particular area. Healthcare-related tweets are considered as source of information related to patient's awareness about the disease, choice of healthcare service providers and feedback about the services received. All this data is usually collected real time as people tweet immediately after they experience something. This analysis provides a rich knowledge to healthcare service providers to improve their service and thereby contribute to the welfare of the society.

The work presented in this paper is continuation of the work presented in [3] where a model is proposed based on clustering for automatic summarization of tweets. Here, the model is improved by including another step in pre-processing the data to overcome the problems with increase in noise as the amount of data increases. We have proposed combination methods of the feature selection techniques used earlier. We have also explored two popular feature extraction methods—principal component analysis (PCA) and singular value decomposition (SVD) to improve the efficiency of clustering. The extracted summary is compared with the original document and evaluated using cosine similarity which is a content-based evaluation measure. In

Sect. 2, the related literature is discussed. The proposed method and techniques used are presented in Sect. 3. Section 4 covers experimental results and Sect. 5 gives the conclusion.

2 Related Work

As social media websites like Facebook, Twitter and Instagram have become an integral part of our daily life, people have also become more open about sharing their feelings and experiences on social media. They actively discuss about their symptoms, service provided and also about their recovery with other patients and healthcare providers. This is more common in people with chronic diseases. In a recent survey, it has been shown that 90% of the physicians who were interviewed also used social media to study and discuss about latest developments in health care. A review of different clustering algorithms with respect to Twitter data is presented in [4] which emphasizes the need of experimenting unsupervised learning with larger data sets to prove its effectiveness. A generic architecture is provided in [5] which uses Apache Hadoop and Mahout to collect and analyse health Twitter data. The importance of analysis of healthcare data is stressed upon as it leads to a transformation in the service provided by healthcare professionals. Sentiment analysis on healthcare-related twitter data in a given area is proposed in [6] which deals with spatio-temporal evolution of a certain disease in a particular geographical area. A novel method of clustering Twitter data based on themes is presented in [7]. The themes are identified with the help of Wikipedia and a graph-based distance measure to cluster tweets. A clustering algorithm which can identify the topics in the twitter data is given in [8]. It also presents a visualization technique to trace the density of the Twitter activity with respect to a geographical location which serves as a source of spatio-temporal information. A graph-based technique to summarize tweets using similarity between tweets and community detection is presented in [9]. A study on topic summarization of tweets in Bahasa Indonesia language is given in [10]. The Phrase Reinforcement Algorithm and semi-abstractive approach are used to construct the summary. A topic preserving summarization model of tweets is reported in [11] which provides summaries using social contexts. The social context is based on user influence and tweet influence. A summarization framework called Sumblr which uses a stream clustering approach to cluster Twitter data is given in [12]. This approach uses a topic evolution detection algorithm and generates online and historical summaries.

Though social media is used by the healthcare industry, it is mostly restricted to marketing, providing latest technical updates, recommending physicians and hospitals, feedback collection and to improve the service provided by the hospitals. There are also instances of Twitter being used to set up a media to exchange information between students, doctors and in medical conferences to tweet the latest information [13]. We observe there is a lot of gap in the analytical tasks which can be carried out in this area. It is the right time to extract information from the available large amount of data and use it in improving the facilities and treatments provided to the patients.

3 Proposed Methodology

In this work, we present the observations of continuation of the work which is presented in [3]. In the previous work, Twitter data was collected for a period of four months. As the data collected is noisy, several pre-processing techniques had been implemented to refine the data. Initial analysis of the data had shown that K-means clustering was beneficial in extracting summarized information from the Twitter data. Two feature selection methods—variance threshold and Laplacian score—were experimented upon to improve the clustering performance. The conclusion was that the Laplacian score feature selection method performed better over variance threshold and clustering performance improved with feature selection. This was consistent with different cluster sizes and different batch sizes.

As continuation, we collected data for another six months and the same experiments were repeated on the new data which was larger in size. The results obtained were intriguing as it was observed that with the increase in the data, the same methods did not perform consistently for the new data. To put it simply, Laplacian score-based feature selection did not improve the clustering performance. This led to further analysis of the raw data which showed heavy noise for which additional pre-processing was done. After pre-processing, the earlier experiments were repeated and summaries extracted using K-means and two feature selection methods for both sets of data. The results thus obtained were quite different to the results obtained earlier. To further strengthen our observations, we explored the feasibility of combining the two feature selection methods in different ways and also experimented upon two popular feature extraction methods—PCA and SVD. Fuzzy C-means clustering was implemented to evaluate the performance of soft clustering algorithm. The extracted summaries are compared with the complete data using cosine similarity which is a content-based evaluation measure. The overall flow of the proposed methodology is given in Fig. 1.

3.1 Data Collection and Pre-processing

In the previous work, data collected for a period of four months was used. Here, we are using an extended data set which is collected over a period of one year. Though the time duration looks quite large enough to collect a huge number of tweets, it is not as expected as we have put filter terms for only cancer related tweets. During one stint of data collection, we learnt that hardly 10% of the tweets collected satisfied our criteria. The reason behind this is though a filter is applied during collection of tweets, Twitter cannot handle filters with more than one word. For example, if

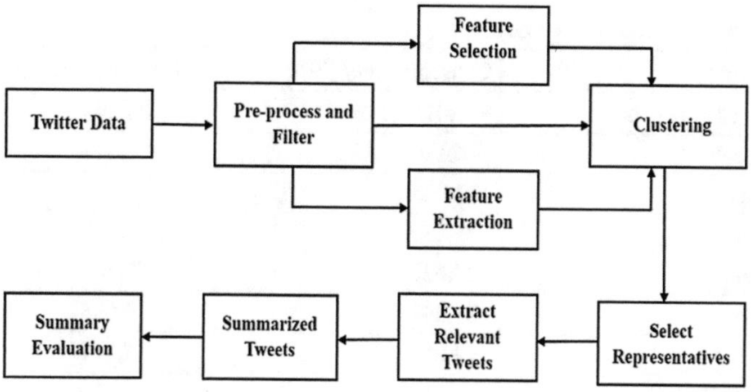

Fig. 1 Overall flow of the proposed method

the filter is set to 'Skin Cancer', it collects all the tweets containing either 'Skin' or 'Cancer'.

Hence after collecting the tweets, it has to be further filtered using Natural Language Processing techniques and only relevant tweets have to be collected. Also there is a lot of repetition in the tweets and have to be removed. Though this part of the pre-processing was applied in the previous work, when we continued our work with more tweets, we observed that feature selection did not work for many of the batches especially with the batch size 100. Upon detailed analysis of the tweets, we found that lots of tweets were repeated and it was observed that sometimes the entire batch of 100 tweets contained the same tweet. This caused a serious problem with feature selection methods as all the features of all the samples were same. We noticed that though we had applied a method to drop duplicates earlier, these tweets escaped the same. The reason was the tweets had the same content but differed in the hyperlinks presented with them. Hence, we applied further pre-processing which is based on Natural Language Processing (NLP) where a regular expression is used to search for hyperlinks in the tweets and remove the same. After this, we removed duplicate tweets which reduced the number of tweets significantly but made the tweet corpus more meaningful. The details of the tweets collected in two phases are given in Table 1.

3.2 Clustering

Twitter data has to be first converted to a vector space form in order to apply learning algorithms. The popular TF-IDF model is used in our work to convert Twitter data to a matrix. The Twitter data set is processed in batches of 100, 200, 300, 400 and 500 tweets. To extract summary from the tweets, we use the well-known unsupervised learning method, i.e. clustering. We have used two types of partition-based clustering techniques—hard and soft. Partition-based methods are used since a representative

Table 1 Tweets' data collection details

Data collection	Cancer_tweets_2017	Cancer_tweets_2018
Total number of tweets	326,412	1,129,549
English tweets	326,114	1,127,727
Breast cancer	117,935	378,812
Skin cancer	25,458	50,742
Prostate cancer	23,627	67,569
Lung cancer	71,244	180,635
Colon cancer	14,051	33,574
Pancreas cancer	118	316
After pre-processing and filtering	252,433	711,648
After removing duplicates	152,603	359,481
After removing hyperlinks	93,475	246,395

can be chosen for each cluster which can be considered as the summary of the underlying data. The algorithms used are *K*-means which is a hard partition-based algorithm and fuzzy *C*-means which is soft partition-based algorithm. *K*-means is used as a continuation of the earlier work, and fuzzy *C*-means is used to evaluate the performance of a soft clustering algorithm with our data. Clustering is performed on the TF-IDF matrix. The tweets nearest to the cluster centroid are collected to form a summary of the twitter data set.

Let $X = \{x_i\}$, $i = 1,\ldots n$, be the set of *n* d-dimensional points to be clustered into a set of *K* clusters, $C = \{c_k, k = 1,\ldots K\}$. *K*-means algorithm finds a partition such that the squared error between the empirical mean of a cluster and the points in the cluster is minimized. Let μ_k be the mean of cluster C_k. The squared error between μ_k and the points in cluster C_k is defined as in Eq. 1.

$$J(c_k) = \sum_{x_i \in c_k} \|x_i - \mu_k\|^2 \quad (1)$$

The goal of *K*-means is to minimize the sum of the squared error over all the *K* clusters [14].

Fuzzy *C*-Means

Fuzzy clustering allows data points the liberty to be organized into overlapping clusters which is similar to natural grouping. The fuzzy *C*-means is also based on minimizing the objective function given in Eq. 2, which is applied for a given set of data with *n* points which have to be grouped into *k* clusters [15].

$$J_m = \sum_{i=1}^{N} \sum_{j=1}^{C} U_{ij}^{m} \|x_i - C_j\|^2, 1 \leq m < \infty \quad (2)$$

3.3 Feature Selection

Feature selection plays a major role in data mining and machine learning tasks. In the present Big Data era, we are dealing with large number of data and the role of feature selection techniques which reduce the dimensionality of the data cannot be undermined. Majority of the feature selection techniques are related to supervised learning tasks. In unsupervised learning, as we do not have class labels which tell us about the contribution of each feature towards the learning algorithm, we have very limited techniques for feature selection [16]. We have used variance threshold and Laplacian score feature selection techniques as a factor of further evaluation. We select features based on a threshold using both the methods. The threshold selected is mean of the values in both the cases. Both the feature selection methods had improved the quality of the results with Cancer_tweets_2017 data. But these feature selection methods became inconsistent with the growth in the data, and hence, we explored the feasibility of combination of the features selected using both the feature selection methods. The combination of the features was explored in three ways:

(1) Union: Set (features selected using variance threshold) Union Set (features selected using Laplacian score)
(2) Intersection: Set (features selected using variance threshold) Intersection Set (features selected using Laplacian score)
(3) Proportional combination: Intersection of the features along with different percentage of features in the Union of features. Along with all the features of intersection, combine 10, 20, 30, 40 and 50% of the features present in the Union.

3.4 Feature Extraction

As there are limited number of feature selection techniques for unsupervised learning, feature extraction becomes the obvious alternative. Hence, we have experimented with two most appreciated feature extraction techniques—principal component analysis (PCA) and singular value decomposition (SVD).

PCA

Principal component analysis (PCA) is a powerful and most commonly used dimensionality reduction technique [17]. It is a statistical procedure which uses linear transformations to map data from high-dimensional space to low-dimensional space. The low-dimensional space is determined by eigenvectors of the covariance matrix. PCA extracts the most discriminating features with higher variances. Consider a data matrix A of $m\ x\ n$ size which is centred where m is the number of samples and n is the number of features. The covariance matric C is given by Eq. 3.

$$C = \frac{A^T A}{n - 1} \quad (3)$$

As it is a symmetric matrix, it can be diagonalized as in Eq. 4.

$$C = VLV^T \tag{4}$$

where V is a matrix in which each column is an eigenvector and L is a diagonal matrix with eigenvalues λ_i in the descending order on the diagonal.

SVD

Singular value decomposition (SVD) is a matrix factorization method used in many applications of linear algebra. SVD is more stable than PCA as it uses a divide-and-conquer approach when compared to eigen decomposition algorithm [18].

Consider a data matrix A of $m \times n$ size which is centred where m is the number of samples and n is the number of features. If SVD is performed on A, we get Eq. 5.

$$A = USV^T \tag{5}$$

where U is a unitary matrix and S is the diagonal matrix of singular values S_i.

3.5 Summary Evaluation

Summary evaluation is a non-trivial task as we have to depend upon humans for both generating an equivalent summary of the data and evaluating it. This becomes much more complicated for our data set as it is continuously growing and also very informal. As an alternative, we have content-based evaluation measures. Cosine similarity measure is one such content-based evaluation measure used in this work to evaluate the summaries generated by clustering techniques [19]. It is given by the formula as in Eq. 6.

$$\cos(X, Y) = \frac{\sum x_i * y_i}{\sqrt{\sum (x_i)^2 * \sum (y_i)^2}} \tag{6}$$

4 Experiments and Results

The experiments were executed on the tweets data as explained in Sect. 3. The system used to carry out the experiments is an Intel Core i5 4210 @ 1.70 GHz machine with 8 GB RAM. Twitter data was collected using Twitter streaming API, and Python is used to implement the experiments.

Experiments were conducted using K-means clustering technique first without feature selection, later with feature selection methods—variance threshold (FSVT) and Laplacian score (FSLS) and two feature extraction methods—PCA and SVD for the old Twitter data (Cancer_tweets_2017) and the new data (Cancer_tweets_2018) which is larger in size. Experiments are carried out for different K values 3, 4 and 5. The number of clusters chosen was based on the elbow method. All the above set of experiments were also repeated with fuzzy C-means technique. The results are given below in the tables.

It is observed that the performance of K-means after further pre-processing goes down for the Cancer_tweets_2017 but improves for the Cancer_tweets_2018. But it is observed that Laplacian score performs slightly better than variance threshold in feature selection. The cosine similarity measures for summaries extracted using K-means without feature selection (WFS), with variance threshold feature selection (FSVT) and with Laplacian score feature selection (FSLS) for both the data sets are given in Tables 2 and 3.

Experiments were carried out with different combinations of two feature selection methods as described in Sect. 3.3. The results for Union and Intersection of the features selected are given in Tables 4 and 5, and the similarity measures for combination of Intersection and a percentage of Union is given in Tables 6 and 7. It is observed that Intersection of the features seems to perform well and with a small feature set.

The results for the cosine similarity evaluation technique for feature extraction techniques PCA and SVD along with the data without feature selection (WFS) are given in Tables 8 and 9. The PCA was performed for 95% of the variance considered and SVD for five components. It can be noted that SVD performs better than PCA for all values of 'k' and in both the sets of data.

The results for the K-means clustering technique and fuzzy C-means technique for the data without feature selection are given in Tables 10 and 11. It can be seen from the results that fuzzy C-means clustering method did not give promising results as the algorithm is very much sensitive to the initial seed and can be easily drawn into local optima [4].

The number of features selected with different methods for the first ten batches with 100 tweets in a batch is given in Table 12. It can be observed that the Intersection of the two feature selection methods obviously selects less number of features with a good similarity measure as shown earlier. SVD performs equally better considering the fact that the components selected are only five.

The time taken for different techniques for three clusters and a batch size of 100 is shown in Fig. 2.

A sample of the summary extracted from our techniques is given in Table 13. Some tweets definitely give some crucial information which may help healthcare professionals and patients or their relatives.

Table 2 Cosine similarity measure for feature selection methods (Cancer_tweets_2017)

No. of tweets in batch	$K = 3$			$K = 4$			$K = 5$		
	WFS	FSVT	FSLS	WFS	FSVT	FSLS	WFS	FSVT	FSLS
100	0.1310	0.1490	0.1585	0.1341	0.1539	0.1613	0.1515	0.1645	0.1618
200	0.0964	0.1396	0.1301	0.1166	0.1333	0.1328	0.1018	0.1373	0.1460
300	0.0756	0.1257	0.0912	0.0739	0.1157	0.1162	0.0906	0.1212	0.1233
400	0.0920	0.1210	0.1077	0.0685	0.1021	0.1168	0.0703	0.1165	0.1206
500	0.0562	0.0854	0.1182	0.0823	0.1167	0.1260	0.0955	0.1195	0.1238

Table 3 Cosine similarity measure for feature selection methods (Cancer_tweets_2018)

No. of tweets in batch	$K = 3$			$K = 4$			$K = 5$		
	WFS	FSVT	FSLS	WFS	FSVT	FSLS	WFS	FSVT	FSLS
100	0.4229	0.4626	0.4661	0.4338	0.4640	0.4661	0.4491	0.4734	0.4722
200	0.3882	0.4542	0.4279	0.4201	0.4492	0.4279	0.4173	0.4637	0.4607
300	0.3105	0.4294	0.4496	0.3154	0.4432	0.4496	0.3467	0.4563	0.4456
400	0.2983	0.3434	0.4139	0.3259	0.3975	0.4139	0.3449	0.4112	0.4468
500	0.2561	0.4135	0.3979	0.3464	0.4070	0.3979	0.2337	0.4382	0.4378

Table 4 Cosine similarity measure for combination of feature selection methods (Cancer_tweets_2017)

No. of tweets in batch	$K = 3$			$K = 4$			$K = 5$		
	WFS	Union	Intersection	WFS	Union	Intersection	WFS	Union	Intersection
100	0.1310	0.1443	0.1495	0.1341	0.1542	0.1588	0.1515	0.1572	0.1611
200	0.0964	0.1256	0.1265	0.1166	0.1394	0.1317	0.1018	0.1504	0.1377
300	0.0756	0.0840	0.1098	0.0739	0.1328	0.1349	0.0906	0.0968	0.1348
400	0.0920	0.0910	0.1067	0.0685	0.1157	0.1169	0.0703	0.1032	0.1182
500	0.0562	0.0943	0.1079	0.0823	0.1186	0.1219	0.0955	0.1154	0.1102

Table 5 Cosine similarity measure for combination of feature selection methods (Cancer_tweets_2018)

No. of tweets in batch	$K = 3$			$K = 4$			$K = 5$		
	WFS	Union	Intersection	WFS	Union	Intersection	WFS	Union	Intersection
100	0.4229	0.4473	0.4705	0.4338	0.4606	0.4785	0.4491	0.4725	0.4789
200	0.3882	0.4316	0.4517	0.4201	0.4626	0.4540	0.4173	0.4670	0.4585
300	0.3105	0.4075	0.4383	0.3154	0.4267	0.4620	0.3467	0.4340	0.4514
400	0.2983	0.3927	0.4009	0.3259	0.3806	0.4403	0.3449	0.4246	0.4451
500	0.2561	0.3551	0.3869	0.3464	0.3966	0.4215	0.2337	0.4076	0.4376

Table 6 Cosine similarity measure for different percentage of combination of features selected (Cancer_tweets_2017)

No. of tweets in batch	$K = 4$					$K = 5$				
	10%	20%	30%	40%	50%	10%	20%	30%	40%	50%
100	0.1579	0.1612	0.1625	0.1604	0.1641	0.1607	0.1633	0.1642	0.1650	0.1661
200	0.1331	0.1315	0.1416	0.1372	0.1424	0.1426	0.1396	0.1481	0.1439	0.1492
300	0.1368	0.1226	0.1063	0.1252	0.1111	0.1338	0.1379	0.1421	0.1425	0.1389
400	0.1017	0.1112	0.0995	0.1350	0.0969	0.1015	0.1308	0.1340	0.1286	0.1144
500	0.0976	0.0942	0.1223	0.0981	0.0826	0.1193	0.0888	0.1287	0.1149	0.0946

Table 7 Cosine similarity measure for different percentage of combination of features selected (Cancer_tweets_2018)

No. of tweets in batch	$K = 4$					$K = 5$				
	10%	20%	30%	40%	50%	10%	20%	30%	40%	50%
100	0.4729	0.4728	0.4761	0.4766	0.4668	0.4784	0.4799	0.4773	0.4745	0.4668
200	0.4544	0.4559	0.4649	0.4568	0.4599	0.4613	0.4664	0.4648	0.4633	0.4599
300	0.4460	0.4456	0.4437	0.4530	0.4415	0.4538	0.4494	0.4522	0.4692	0.4415
400	0.4263	0.4378	0.4060	0.4232	0.4285	0.4316	0.4206	0.4253	0.4087	0.4285
500	0.4195	0.4232	0.4254	0.4385	0.4179	0.4188	0.4503	0.4498	0.4345	0.4179

Table 8 Cosine similarity measure for feature extraction methods (Cancer_tweets_2017)

No. of tweets in batch	$K = 3$			$K = 4$			$K = 5$		
	WFS	PCA	SVD	WFS	PCA	SVD	WFS	PCA	SVD
100	0.1310	0.1326	0.1454	0.1341	0.1515	0.1568	0.1515	0.1418	0.1608
200	0.0964	0.1152	0.1324	0.1166	0.1018	0.1371	0.1018	0.1319	0.1543
300	0.0756	0.0964	0.1140	0.0739	0.0906	0.1375	0.0906	0.1211	0.1319
400	0.0920	0.0675	0.1173	0.0685	0.0703	0.1299	0.0703	0.0804	0.1158
500	0.0562	0.0746	0.0920	0.0823	0.0955	0.1095	0.0955	0.0788	0.1327

Table 9 Cosine similarity measure for feature extraction methods (Cancer_tweets_2018)

No. of tweets in batch	$K = 3$			$K = 4$			$K = 5$		
	WFS	PCA	SVD	WFS	PCA	SVD	WFS	PCA	SVD
100	0.4229	0.4292	0.4431	0.4338	0.4418	0.4659	0.4491	0.4480	0.4663
200	0.3882	0.4082	0.4562	0.4201	0.4328	0.4567	0.4173	0.4320	0.4544
300	0.3105	0.3486	0.4356	0.3154	0.3997	0.4583	0.3467	0.3929	0.4367
400	0.2983	0.3016	0.4248	0.3259	0.3359	0.4083	0.3449	0.3329	0.4223
500	0.2561	0.2062	0.3960	0.3464	0.3119	0.4190	0.2337	0.3414	0.4103

Table 10 Cosine similarity measure for K-means and fuzzy C-means (Cancer_tweets_2017)

No. of tweets in batch	$K = 3$		$K = 4$		$K = 5$	
	K-means	FCM	K-means	FCM	K-means	FCM
100	0.1310	0.0863	0.1341	0.0902	0.1515	0.0860
200	0.0964	0.0553	0.1166	0.0552	0.1018	0.0552
300	0.0756	0.0496	0.0739	0.0497	0.0906	0.0496
400	0.0920	0.0348	0.0685	0.0348	0.0703	0.0348
500	0.0562	0.0391	0.0823	0.0391	0.0955	0.0391

Table 11 Cosine similarity measure for K-means and fuzzy C-means (Cancer_tweets_2018)

No. of tweets in batch	$K = 3$		$K = 4$		$K = 5$	
	K-means	FCM	K-means	FCM	K-means	FCM
100	0.4229	0.3596	0.4338	0.3592	0.4491	0.3621
200	0.3882	0.3290	0.4201	0.3290	0.4173	0.3290
300	0.3105	0.2170	0.3154	0.2172	0.3467	0.2169
400	0.2983	0.1324	0.3259	0.1324	0.3449	0.1324
500	0.2561	0.1615	0.3464	0.1615	0.2337	0.1615

Table 12 Number of features selected/extracted

No. of features	No. of features selected			
	VT	LS	Intersection	PCA
101	34	58	19	48
130	53	63	17	55
126	40	73	18	53
103	38	60	12	47
125	48	76	15	49
125	52	71	21	53
122	44	65	24	55
139	44	80	16	52
137	46	78	18	54
152	57	102	25	52

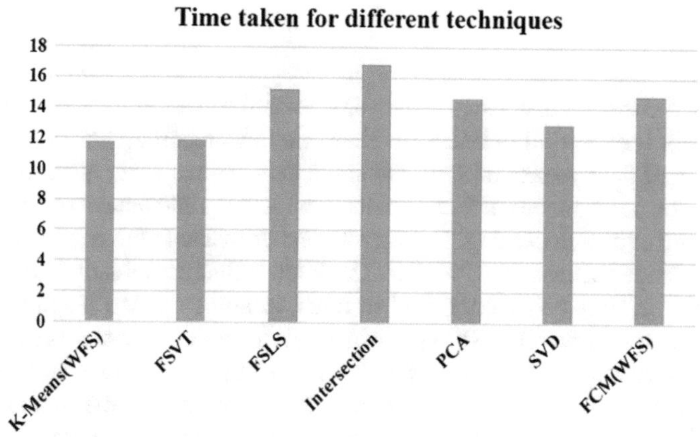

Fig. 2 Time taken for different techniques in seconds

Table 13 Sample tweets in the summary

RT @LoganJames: @SenPatRoberts Your flippant comment was a direct insult to anyone who has lost a loved one to breast cancer or is fighting…
RT @RepBarbaraLee: 40,000 US women will die from breast cancer this year. But Republicans think taking away coverage for mammograms is…
RT @BestScienceFact: Glycoprotein and selenium is found in human semen which helps reduce breast cancer in women by almost 50%.
“The Breast Cancer Awareness Show” this morning at 11:00 AM ET. Go to: and click “TUNE IN”
#endbreastcancernow dont be afraid to attend to early breast cancer detection
#surprised Dogs can smell breast cancer?
whenever reps. talk ab defunding PP they always talk ab defunding abortion & fail to mention breast cancer screenings, pap smears & etc.
RT @AndrewMukose: Once early detected breast cancer can be cured,join us tomorrow for breast cancer free Uganda Campaign at Pride Col…
Mammography NOT the best screening tool for breast cancer. Thermography? Read on. Clinics coming to my New…
Dogs can detect breast cancer by smelling a bandage

5 Conclusion

This work focuses on analysing twitter data related to cancer as we collected data continuously for a period of one year. Here, we have recorded our observations in comparison of the two sets of data. The large data set had more noise which interfered with the performance of feature selection methods and hence we implemented a further step of pre-processing, and later, we experimented with combination of the feature selection methods used earlier. The Intersection combination with the least feature set performs better compared to other techniques. Two feature extraction methods were also explored. SVD performs well for both the data sets and different cluster sizes with only five features which is remarkable. A soft clustering algorithm fuzzy C-means was also experimented upon. It is observed that the summary generated from the larger data set was more meaningful than the smaller data which gave higher measures in the earlier experiments. This shows the significance of the additional pre-processing carried out. Further, the summaries generated from the Twitter dataset are certainly useful for the healthcare fraternity to understand the genuine feelings and opinions of the patients or their relatives and thereby to improve the services and care provided to them which is a part of the holistic approach required in the healthcare industry. In future, we would like to summarize the data based on specific keywords to garner key information from the data.

References

1. www.internetlivestats.com/twitter-statistics
2. www.similarweb.com/website/twitter.com#overview
3. Lavanya, P.G., Mallappa, S.: Automatic summarization and visualisation of healthcare tweets. In: Proceedings of the International Conference on 2017Advances in Computing, Communications and Informatics (ICACCI), pp. 1557–1563 (2017). https://doi.org/10.1109/icacci.2017.8126063
4. Crockett, K., Mclean, D., Latham, A., Alnajran, N.: Cluster Analysis of twitter data: a review of algorithms. In: Proceedings of the 9th International Conference on Agents and Artificial Intelligence, pp. 239–249 (2017). https://doi.org/10.5220/0006202802390249
5. Cunha, J., Silva, C., Antunes, M.: Health twitter big d ata management with hadoop framework. Procedia Comput. Sci. **64**, 425–431 (2015). https://doi.org/10.1016/j.procs.2015.08.536
6. Carchiolo, V., Longheu, A., Malgeri, M.: Using twitter data and sentiment analysis to study diseases dynamics. In: Proceedings of the International Conference on Information Technology in Bio-and Medical Informatics, pp. 16–24 (2015). https://doi.org/10.1007/978-3-319-22741-2_2
7. Tripathy, R.M., Sharma, S., Joshi, S., Mehta, S., Bagchi, A.: Theme based clustering of tweets. In: Proceedings of the 1st IKDD Conference on Data Sciences, pp. 1–5 (2014). https://doi.org/10.1145/2567688.2567694
8. Sechelea, A., Do Huu, T., Zimos, E., Deligiannis, N.: Twitter data clustering and visualization. ICT, pp. 1–5 (2016). https://doi.org/10.1109/ict.2016.7500379
9. Dutta, S., Ghatak, S., Roy, M., Ghosh, S., Das, A.K.: A graph based clustering technique for tweet summarization. In: Proceedings of the 2015 4th International Conference on Reliability, Infocom Technologies and Optimization (ICRITO)(Trends and Future Directions), pp. 1–6 (2015). https://doi.org/10.1109/icrito.2015.7359276

10. Jiwanggi, M.A., Adriani, M.: Topic summarization of microblog document in Bahasa Indonesia using the phrase reinforcement algorithm. Procedia Comput. Sci. **81**, 229–236 (2016). https://doi.org/10.1016/j.procs.2016.04.054
11. Zhuang, H., Rahman, R., Hu, X., Guo, T., Hui, P., Aberer, K.: Data summarization with social contexts. In: Proceedings of the 25th ACM International on Conference on Information and Knowledge Management, pp. 397–406 (2016). https://doi.org/10.1145/2983323.2983736
12. Sindhuja, P., Suneetha, J.: An Advanced approach for summarization and timeline generation of evolutionary tweet streams
13. Ventola, C.L.: Social media and health care professionals: benefits, risks, and best practices. Pharm. Ther. **39**, 491 (2014)
14. Jain, A.K.: Data clustering: 50 years beyond K-means. Pattern Recogn. Lett. **31**, 651–666 (2010). https://doi.org/10.1016/j.patrec.2009.09.011
15. Panda, S., Sanat, S., Jena, P., Chattopadhyay, S.: Comparing fuzzy-C means and K-means clustering techniques: a comprehensive study. In: Advances in Computer Science, Engineering & Applications, pp. 451–460. Springer, Berlin, Heidelberg (2012). https://doi.org/10.1007/978-3-642-30157-5_45
16. Dash, M., Liu, H.: Feature selection for clustering. In: Pacific-Asia Conference on knowledge discovery and data mining, pp. 110–121 (2000)
17. Vasan, K.K., Surendiran, B.: Dimensionality reduction using principal component analysis for network intrusion detection. Perspect. Sci. **8**, 510–512 (2016). https://doi.org/10.1016/j.pisc.2016.05.010
18. Wang, Y., Zhu, L.: Research and implementation of SVD in machine learning. In: Proceedings of the 2017 IEEE/ACIS 16th International Conference on Computer and Information Science (ICIS), pp. 471–475 (2017). https://doi.org/10.1109/icis.2017.7960038
19. Steinberger, J., Jevzek, K.: Evaluation measures for text summarization. Comput. Inform. **28**, 251–275 (2012)

Intrinsic Evaluation for English–Tamil Bilingual Word Embeddings

J. P. Sanjanasri, Vijay Krishna Menon, S. Rajendran, K. P. Soman and M. Anand Kumar

Abstract Despite the growth of bilingual word embeddings, there is no work done so far, for directly evaluating them for English–Tamil language pair. In this paper, we present a data resource and evaluation for the English–Tamil bilingual word vector model. In this paper, we present dataset and the evaluation paradigm for English–Tamil bilingual language pair. This dataset contains words that covers a range of concepts that occur in natural language. The dataset is scored based on the similarity rather than association or relatedness. Hence, the word pairs that are associated but not literally similar have a low rating. The measures are quantified further to ensure consistency in the dataset, mimicking the cognitive phenomena. Henceforth, the dataset can be used by non-native speakers, with minimal effort. We also present some inferences and insights into the semantics captured by word vectors and human cognition.

Keywords Intrinsic evaluation · Bilingual word embeddings · English-Tamil · Lexical gaps · Semantic similarity · Semantic relatedness

1 Introduction

To do any semantic tasks in natural language processing (NLP) such as part-of-speech (POS) tagging, sentiment analysis and semantic role labelling, semantics of

J. P. Sanjanasri (✉) · V. K. Menon · S. Rajendran · K. P. Soman
Center for Computational Engineering & Networking (CEN), Amrita School of Engineering, Amrita Vishwa Vidyapeetham, Coimbatore, India
e-mail: p_sanjanashree@cb.amrita.edu

V. K. Menon
e-mail: m_vijaykrishna@cb.amrita.edu

S. Rajendran
e-mail: s_rajendran@cb.amrita.edu

K. P. Soman
e-mail: kp_soman@amrita.edu

M. Anand Kumar
Department of Information Technology, National Institute of Technology, Surathkal, Karnataka, India
e-mail: m_anandkumar@nitk.edu.in

S. M. Thampi et al. (eds.), *Intelligent Systems, Technologies and Applications*, Advances in Intelligent Systems and Computing 910,
https://doi.org/10.1007/978-981-13-6095-4_3

the word needs to be measured. With the evolution of word embeddings, quantitative representation of the relative semantics of words is now possible. The word embeddings are obtained by distributed representation; in the embedding vector space, the semantic relations of words are translated to spatial positions. The word embeddings are particularly attractive because it can be trained from an unannotated monolingual corpus.

Word embeddings offer ready-made features for machine learning problems in NLP compared to the conventional ones (LSA, *n*-grams, one hot representation) [10, 11]. Most of the NLP problems can now be rebooted using word embeddings [5, 12].

Since word embeddings have no ground truth data available, this needs to be collected from human evaluators; by scoring given word pairs on the degree of their similarity, using this core model we evaluate the semantics captured by word embeddings for word pairs with varying degree of association. The human evaluators themselves do not agree upon certain word pairs, so it was necessary to take the weighted average to get the absolute score for each word pair. The method of computing weights will be detailed in Sect. 5.2.

2 Evaluating Word Embeddings

There are generally two types of evaluations, (a) intrinsic evaluation and (b) extrinsic evaluation [14, 15]. The popular intrinsic evaluation is done in two ways; one evaluates by correlating human evaluation with the word vector similarity (like cosine similarity), and the other one uses analogy and reasoning [6, 12]. Extrinsic evaluation is done by applying the obtained word embedding as feature for performing any supervised tasks like POS tagging and text classification and checking the improvement in accuracy iteratively. Since the word embedding is opaque, i.e. we still do not know what those numerals in the vector for a word represent, a good performance in one task cannot imply good performance in another. After all, this cannot be generalised since it is too laborious.

There are lots of work done in the area of intrinsic evaluation for English. A conventional method is based on an idea that the distances between words in the embedding space reflect the human judgments on actual semantic relations between these words [6, 8]. For example, the cosine distance between the word vectors, "*baby*" and "*child*", would be 0.8 since these words are synonymous, but not really the same thing. Similarly, an assessor is given a set word pairs and asked to assess the degree of similarity for each pair. The cosine similarity between these pairs is also computed using word embeddings, and the two obtained distance sets are compared. The more similar they are, the better is the quality of the embeddings. WordSim-353 [6] is the conventionally used database. It has 353 noun pairs. This is divided into two subsets of file. The set-1 (153 words) is scored based on similarity. The set-2 (200 words) is scored based on association. SimLex-999 [8], developed by stanford, has 999 word pairs. In this, the most similar words are given higher rating compared to associated word pairs (e.g. word pairs such as "coast" and "shore" have similarity score of 9.00, and "clothes" and "closet" have 1.96). The other works include datasets such as MEN [2] and SCWS [9]. Akhtar et al. [1] created word similarity datasets for six Indian languages, namely Urdu, Telugu, Marathi, Punjabi, Tamil and Gujarati.

All the above-mentioned works are done for monolingual evaluation. To our knowledge, no datasets are available for validating semantic relatedness for English–Tamil bilingual embeddings. In this paper, we focus on English–Tamil bilingual word pair dataset that contains 750 word pairs scored by 13 human subjects. The dataset is scored based on similarity rather than association. It is not focused on any particular relation, but on the reader's cognitive perception. We use this dataset for evaluating the English–Tamil bilingual word vectors obtained using BilBOWA algorithm [7].

3 Experimental Design

We can perceive a single word as a combination of a large number of attributes such as morphological, lexical, syntactic, semantic, discourse and other features. Its embedding should accurately and consistently represent these ideas; ideally, this is the verification we use, to assess the embedding's quality. In the following subsections, the main experiments conducted towards evaluating the embeddings are described in detail.

3.1 Semantic Similarity and Relatedness

To measure similarity between word pairs in a bilingual setting, it is crucial to understand the discrepancy between the notion of **semantic similarity** and **semantic relatedness**. The semantic similarity is more specific (synonyms), whereas the latter includes any kind of relation such as meronymy, hypernymy and hyponymy [3]. For example, [*bird*, *paRavai* (*bird in Tamil*)] and [*bird*, *charanNaalayam* (*a nature reserve in Tamil*)] exemplify the difference between relatedness or association and semantic similarity. The word "*bird*" is semantically similar to "*paRavai*" (refer to the same thing bilingually) and associated with "*charanNaalayam*". Instinctively, it can be presumed that "*bird*" and "*paRavai*" are translational equivalence and more similar. In contrast, "*bird*" and "*charanNaalayam*" are firmly related but not similar; this relation comes from the cognitive understanding of humans, in this case as a result of frequent association.

Similarity and association are neither mutually exclusive nor independent [8]. For instance, "*paeruNthu*" (*bus in Tamil*) and "*car*" are related to some degree because they share similar attributes (*vehicle, transportation*). The word pairs such as "*car*" and "*kiLampukiRathu*" (*to commence in Tamil*) is common to occur together in both real world and bilingual setting, and it is relatively easy to perceive the relation of these word pairs. Identifying the word pairs which are contrary to the above statement is comparatively more difficult. For example, consider the pairs [car, Hoontaa]. "*Honda*" is a company which manufactures the multitude of machines and not just cars. The term "*car*" need not necessarily make one think of "*Honda*", yet they are similar because the term "*Honda*" can be used interchangeably with car manufactured by them. The word pairs with low frequency association, such as "*bird*" and "*teel*" (*a type of bird in Tamil*). These pairs can seem, at least intuitively, to be similar but not strongly associated.

3.2 Lexical Gaps in English–Tamil Bilingual domain

Lexical gaps have an impact on the correlation score calculated between human judgment and word embedding similarity measure (cosine similarity). According to Chomsky, there are no gaps in a language; gaps appear only when the languages are compared [4]. Lexical gap indicates the absence of exact matching words in the corresponding parallel language. It is natural that languages show cultural gaps. As one language is culturally different from another language, it is likely that the cultural items of one language may not be found in the other language. So, it is needless to say that lexical gaps are inevitable in the vocabulary structure of a language. *Hand* and *arm* in English are mapped to a single entity "*kai*" in Tamil, though *hand* and *arm* are meronyms and carry different meaning in English. In Hindi, however, this particular gap does not exist; we have separate words *hAT* and *baAH*.

Influence and adoption of foreign words will minimise the lexical gaps. For example, the word *bicycle* has been introduced into the Tamil lexicon. Later on, an indigenous name is coined from its own units of meaning, thus "*bicycle*" such as "*methivanNti*" which literally means *vehicle which needs to be peddled*' and "*uNthuvanNti*", "*vehicle which needs to be pushed*". Similarly, *car* is taken with its foreign name car ("*kaar*"). Later on, "*chiRRuNthu*" which is means small vehicle and *Bus* which came from the same foreign word was coined as "*paeruNthu*".

3.3 Morphological Gaps

Lexical gaps can be due to morphological differences between word pairs in the two languages. Tamil being agglutinative language, the word features are always appended to the root word. Prepositions does not exist in Tamil; rather, postpositions are used [13]. For instance, the translation of the prepositional phrase "to bird" in English is equivalent to "*paRavaikku*" (*paRavai +kku* = bird + to), a case-inflected noun in Tamil.

Consider a document having following sentences,[1]

put:0 the:1 bird:1 in:2 the:2 cage:2

---

[1]Each word is appended with an id (:id) to understand the mapping between the sentences.

paRavaiyai:1 koontil:2 itu:0
that:0 bird:1 is:2 beautiful:3
aNtha:0 paRavai:1 azhakaaka:3 irukkiRathu:2
bird:0 's:1 home:2
paRavaiyin:0,1 veetu:2

In embedding space, the word "*bird*" in English would be mapped not just to its translational equivalent "*paRavai*" but also to the varied inflected forms of its translation. Henceforth, the word embedding similarity metric such as cosine similarity between word pairs such as [bird, "*paRAvai*"], [*bird*, "*paRavaiyin*"], [bird, "paRavaiyai"] might be very close in the embedding space. Though, we know cognitively they are not one and the same.

4 Dataset Description

The bilingual model is trained using the corpus created in house. The bilingual model is trained using the popular BilBOWA toolkit [7]. The monolingual corpus for Tamil and English was created by crawling the web pages. The parallel corpus is created from the bilingual novels, story books, etc. The embeddings are trained with 300 dimensions. Table 1 shows the statistics of the data.

After training, the obtained bilingual embeddings are evaluated using the newly created dataset. The most frequent nouns are chosen. Most of the words are taken from WordSim-353 and SimLex-999 datasets. The bigram translational equivalence for the English word is avoided. For example, "lunch" in English has bigram translational equivalent "*maththiya chaappaatu*". For each source word, a set of target word (translational equivalence, inflected forms, associated words) is chosen. This dataset contains a considerable number of pairs that are strongly associated such as [*bird*, *charanNaalayam*] and [*book*, *patikka* (*read*)]. The dataset is provided as a set of two files. First two columns in each file represent the source word and target word. File1 contains word pairs followed by the participants' mean score. File2 contains the raw scores of each participant for every word pair. In the general case, all scores are floating points, although many appear as integers. The sample of dataset is shown in Table 2.

Table 1 Data statistics

	#. of sentences
Parallel (English–Tamil)	51,000
Monolingual (English)	2,278,564
Monolingual (Tamil)	1,765,499

Table 2 Sample data. Here, the superscript **, *, +, # in target word indicates the actual translation of the source word, inflected form of the actual translation, plural form of the translation and the associated term

Source	Target
Computer	kanNiniyil *
Computer	kanNippoRi **
Computer	vaLaiththaLam #
Computer	kanNippoRiyai *
Coast	kataRkarai **
Coast	kataRkaraikaL +
Coast	kataRkaraiyil *
Coast	theRkae #
Coast	theevin #

5 Scoring Scheme

5.1 Evaluator Instructions

The dataset is evaluated by 13 participants.[2] All the participants have commandable understanding of English and native Tamil speakers. The following are the instructions given to the participants (in the guidance of senior linguist).

- The score has to be given on a scale 0 to 10 based on similarity rather than association.
- Associated words should be given little lower rating than similar words.
- The association between the word pair is not bound to any specific lexical relation, but rather to intellectual knowledge.
- Scores can be given in decimal form too.

5.2 Statistical Pruning of Human Evaluation Scores

The interpretation of the meaning of the words differs from person to person based on their state of mind and their linguistic understanding. Inter-assessor agreement denotes the trustworthiness of the annotation provided. We derive statistical confidence on the score so that we will know how correct they are. This measure suggests the correlation between intuitive understanding of an assessor and other assessors towards the similarity of word pairs in the dataset. The score also shows some degree of homogeneity in the ratings given by human evaluators.

[2]All the participants were informed about this study, and they have provided their consent to be part of this.

Once the evaluator completed scoring, we estimate some general measures of central tendency (mean and standard deviation) from them. Using these, we can construct weighting coefficient for each assessor. The final human assessment is obtained as a weighted average of all assessors' scores. We have formalised the methodology we used below in Algorithm 1. We represent the word pair as $w_i^p, i \in \{1, 2, 3 \ldots 750\}$; human evaluator is represented as $h_j, j \in \{1, 2, 3, \ldots 13\}$; x_{ij} represents the score for word pair w_i^p, evaluated by assessor h_j; μ_i^H is the (horizontal) mean of scores assigned by all human evaluators for the word pair, w_i^p; N^p indicates the total number of word pairs, and N^h indicates the total number of human evaluators.

Algorithm 1: Algorithm for calculating weighting coefficient

Data: x_{ij}

Result: weighting coefficient for each assesor, θ_j

Step 1 : compute the (horizontal) deviation, d_{ij}^H, of every word pair,

$$d_{ij}^H = (x_{ij} - \mu_i^H) \tag{1}$$

Step 2: compute the standard deviation for each assessor (σ_{h_j})

$$\sigma_{h_j} = \sqrt{\frac{1}{N^p}\sum_{i=1}^{N^p}(x_{ij} - \mu_i^H)^2} \tag{2}$$

Step 3: compute pairwise correlation coefficient between assessors, $\rho_{h_x h_y}$

$$\rho_{h_x,h_y} = \frac{cov(h_x, h_y)}{\sigma_{h_x}.\sigma_{h_y}}; x, y \in 1, 2, 3, ...13 \tag{3}$$

$$cov(h_x, h_y) = \frac{1}{N^p}\sum_{i=1}^{N^p} d_{ix}d_{iy}$$

Step 4 : compute standard error, $s.e(h_j)$ for each human assessor (h_j)

$$s.e(h_j) = \left|\frac{\mu_{\rho_j}}{\sigma_{\rho_j}}\right| \tag{4}$$

$$\mu_{\rho_j} = \frac{1}{N^h - 1}\sum_{x,x\neq j}^{N^h} \rho_{xj}$$

$$\sigma_{\rho_j} = \sqrt{\frac{1}{N^h - 1}\sum_{x,x\neq j}^{N^h}(\rho_{xj} - \mu_{\rho_j})^2}$$

The absolute value of standard error is taken as weighting coefficient (θ_j). The average score is calculated as given below

$$s_i^{\mu} = \frac{\sum_{j=1}^{N^p} x_{ij}.\theta_j}{\sum_{j=1}^{N^p} \theta_j} \tag{5}$$

The correlation between the weighted average score and the cosine similarity obtained from the bilingual word embedding model built using [7] is calculated.

6 Result

The results that we obtained indicate strong correlation between human evaluation and word embedding similarity for *similar words* and lesser correlation for *associated words*.This can be interpreted as the difference between how the concept of (*weak*) *association* is perceived by people and by embedding vectors. The vectors apparently consider both these semantic relations (similarity and association) as the same. However, the human cognition is able to differentiate more effectively. Table 3 gives sample word with mean human evaluation scores as mentioned in Sect. 5.2. The first two columns represent the word pairs, and third column is the weighted mean score of the evaluator's individual assessment.

The correlation between the weighted average score, s_i^{μ}, and the word similarity metric is obtained through the Algorithm 2. The word pairs are represented as w_i^p, $i \in \{1, 2, 3 \ldots 750\}$; the mean weighed score obtained from Sect. 5.2 is represented as s_i^{μ} and the cosine similarity of the embedding as c_j.

Table 3 A sample of mean word pair score obtained through weighted average of individual human evaluation as described in Sect. 5.2

Source	Target	Score
Computer	kanNiniyil	8.82
Computer	kanNippoRi	8.88
Computer	kanNini	10
Computer	vaLaiththaLam	3.71
Computer	kanNippoRiyai	7.89
Coast	kataRkarai	10
Coast	kataRkaraikaL	8.76
Coast	kataRkaraiyil	8.81
Coast	theRkae	3.74
Coast	theevin	4.36
Coast	kizhakku	3.62

Standard deviation among human evaluators is estimated. This shows how much the score of each assessor deviates from the mean score of word pairs. Tables 4, 5, 6, 7 and 8 show the sample of the data that is segregated based on standard deviation. SD (score) means the standard deviation over weighed average score by the assessors for each word pair. Low values indicate good agreement between the 13 assessors on the similarity value. Higher scores indicate less certainty. Table 4 indicates that there is not much deviation among assessors on score for exact translational equivalent pairs of words; these represent *most similar words* among all word pairs. Table 5 shows that the most of the inflected terms' score is within one standard deviation. Table 6 shows

Algorithm 2: Algorithm for calculating correlation between human judgment and cosine similarity

Data: Set of all $s_i^{\mu}, c_i, i \in 1, 2, 3, \ldots 750$

Result: correlation between the human judgment and cosine similarity, s_corr

Step 1 : compute the deviation,d_{s_i}, over every word pair's average human score (s_i^{μ}) from its mean (μ_s),

$$d_{s_i} = (s_i^{\mu} - \mu_s)$$

$$\mu_s = \frac{1}{N^p} \sum_{i=1}^{N^p} s_i^{\mu}$$

Step 2: compute the deviation of every word pair's cosine similarity (c_i) from its mean (μ_c), d_c

$$d_{c_i} = (c_i - \mu_c) \tag{6}$$

$$\mu_c = \frac{1}{N^p} \sum_{i=1}^{N^p} c_i$$

Step 3: compute the standard deviation of human judgment

$$\sigma_s = \sqrt{\frac{1}{N^p} \sum_{i=1}^{N^p} (d_{s_i})^2} \tag{7}$$

Step 4 : compute standard deviation of cosine similarity

$$\sigma_c = \sqrt{\frac{1}{N^p} \sum_{i=1}^{N^p} (d_{c_i})^2} \tag{8}$$

Step 5: compute the correlation between human evaluation and cosine similarity

$$s\_corr = \frac{cov(s, c)}{\sigma_s . \sigma_c} \tag{9}$$

$$cov(s, c) = \frac{1}{N^p} \sum_{i=1}^{N^p} d_{c_i} . d_{s_i}$$

Table 4 Sample data (SD in range [0, 0.4])

Source	Target	Avg. score	SD (score)
Baby	kuzhaNthai	10	0
Forest	kaatu	10	0
Sun	suuriyan	9.97	0.05
Hat	thoppi	9.92	0.13
Salary	sampaLm	9.87	0.27
Island	theevu	9.74	0.37

Table 5 Sample data (SD in range [0.4, 1.4])

Source	Target	Avg. score	SD (score)
Forest	kaatukaLai	8.58	0.55
Forest	kaatukaL	8.86	0.61
Forest	kaatukaLin	8.73	0.71
Forest	vana	7.64	1.18
Money	paNaththai	8.16	1.36
West	meeRkil	8.45	1.4

Table 6 Sample data (SD in range [1.4, 2.6])

Source	Target	Avg. score	SD (score)
Music	paatukiRaarkaL	6.38	1.56
Shore	katalil	6.43	1.84
Forest	atarNtha	5.26	1.66
Car	kiLampukirathu	3.38	1.89
King	pataiyetuppinaal	4.17	2.06

that the strongly associated terms' score is within two standard deviations. Table 7 shows that the weakly associated terms' score is within three standard deviations. It is evident that for the words having strong relations, the evaluators agree on their score; the agreement decreases when the words become semantically decoupled. This is an empirical proof for an aspect of the human cognition that it is very clear on the semantic relation of actual related words and becomes fuzzy as and when the association between the words decreases.

As mentioned in Sect. 3.2, foreign-influenced lexemes cannot be avoided in Tamil; certain Tamil documents still use it. Hence, it is also been considered; in Table 8, its score falls above three standard deviations.

Table 9 shows the correlation of a subjects' assessment with the mean of all the other subjects' scores. The assessors are tagged with ids instead of names. The average pairwise standard correlation between 13 assessors is 0.92. This shows that the

Table 7 Sample data (SD in range [2.6, 2.9])

Source	Target	Avg. score	SD (score)
Air	vazhiyae	2.16	2.70
Bird	teel	2.94	2.56
Forest	maachillaatha	2.66	2.98
Forest	kaattu	5.90	2.89

Table 8 Sample data (SD in range [3.15, 4.00])

Source	Target	Avg. score	SD (score)
Hat	haat	6.94	3.15
Boat	booat	5.33	3.66
Gate	gaet	6.24	3.72
Hall	haal	6.13	3.84
Air	aeer	4.45	4.16

Table 9 Standard correlation score of each assessor with mean score of the other assessors

Subject_id	Correlation score
Subject_a	0.94
Subject_b	0.91
Subject_c	0.93
Subject_d	0.93
Subject_e	0.91
Subject_f	0.89
Subject_g	0.89
Subject_h	0.87
Subject_i	0.92
Subject_j	0.93
Subject_k	0.91
Subject_l	0.90
Subject_m	0.94

average response of each assessor was within one standard deviation from all other assessors. The standard correlation between the mean of assessor scores and cosine similarity of the word pairs is 0.61. The standard correlation between human judgment and automated similarity is low, though the correlation shows good agreement among the assessors; this is because of the lexical gap between the languages, which is discussed in Sect. 3.2.

7 Conclusion

A gold standard dataset for English–Tamil bilingual language pair is created and assessed by 13 evaluators. The weight value showing the statistical confidence on each evaluator is computed using algorithm in Sect. 5.2. The weighted mean of the human evaluation scores is taken and compared with the cosine similarity of bilingual word embeddings built by BilBOWA algorithm [7]; the correlation score obtained is 0.61. The correlation for similar words is as high as 0.92 and for dissimilar words is 0.38. From this, we can empirically infer that the human scores for dissimilar words have more randomness in them. However, human cognition is able to distinguish between similar and dissimilar words. Human evaluation for dissimilar words tends to become more fuzzy as the degree of dissimilarity increases.

Word embeddings see both similar and associated words as having the same semantic relations. They are incapable of distinguishing it.

References

1. Akhtar, S.S., Gupta, A., Vajpayee, A., Srivastava, A., Shrivastava, M.: Word similarity datasets for indian languages: Annotation and baseline systems. In: LAW@ACL (2017)
2. Bruni, E., Tran, N.K., Baroni, M.: Multimodal distributional semantics. J. Artif. Int. Res. **49**(1), 1–47 (2014). URL http://dl.acm.org/citation.cfm?id=2655713.2655714
3. Budanitsky, A., Hirst, G.: Semantic distance in wordnet: An experimental, application-oriented evaluation of five measures. In: IN WORKSHOP ON WORDNET AND OTHER LEXICAL RESOURCES, SECOND MEETING OF THE NORTH AMERICAN CHAPTER OF THE ASSOCIATION FOR COMPUTATIONAL LINGUISTICS (2001)
4. Chomsky, N.: Aspects of the Theory of Syntax. The MIT Press, Cambridge (1965). URL http://www.amazon.com/Aspects-Theory-Syntax-Noam-Chomsky/dp/0262530074
5. Collobert, R., Weston, J., Bottou, L., Karlen, M., Kavukcuoglu, K., Kuksa, P.: Natural language processing (almost) from scratch. J. Mach. Learn. Res. **12**, 2493–2537 (2011). URL http://dl.acm.org/citation.cfm?id=1953048.2078186
6. Finkelstein, L., Gabrilovich, E., Matias, Y., Rivlin, E., Solan, Z., Wolfman, G., Ruppin, E.: Placing search in context: The concept revisited. In: Proceedings of the 10th International Conference on World Wide Web, WWW '01, pp. 406–414. ACM, New York, NY, USA (2001). 10.1145/371920.372094. http://doi.acm.org/10.1145/371920.372094
7. Gouws, S., Bengio, Y., Corrado, G.: Bilbowa: Fast bilingual distributed representations without word alignments. In: F. Bach, D. Blei (eds.) Proceedings of the 32nd International Conference on Machine Learning, *Proceedings of Machine Learning Research*, vol. 37, pp. 748–756. PMLR, Lille, France (2015)
8. Hill, F., Reichart, R., Korhonen, A.: Simlex-999: Evaluating semantic models with (genuine) similarity estimation. CoRR **abs/1408.3456** (2014). URL http://arxiv.org/abs/1408.3456
9. Huang, E.H., Socher, R., Manning, C.D., Ng, A.Y.: Improving word representations via global context and multiple word prototypes. In: Proceedings of the 50th Annual Meeting of the Association for Computational Linguistics: Long Papers - Volume 1, ACL '12, pp. 873–882. Association for Computational Linguistics, Stroudsburg, PA, USA (2012). URL http://dl.acm.org/citation.cfm?id=2390524.2390645
10. Li, Q., Shah, S., Nourbakhsh, A., Liu, X., Fang, R.: Hashtag recommendation based on topic enhanced embedding, tweet entity data and learning to rank. In: Proceedings of the 25th ACM

International on Conference on Information and Knowledge Management, CIKM '16, pp. 2085–2088. ACM, New York, NY, USA (2016). http://doi.acm.org/10.1145/2983323.2983915
11. Liu, Y., Liu, Z., Chua, T.S., Sun, M.: Topical word embeddings. In: Proceedings of the Twenty-Ninth AAAI Conference on Artificial Intelligence, AAAI'15, pp. 2418–2424. AAAI Press (2015). URL http://dl.acm.org/citation.cfm?id=2886521.2886657
12. Mikolov, T., Sutskever, I., Chen, K., Corrado, G., Dean, J.: Distributed representations of words and phrases and their compositionality. In: Proceedings of the 26th International Conference on Neural Information Processing Systems - Volume 2, NIPS'13, pp. 3111–3119. Curran Associates Inc., USA (2013). URL http://dl.acm.org/citation.cfm?id=2999792.2999959
13. Rekha, R.U., Anand Kumar, M., Dhanalakshmi, V., Soman, K.P., Rajendran, S.: A novel approach to morphological generator for tamil. In: Kannan, R., Andres, F. (eds.) Data Engineering and Management, pp. 249–251. Springer, Berlin Heidelberg, Berlin, Heidelberg (2012)
14. Tsvetkov, Y., Faruqui, M., Ling, W., Lample, G., Dyer, C.: Evaluation of word vector representations by subspace alignment. In: EMNLP (2015)
15. Zahran, M.A., Magooda, A., Mahgoub, A.Y., Raafat, H., Rashwan, M., Atyia, A.: Word representations in vector space and their applications for arabic. In: Gelbukh, A. (ed.) Computational Linguistics and Intelligent Text Processing, pp. 430–443. Springer International Publishing, Cham (2015)

A Novel Approach of Augmenting Training Data for Legal Text Segmentation by Leveraging Domain Knowledge

Rupali Sunil Wagh and Deepa Anand

Abstract In this era of information overload, text segmentation can be used effectively to locate and extract information specific to users' need within the huge collection of documents. Text segmentation refers to the task of dividing a document into smaller labeled text fragments according to the semantic commonality of the contents. Due to the presence of rich semantic information in legal text, text segmentation becomes very crucial in legal domain for information retrieval. But such supervised classification requires huge training data for building efficient classifier. Collecting and manually annotating gold standards in NLP is very expensive. In recent past the question of whether we can satisfactorily replace them with automatically annotated data is arising more and more interest. This work presents two approaches entirely based in domain knowledge for automatic generation of training data which can further be used for segmentation of court judgments.

Keywords Natural language processing · Legal text segmentation · Legal information retrieval · Supervised learning · Generation of training dataset

1 Introduction

Information retrieval in legal domain, ever since its emergence in 1960s has been posing challenges to information scientist community. Legal information retrieval is complicated by complex nature of legal language used for interpretations of law, open texture of legal documents and multifaceted information needs of legal professionals [1]. A court judgment is a document that contains decision pronounced in the court along with the details of the legal dispute in question. It also provides the

R. S. Wagh (✉)
Jain University, Bangalore, India
e-mail: rupali.wagh@christuniversity.in

D. Anand
CMR Institute of Technology, Bangalore, India
e-mail: deepa.a@cmrit.ac.in

S. M. Thampi et al. (eds.), *Intelligent Systems, Technologies and Applications*, Advances in Intelligent Systems and Computing 910,
https://doi.org/10.1007/978-981-13-6095-4_4

court's explanation of why it has arrived at the given decision. In common law system relevant court judgments or precedents are the principal sources of knowledge for a legal professional. Information queried by the user is available implicitly in these judgment documents. Thus, locating specific semantic information units within the document is very crucial for high precision legal information retrieval systems. While identification of semantic constructs like case citations and arguments for establishing relevance between cases [2–4] is very common in legal information retrieval, text segmentation is used to identify and locate semantically similar text fragments. Automatic text segmentation of a court judgment is one of the widely studied problems in legal domain [5] and its applications are manifold. Finding case similarity with respect to specific legal dimension labeled (for example facts or decision) during automatic text segmentation can be useful for legal professionals [6] for deeper analysis of precedents. Legal text summarization which aims at generating concise summaries of lengthy court judgments is generally performed as antecedent activity of text segmentation [7–9]. Though a judge follows a set of prescribed guidelines while drafting a judgment, due to different writing styles and multiple dimensions of legal issues discussed during case proceedings, segmentation of a judgment is a very complex task. Non-availability of sufficient training data and the cost involved in building new labeled dataset aggravates the challenges further. In this work, we address the problem of insufficient labeled data by proposing approaches for automatic training data construction by leveraging domain knowledge. The results presented in the paper are preliminary findings of court judgment segmentation by using the proposed approach.

2 Literature Review

Legal knowledge and its interpretation is implicitly available Legal documents like judgments, court orders transcripts. Though the finer details of judiciary system and court proceedings are country specific, a court judgment always has paragraphs for case description, facts, arguments by counsels and judgment pronounced by the judge [10] and these semantic categories have mainly been used by information scientist in their work for legal document segmentation worldwide. Legal text segmentation is studied at statement as well at paragraph levels [11]. Defining role categories [12, 13] for sentences is a common approach for text segmentation in legal domain across the globe. Yamada et al. [14] has presented a study on annotation of legal text for Japanese Language court decisions. This study relies on comparison of annotation results with human annotators for evaluation of the scheme. Legal text annotation with finer level of granularity to be used for semantic web and ontologies is presented in [15] for Scottish court data. This study also uses data labeled by human annotators for training and test data generation. A vector space model for classifying sentences in different stereotypical fact category was proposed in [5]. Identification of sentences stating facts and principles discussed in the case [6] using machine learning is proposed. This approach used features like sentence length, its position in the text, part of speech tags and multinomial Naïve Bayes classifier to classify a sentence as a fact, principle or neither. Identification of rhetorical roles for generating summaries of case documents is demonstrated in [13]. Corpus of Cases from Indian courts related to rent control act was used for this segmentation task. This study used Conditional random fields for sentence classification into seven categories. Importance of Gold standard annotated dataset for researchers for computational legal reasoning is emphasized [11] where authors in their previous work have used sentence classification and paragraph annotation from extracting semantic information from a legal document.

Thus studies suggest that legal text segmentation is one of the most important tasks in intelligent management of legal data. Automatic labeling of text in legal domain is very complex and different from other domains due to the open texture of processes in this domain. Supervised machine learning has been used extensively in all previous work on legal text segmentation. All these above mentioned studies used gold standard human annotated data and have focused primarily on a specific sub-domain of law like claims, IPR, rent control etc. Obtaining sufficient gold standard training data for all domains of law is a very expensive task and can be highlighted as one of the major challenges in generalizing legal text segmentation across all domains of law. Semi supervised learning methods pave their way in such situations of non-availability of sufficient training data. Studies indicate that semi-supervised methods for sentence classification have been used effectively in many domains to overcome this limitation. Semi-supervised method for text segmentation for identifying legal facts is proposed in [16] for more contextual semantic search. These approaches can accelerate the effective application of deep learning methods in providing supervised solutions in complex scenarios of legal domain.

In the proposed work we leverage on domain knowledge and present two approaches for training data construction—(1) Based on the position of the paragraph in the judgment on which we apply evolutionary algorithm to get optimum positions of paragraphs for most accurate segmentation. (2) Based on keywords which labels paragraphs based on the presence of certain cue words.

3 Proposed Approach

In this section we explore various approaches based on leveraging existing legal domain knowledge to perform the task of Legal Case Segmentation. The various segments we wish to identify are described in the Table 1.

3.1 Position Based Segmentation

It is observed that in most of the case judgments the segments occur in a sequence though there are a fair number of exceptions. For instance, most case judgments start with a case description, followed by the arguments advanced by the counsel for

Table 1 Segmentation Categories

S. No.	Segment name	Description
1	Case description	Paragraphs in the judgment which describe the sequence of events that gave rise to the case.
2	Attorney argument	The arguments offered by the attorneys for either the appellant or respondent for the case.
3	Discussion of case by judge(s)	The segment in the judgement where the judges express their responses to the case history and the arguments offered by the attorneys for either sides
4	Judgment	The final decision regarding the case

both sides. The discussion by the judge follows this and the last segment is usually the decision on the case. However, there are many exceptions to this sequence. For instance, in judgments examining several issues, each issue is discussed separately and the judgment is given. In such a case for instance the sequence followed may be: <Case Description>, <Argument on Issue No. 1>, <Discussion of Issue No. 1> <Decision on Issue No. 1> <Argument on Issue No. 2>, <Discussion of Issue No. 2> <Decision on Issue No. 2> … etc.

It is also to be noted that in many instances the order may change, for instance the Judges discuss their views in their case first before stating the arguments by the counsel. Several paragraphs may be categorized in more than one category, for e.g. Facts of the case as well as judge's discussion on the case. Many a times paragraphs do not fall neatly into any one of the categories above.

Due to the above issues, segmentation remains a non-so straightforward process. We propose to leverage on this sequence, generally observed in case judgments, to construct our training set. More specifically, we aim to learn a set of limits $L = \{(s_i, \text{sp}_i)|0 \leq s_i, \text{sp}_i \leq 1,\ 1 \leq i \leq 4\}$.

Where s_i and sp_i (sp-stands for span) can be used to compute the start point and the end point of the probable location of the segment in a case judgment. Once the set of limits is learnt, to construct the training data, we take each case judgment C_i, having length L_i (in terms of the number of paragraphs in the judgment) and compute the first paragraph number where segment j starts as start $= \lfloor s_j \times L_i \rfloor$ and similarly the paragraph number where segment j ends as end $= \lfloor (s_j + \text{sp}_j) \times L_i \rfloor$. Once the start and end paragraph numbers are determined these paragraphs from the case C_i ranging from start to end are added to the training set with the segment label j. This process is carried out for all segments across all the cases in order to construct the training set. Note that here s_j is a fraction which denotes the relative position in most judgments where the text corresponding to that segment would start and sp_j denotes the relative fraction of contiguous paragraphs after the starting line which belongs to the same segment category. For instance, if for a segment [0.1, 0.25] are the learned limits then from each judgment document the paragraphs starting from after the first

10% of the document till 35% of the document are added to the training set labeled with that segment as the category. Though the training set, thus constructed, would be noisy due to most judgments not adhering to any fixed pattern, it has the advantage of not relying on a labeled dataset while offering a reasonable performance.

To learn the limits, we employ EA (Evolutionary Algorithm). An individual in the population is composed of a list of eight real numbers $\{s_1, sp_1, s_2, sp_2, s_3, sp_3, s_4, sp_4\}$ corresponding to the start point and span of each of the four segments. The population comprises of 30 individuals and the algorithm runs for 100 iterations. The iterations halt when there is no change in the max fitness of the population for five consecutive generations. The selection is elitist, the crossover and mutation probability is set to 0.5 and 0.1 respectively and a two-point crossover is adopted. An individual in the population would result in construction of a training set. The fitness of the individual is evaluated by the accuracy of the model trained using the training set using logistic regression, generated based on the individual, when tested on a validation set. Steps of proposed positioned based approach are depicted in following algorithm.

Algorithm 1: **Position Based Generation of Training Data using Evolutionary Algorithm (PBEA)**

```
Algorithm labelParagraphs(C,S)
      # Input: C= {C1,C2,…Cn} Set of unlabeled Case Judgments
      # S={(sj,spj);1<=j<=4} -limits guiding the labeling of paragraphs
      #Output :C_labeled - Judgments with paragraphs labeled
      For every judgment Ci in C
            Li = number_of_paragraphs(Ci)
            For jin[1,4]
                  Label paragraphs from floor(sj*Li) to floor((sj+spj)*Li as segment j
                  Append the labeled paragraphs to C_labeled
   returnC_labeled
```

```
Algorithm Fitness(C, S, Eval_data)
      # Input: C= {C1,C2,…Cn} Set of unlabeled Case Ju dgments
      # S={(sj,spj);1<=j<=4} -limits guiding the labeling of paragraphs
      #Output :Fitness of the individual - contained in S
      C_labeled = labelParagraphs(C,S)
      Model = Train_model(C_labeled)
      accuracy = Evaluate_model(Model, Eval_data)
      return accuracy
```

```
Algorithm EA(C,Eval_data)
    # Chromosome in each population is a set of 8 numbers in
the range[0,1]
    population = InitializePop() # random initialization of
population
    for each individual in population
        individual.fitness      =      Fitness(C,individ-
ual,Eval_data)
    Repeat till convergence
        offspring = breed individuals through crossover and
mutation according to the crossover and mutation rates
        for individual in offspring
            individual.fitness   =   Fitness(C,individ-
ual,Eval_data)
        population = best n individuals from population U
offspring
    return best(population)
```

3.2 *Keyword Based Segmentation*

The existence of certain keywords in paragraphs is strong indicator of the type of segments. However, as shown in Sect. 4 use of keywords alone is not able to offer prediction in all cases. For some cases, because of the keyword patterns not being universally applicable, a prediction as to their segment type cannot be made.

The proposed approach, therefore, utilizes the clues offered by the presence of keywords to populate the training set. Some case documents were scanned and a set of clue words for each segment were identified manually. The set of case judgments available for training were taken. Each paragraph in the document is examined in turn for the presence of keywords indicators for each of the segments. Each paragraph is then labeled according to the match in keyword indicators. Following algorithm depicts the steps of keyword based approach.

Algorithm 2: Keyword Based Generation of Training Set (KBTS)

```
Algorithm labelParagraphs(C,S,KW)
    # Input: C= {C1,C2,…Cn} Set of unlabeled Case Judgments
    # S={(sj);1<=j<=4} Paragraph Labels
    #KW={(KWK);1<=k<=4} Set of keywords for paragraph extrac-
tion
    #Output :C_labeled - Judgments with paragraphs labeled
    For every judgment Ci in C
            For jin[1,4]
                For w in KWj
                    Select all paragraphs contain-
    ing w from Ci
                Label paragraphs as segment j
            Append the labeled paragraphs to C_labeled
     returnC_labeled
```

In case of a paragraph containing keywords from multiple categories a resolution process is carried out depending on the paragraph's position in the text and the labels given to the previous paragraphs. For instance, a paragraph which appears towards the end would most probably be the court decision. Similarly, if the previous few paragraphs were related to the judge's reasoning about the case then the current paragraph could also be assigned the same label. In case the paragraph does not contain any keywords matching with segment categories, no label is assigned and the paragraph is not included in the training set.

Once a training set using any of the above proposed method is constructed, it is converted to a vector representation using tf-idf with 'n' features. The number of features we tested with is 5000. We did not do stop-word removal since even common words such as 'we' and 'our' are strong indicators for the type of segment a paragraph can be categorized into. Subsequently a model is learnt, and a prediction is made using prediction algorithms (details in Sect. 4). The proposed approaches along with modifications of the same were tested on a small test set. The next section gives the relevant details.

4 Experimental Evaluation

The case judgments for our experiments were obtained by crawling judic.nic.in [17] website.

4.1 Experimental Setup

We extracted around 21,560 cases for years ranging from 1947 to 2014. For various experiments a random fraction of these cases were selected to participate in the training data generation. The cases were cleaned up by removing the initial part of the judgment which lists the title, judge names, order number etc. We also replaced out the names in the judgment by a token <name>. This was done because names of people involved in case will be repeated a large number of times in the case but would be fairly unique across judgments. Using a tf-idf to represent the features in the documents would therefore give high importance to names, which are actually unimportant for the sake of the classification. All the cases were converted to lower case and all punctuation other than '-' and '_' were removed.

We retained around 20 cases for testing and manually labeled the paragraphs in these cases assigning one of the four labels. Some paragraphs, where a unique label could not be assigned were not included. These manually labeled paragraphs were set aside for creating the test and validation sets. All the methods were tested using three learning methods—Logistic Regression (LR), Multinomial Bayes Classifier (BC) and Support Vector Machine (SVM).

4.2 Experimental Results

We compare the performance of the position based generation of training set using Evolutionary Algorithm (Sect. 3.1—henceforth referred to as PBEA), Keyword based generation of Training Set (Sect. 3.2—henceforth referred to as KBTS), assigning labels only using keywords (henceforth referred to as KB) and a supervised technique using part of the small manually labeled set as training and using it to test on the rest. Note that for PBEA and KBTS methods the results in Table 1 used cases which had 30 or more paragraphs. Moreover, the training rate was set to 30% i.e. only a selected 30% of the cases would be considered while building the training set. This was done to avoid long training time and we did not observe much benefit in increasing the training rate. The small set of manually labeled data was split into a validation and test set. The tuning of parameters for the proposed methods was performed using the validation set and the results reported are on the test set using the parameters learnt. In our case the split between validation and test was 50–50%. The results of running these experiments are presented in Table 2.

The methods were compared using classification accuracy as the metric. We observe that among the various methods the keyword based ones gave the best accuracy. Here the most effective prediction was performed using Bayesian classifier. However, because the method could not be applied to all paragraphs it could not cover all the test cases. Hence its coverage was around 60%. We also observe that the simple supervised technique gives the next best result but is closely matched by the Keyword Based Training Set generation method. The SVM and the Bayesian

Table 2 Comparison of accuracy of classifying into segment categories

	LR	BC	SVM	Coverage (%)
PBEA	0.602	0.572	0.501	100
KBTS	0.51	0.707	0.518	100
KB	0.77	0.82	0.741	62
SL	0.685	0.675	0.712	100

Table 3 Comparison of accuracy on varying input file sizes (columns heading indicate number of paragraphs used for developing the training set)

	<10	10–20	20–30	>30
PBEA	0.52	0.602	0.56	0.47
KBTS	0.57	0.592	0.661	0.707

classifier performed best in the methods respectively. This shows that employing domain knowledge in the form clue words indicating case segments to create a training set is a promising method to mitigate nonexistent labeled set for this task. In this work we used simple single word keywords to construct the training set. A more sophisticated scheme maybe using phrases and complex rules if devised, may push the accuracy further and may exceed that of the supervised techniques. The position based method could not match up with the other methods in terms of accuracy.

We also experimented with using the effect of the length of cases in improving the accuracy of the PBEA and KBTS methods. The results are presented in Table 3.

We notice here that for the PBEA scheme the best result is obtained when the input cases have number of paragraphs between 10 and 20. Similarly KBTS scheme performs best when cases with number of paragraphs is greater than 30.

In addition, we also experimented with turning the multi-class classification problem into a binary classification for each of the segment types. For this we labeled any paragraph not having the desired label as 'negative' and the paragraphs having the desired label as positive. We followed a similar approach as PBEA to finding the limits for just this single class we wanted to classify correctly. Table 4 lists the best accuracy found using any of the three prediction methods i.e. LC, BC and SVM (we only report the best result obtained). The table also lists the limits we found the best in order to build a good training set for predicting that particular class. For instance, for predicting whether a given paragraph is describing the judge's discussion of the case, we found that if the training set is constructed using the paragraphs after the first 65% of the paragraphs and up to 70% of the paragraphs as positive example and the rest of the paragraphs as negative example, then we obtained the best result, i.e. 74% accuracy. We find that this method is more effective for finding final decisions and case history. This is intuitive since the decision and case history normally appear towards the end and the at the beginning respectively of the case judgment.

Table 4 Segment category-wise prediction accuracy

	Best accuracy	Range of paragraphs (based on length of case)
Case history	0.771	5–45%
Advocate argument	0.72	35–55%
Judge's discussion	0.74	65–70%
Final decision	0.864	80–100%

5 Conclusion

Legal case segmentation is a problem which can be utilized in several applications from case summarization to finding relevant cases. However, labeled data for the same is unavailable. Even if labeled by experts, creation of such a labeled data is cumbersome, prone to ambiguity and becomes very complicated due to different ways of interpreting the text. Legal documents, moreover, do not follow sentence structures similar to normal texts in English. Hence, we explore methods to leverage on domain knowledge to create training data which could be used in the classification of paragraphs in a case into four categories. We found that the training data so generated albeit noisy was able to achieve a reasonable level of accuracy almost as good as a supervised technique. We believe that these techniques can be fine tuned further using more sophisticated approaches to achieve higher accuracy. The proposed methods can be used for segmentation for a set of documents in the legal domain to augment a small training data set to improve the accuracy. In the future, we would attempt to improve these methods of generating training documents and test their effectiveness in augmenting an already existing training data set. Moreover, we intend to test the effectiveness of using paragraphs belonging to different categories to aid finding similar judgments to a given judgment.

References

1. A History of Artificial Intelligence and Law: 25 year of the international conference on AI and Law. Artif. Intell. Law. https://doi.org/10.1007/s10506-012-9131-x. Springer Science+Business Media B.V. (2012)
2. Kumar, S., Reddy, P.K., Reddy, V.B., Singh, A.: Similarity analysis of legal judgments. In: Proceedings of the Fourth Annual ACM Bangalore Conference on—COMPUTE 11 (2011)
3. AI and Justice/Legal information systems
4. Verheij, B.: Formalizing correct evidential reasoning with arguments, scenarios and probabilities. In: Workshop at the 22nd European Conference on Artificial Intelligence (2016)
5. Falakmasir Mohammad, H., Ashley Kevin, D.: Utilizing vector space models for identifying legal factors from text. In: Legal Knowledge and Information Systems. IOS Press (2017)
6. Wyner, A., Shulayeva, O., Siddharthan, A.: Recognizing cited facts and legal principles in judgments. Artif. Intell. Law (2017) (Springer)

7. Kanapala, A., Pal, S., Pamula, R.: Text summarization from legal documents: a survey. Artif. Intell. Rev. (2017). https://doi.org/10.1007/s10462-017-9666-2
8. Jia, J., Miratrix, L., Yu, B., Gawalt, B., El Ghaoui, L., Barnesmoore, L., Clavier, S.: Concise comparative summaries (CCS) of large text corpora with a human experiment. Ann. Appl. Stat. **8**(1), 499–529 (2014)
9. Shastri, L.: System and method for identifying text in legal document for preparing headnote. United States Patent US9058308B2, 2015
10. Judgment Writing, Delivered by the Honourable Justice Roslyn Atkinson, Supreme Court of Queensland, to the AIJA Conference, Brisbane, 13 Sept 2002
11. Walker, V.R., Han, J.H., Ni, X., Yoseda, K.: Semantic types for computational legal reasoning: propositional connectives and sentence roles in the veterans' claims dataset. In Proceedings of the 16th International Conference on Artificial Intelligence and Law, London, UK, June 2017 (ICAIL '17), 10 p (2017)
12. Yamada, H., Teufel, S., Tokunaga, T.: Designing an annotation scheme for summarizing Japanese judgment documents. In: 2017 9th International Conference on Knowledge and Systems Engineering. https://doi.org/10.1109/kse.2017.8119471
13. Saravanan, M., Ravindran, B., Raman, S.: Improving legal document summarization using graphical models. JURIX (2006)
14. Yamada, H., Teufel, S., Tokunaga, T.: Annotation of argument structure in Japanese legal documents. In: Proceedings of the 4th Workshop on Argument Mining, pp. 22–31, Copenhagen, Denmark, 8 Sept 2017. Association for Computational Linguistics
15. Wyner, A., Gough, F., Levy, F., Lynch, M., Nazarenko, A.: On annotation of the textual contents of Scottish legal instruments. In: Legal Knowledge and Information Systems. IOS Press (2017)
16. Nejadgholi, I., Bougueng, R., Witherspoon, S.: A semi-supervised training method for semantic search of legal facts in Canadian immigration cases. In: Legal Knowledge and Information Systems. IOS Press (2017). https://doi.org/10.3233/978-1-61499-838-9-125
17. judic.nic.in

Sarcasm Detection on Twitter: User Behavior Approach

Nitin Malave and Sudhir N. Dhage

Abstract Nowadays, people often use social media platforms to express their feelings and opinions of a certain event in a very informal language. Complex ways of expressing opinions by different people make it difficult to determine the actual sentiments. Different elements that influence these sentiments are briefly discussed in this paper. To do this, only the content of that tweet is not enough; there is an emerging need to find some generic approach for sarcasm detection on Twitter. Our proposed framework concentrate not only on insights of tweets but also shades light on important factor of user behavior and its influence on other users. This factor is studied enormously in the field of psychology, but very few have worked on this in the domain of recognizing sarcasm in texts. This suggested method also considers context-based evaluation, based on data acquired from past experience. Context plays important role in determining user behavior, and it should not be ignored while detecting mockery. Our framework proposes to record user behavior pattern and personality traits along with context information. Accessing this information along with existing sarcasm-detection mechanism would help us to achieve generic approach to detect sarcasm on Twitter.

Keywords Sarcasm detection · Sentiment analysis · Polarity detection · User behavior pattern · Personality traits

1 Introduction

Social networking sites have turned into a prominent stage for clients to express their emotions and sentiments on different topics,for example, any event or any individual.

N. Malave (✉) · S. N. Dhage (✉)
Department of Computer Engineering, Bharatiya Vidya Bhavan's,
Sardar Patel Institute of Technology, Andheri (West), Mumbai 400058, India
e-mail: nitin.malave@spit.ac.in

S. N. Dhage
e-mail: sudhir_dhage@spit.ac.in

S. M. Thampi et al. (eds.), *Intelligent Systems, Technologies and Applications*,
Advances in Intelligent Systems and Computing 910,
https://doi.org/10.1007/978-981-13-6095-4_5

Twitter is a similar web-based social networking platform for individuals to express their emotions, conclusions, and suggestions. People post about 340 million tweets daily. Twitter framework allows users to put short messages, i.e., of 140 characters only per tweet, Due to this limitation, we often notice use of symbolic characters like "emojis," "emoticons," and abbreviations (LOL, i.e, laugh out loud). From research, it has been observed that use of interjections while expressing opinions has tremendously increased over past few years. These behaviors of user are associated with some psychological and behavioral science of humans. This has been studied widely in psychology but still not used extensively in actual design framework for detecting Sarcasm on Social media.

Sentiment analysis is a branch of NLP which deals with detection of polarity of a sentence, i.e, whether it is a positive,negative, or neutral sentence. Twitter is a social media platform where users post their perspectives on everyday life regularly. Many big brands, organizations, and companies have relied on these information to study the opinions of people regarding a political events, brands, or movies to design business strategies. Because of the massive popularity of these social platform, people start tweeting, writing reviews, post comments, etc. on the current trends immediately. Also, people get influenced by other person's review and makes his opinion about a certain brand. So social media also acts as an indirect guide about opinions. Organizations depend on such tweets for recording feedback which would help them to improve their products and plan effective business strategies.

Sentiment analysis is extraction of feelings from any verbal or non-verbal communication. Two ways to express opinions are—direct and indirect way, of which sarcasm is an indirect way. Sarcasm plays a major role as an interfering factor that can flip the polarity of the given text. Example: In a context of late attending a lecture, professor may use the phrase "You have a perfect timing kid!!" In this example, professor has used positive words to express his anger. But overall tweet reflects contradicting sentiment toward the situation. People also find it difficult to recognize sarcasm in a conversation. Normal user face difficulties in recognizing sarcasm in reviews.

Some important features such as hyperbole and interjections, or combination of features like interjection, intensifier, quotes, and punctuation mark in a given text plays a crucial role in identifying sarcasm. The presence of intensifier in a sentence increases the probability of sarcasm. In a given text, adjective and adverb act as an intensifier. Few interjections such as wow, yay, yeah, yeahhhh, oh, nah, and aha are frequently used in tweets. Also the trend of using punctuation marks and quotes such as ?????, !!!!!, , is very popular nowadays. The lexical features, viz uni-gram, bi-gram, tri-gram and n-grams are mostly used by researchers to identify sarcasm.Use of emoticons and "emojis" has become a habit of people nowadays. Peoples often express their feelings through emojis, that also implies some sentiment within sentence, so for better understanding and conclusions, we also have to consider these characters. On Twitter, from research we have claimed that tweets which start with an interjection have high tendency to get labeled as sarcastic [1].

Therefore, the main contributions of this paper are as follows:

- Identify the main purposes behind sarcasm used in social networks.
- How User Behavior could be helpful to detect sarcasm in future tweets, and how to use this information to enhance the accuracy of sentiment analysis.
- Study the added value of the context features,which would solve the insufficient situational database problems on Twitter.

The remainder of this paper is organized as follows: Sect. 2 describes literature survey. Section 3 highlights issues in current systems and approaches. Section 4 proposes a novel approach for sarcasm detection and conclusion in Sect. 5.

2 Literature Survey

Sarcasm detection has become an interesting area of research in sentiment analysis. Most of the research has been done on lexicon-, pragmatic-, and hyperbole-based approaches.

Basic model for sarcasm detection on twitter is stated below as per researchers. The entire process includes following steps [2]:

- Data collection from Twitter using streaming API's.
- Data Preprocessing: removes stop words, URL and references to other users, like '@user'.
- Identify whether Tweet is sarcastic.
- Determine the polarity at sentence and phrase level.
- Polarity detection of the overall result.
- Prediction of analyzed results.

Figure 1 is the diagrammatic representation of current approaches [3]. Different types of sarcasm encountered commonly are stated below [4]:

- Contradistinction between positive sentiment and negative circumstance
- Contradistinction between negative sentiment and positive circumstance
- Fact Negation—text contradicting a fact
- Lexical Analysis—sarcasm hashtag-based
- Likes and Dislikes—Prediction behavior-based.

Important features extracted from the tweets are as follows [5]. We have listed different purposes for which sarcasm is commonly use as follows:

- **Sarcasm as wit**: When used as a wit, sarcasm is used with the purpose of being funny. Mostly, it is used by people who knows each other well. A person employs some special forms of speeches, especially focus on his voice modulation to make it recognizable. On social networking sites, voice tones are replaced by special forms of sentiment expression, use of capital letter words, exclamation and question marks, as well as some sarcasm-related emoticons.

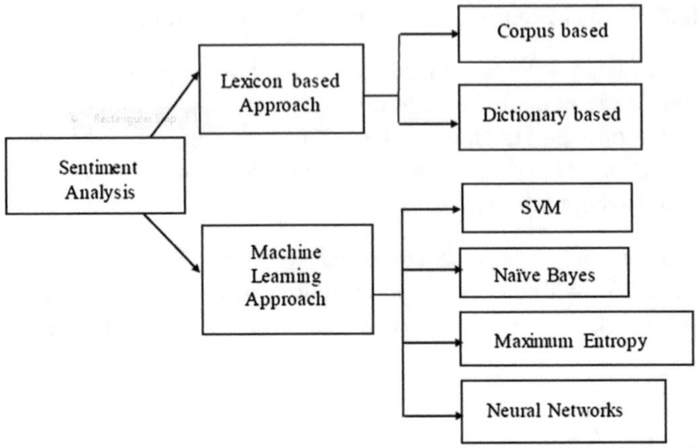

Fig. 1 Sentiment analysis approaches

- **Sarcasm as whimper**: It is used to show how much someone is annoyed or angry on someone. Thus, it tries to highlight a bad situation by making use of positive words for a negative sentiment.
- **Sarcasm as Evasion**: This kind of sarcasm is used by people when they want to avoid something or someone by giving clear straightforward answers.

Important features extracted from the tweets are as follows [5].

2.1 Sentiment-Related Features

A very popular type of sarcasm is used when an emotionally positive expression is used in a negative context. A similar way to express sarcasm is to use expressions having contradictory sentiments.This type of sarcasm can be identified and detected when a positive statement, usually a verb or a phrasal verb, is collocated with a negative situation. However, learning all possible negative situations requires a big and rich source. It is infeasible because negative situations are unpredictable.

In paper [4, 5], they have used POS-Tagging-based approach for classification and sentimental-based detection approach. The process of assigning corresponding part-of-speech to each word in a text, based on its definition and context is referred to POS-Tagging. That is, relationship with adjacent and related words in a phrase, sentence, or paragraph. In this paper,python-based package working on NLTK name TEXTBLOB was deployed to calculate POS tagging of a given text. They used *SentiStrength* database to classify positive and negative words within texts, and their respective scores to classify tweet as sarcastic or not sarcastic.

2.2 Punctuation-Related Features

As specified before, sarcasm is a perplexing-type of speech, it plays with words and meanings. Set of features, i.e., number of exclamation marks, question marks, full stops, all-capital words, and quote also adds some probability toward showing sarcasm. Existence of any vowel with intense expression may alter the polarity of overall observation. (e.g., looooove). The "unreasonable" utilization of exclamation or question marks, or the redundancy of a vowel, especially in an enthusiastic word may mirror a specific tone that the user plans to express, but it is not necessary to be sarcastic always [5]. Also we need to consider that some very short tweets which end with many exclamation marks might show surprise rather than sarcasm. Consider the below case where exclamation marks shows sincere feelings of gratitude.

– Thanks dude, it was awesome !!!

2.3 Syntactic and Semantic Features

According to researchers, Some common expressions has been seen often widely in sarcastic tweets. They often seem related to some punctuation marks. Sometimes, people use conscious efforts to make statements complicated by making use of uncommon words. Thus, they try to give clear answers which can confuse listener. This is normally occur in sarcasm as a "Evasion" situation, where people normally hide their feelings by using sarcasm. Some important features such as use of uncommon words, number of uncommon words, existence of common sarcastic expressions, number of interjections, and number of laughing expressions can be useful for interpreting data [5].

2.4 Pattern-Related Features

"Common sarcastic expressions" often occur in human conversations. They are expressible with human tones, facial expressions. However, in tweets , there are particular word combination structures. Such words and their structural combinations, i.e., n-gram representations are used for determining polarity of sentence.

Mondher Bouazizi et al. and Tomoaki Ohtsuki et al. used POS-Tagging to extract patterns that characterizes the level of sarcasm and ran the classification using classifiers "Random Forest," "Support Vector Machine (SVM)," "K Nearest Neighbors," and "Maximum Entropy" and Random Forest gives accuracy of 83.1% and F1-score upto 81.3%. Large amount of training data is needed to determine exact pattern of classification [5].

Santosh Kumar Bharti et al. worked on contradiction between negative sentiment and positive situation scenario of sarcasm and implemented using PBLGA algorithm. Bharti et al. [4] introduced interjection-word-start-based approach to detect mockery in tweets and claims to get better results. Tweets that starts with interjection word have more tendency to be classified as sarcastic [4].

Novel research of classifying tweets into positive,negative, and neutral feelings have been used by Khan et al. in [6]. They acquired data through streaming API, preprocessed it, and refined tweets are classified using emoticon classifier, Bag-of-words classifier, and Sentiwordnet classifier simultaneously. Approach resolves limitations of existing algorithms and increases classification accuracy with great ease. Further, visual analysis of tweets is done to locate them using GIS-based visual analysis. Visual information helps in elaborating which part of the globe have what type of sentiment orientation toward the search query.

Supervised algorithm is used to predict election poll results by Tayal et al. and Yadav et al. in [7].They used a particular key feature called user retweet. They successfully identified sarcastic tweets, determined polarity through it, and predicted the results of election poll efficiently.

Ms. Payal Yadav et al. highlights the importance of text preprocessing, feature extraction and selection. These techniques plays major role in analyzing sentiments efficiently. Machine learning techniques are domain-specific and work well for specific domains, but they are not contributing toward any generic approach of sarcasm detection on Twitter [3]. They propose that dictionary-based approach is advantageous for all areas as it emphasizes on content present in the lexicon. Machine learning methods together with vocabulary-based approach may bring about a better hybrid approach which may result in better sentiment analysis.

Ashwin Rajadesignam et al. presented SCUBA, called as behavioral-modeling framework for sarcasm detection.They discuss different variations of sarcasm within tweets. Their results shown us SCUBA is quite effective in detecting sarcastic tweets with a novel approach. SCUBA has two advantage of considering psychological and behavioral aspects of sarcasm and retrieve user's historical information to decide whether tweets are sarcastic or not. They claim that even with restricted chronicled data can significantly help enhance the productivity and results of sarcasm identification [8].

Bharti et al. and Vachha et al. present hybrid framework of big data technology for sarcasm detection [9]. Twitter data acquired through flume, for identifying the sentiment of a given tweet, refined data passes through MapReduce functions for sentiment classification. The tweet is classified into either a negative,positive, or neutral,based on the detection engine. Tweets with positive and negative polarities are further checked for actual positive and actual negative. They achieve fast and efficient results by parallelizing the work over HDFS [9].

3 Issues

Field of sentiment analysis deals with finding exact sentiment, attitude, or opinion behind that tweet. As a human, we can easily interpret the true sentiment of a sentence. But a machine is unable to detect true sentiment in a contradicting statement. Following are some of the issues gathered from literature survey.

1. **Language fluidity**: English is one of the most popular language being used all over the world.Social media has become tremendously rich in terms of technology and supports other native languages of users like hindi, marathi and kannada etc. Therefore these languages should be incorporated in these intelligent systems.
2. **Excessive use of emoticons and emojis**: Use of emojis has increased widely since it has become available in chat interfaces. People often use it to express their feelings, which otherwise required typing long sentences. Since there can be some ambiguities among few users about exact meaning of a particular emoticon or emoji. Current sarcasm-detection frameworks mostly focus on understanding meaning meaning of a text. State of the art systems are not integrated with these emoticons feature. These emoticons have high potentials to flip the overall polarity of sentence [10, 11].
3. **Misspells in tweets**: People often make mistakes while typing. An incorrect spelling is effortlessly rectified by humans yet a machine could fail to perceive that incorrectly spelled word. It may happen that it will not match with domain-specific dictionary words and will get removed in preprocessing step by considering it as stop word. One single misspell of word can cause impact on overall polarity of a sentence, as misspells are generally ignored by machines or may replace by any other possible similar word. These are the most trending issues and need to pay attention on it [10].
4. **Lexical Variation**: A few words are written by users in more one style. For example, "awesome" is "awsum," "awesum," "awsom," and many more. Managing these variety of dictionaries is a complex task in sentiment analysis [10].
5. **Insufficient context information**: Human beings normally use some intense words or sarcasm into a conversation based on context level. A machine can automatically detect sarcasm in a text, which have some user-provided lexicon features. Most of the researchers have not considered this context-based information. Presence of context plays an important role in determining whether or not people would react to it sarcastically or not. That is, prior knowledge about context must be known [12].
6. **Generic approach**: Most research done on this topic of "Sarcasm Detection on Twitter" uses machine learning approaches with SVM, naive Bayes, etc. for classification. But all these approaches work toward domain-specific work. Therefore, we need some kind of smart artificial intelligence frameworks which does not follow fixed set of rules and can be use as generic approach for detecting sarcasm in any scenario [10, 13].

4 User Behavior Approach

Given a tweet 't' for user 'u', our objective is to appropriately detect sarcasm in a tweet. We will further try to go beyond this and try to extract some set of features from a tweet which would be helpful in training patterns.We will also focus on studying some sort of user behavior patterns and psychological traits mostly seen in user while posting opinions on social media [14, 15].

There are two types of tweets encountered while detecting sarcasm on twitter. First type of tweet is simply single sentence with suffix as #sarcasm and second type is nothing but conjunction segregated tweets (i.e., and, or, but, if, while).

1. U are looking amazing!!!! **#sarcasm**
2. You look beautiful **and** makeup really brings out your natural beauty ;)

In first example, user has directly expressed sarcasm through the use of #sarcasm marker; so, such type of sarcasm easily gets recognized and we do not really need to extract various patterns or features to detect sarcasm. #sarcasm is enough to detect sarcasm in this tweet. As a result, it would reduce the complexity of existing approach. In second example, we can say that this type of sarcastic tweets are often commonly used by users on twitter. Use of conjunction to express opinion may alter the overall polarity of text. Most of sarcastic tweets contain negative sentiment before conjunction and positive or neutral sentiment in other part of a statement. This is the important parameter we have observed in "contrast between negative situation and positive sentiment." This would help in achieving better accuracy and obtain true results.

Rajadesignam et al. proposed approach of considering user's past tweet records for predicting user behavior pattern shades light on a diversified approach for looking sarcasm on social media. As a study of human psychology plays role as important as lexicon-based and machine-learning-based approach, proposed by most of the researchers. As very less work on behavioral science of humans is implemented in their work, so our aim is to manifest that user behavior pattern in sarcasm detection on Twitter. In our approach, we will record user behavior pattern from their current and past tweets, and also consider context equally important for determining sarcasm in future tweets. Framework for user behavior pattern is define in Part A and framework for context calculation is defined in Part B.

4.1 Part A: User Behavior Pattern

Most of the researchers have used lexicon-based and machine-learning-based approach in their studies and got better results and accuracy. Very few of them looked away for something different and focus on different aspects of this problem. But we forgot that Sarcasm is a special form of speech which originated from individual minds and public get influenced by dominant energies of individuals in everyday life. Our proposed approach will focus on determining user behavior and finding personality traits by analyzing user tweet pattern and classifying these users into clusters based on their behavior. This classification will help us predicting future behavior of individual as per the appropriate context. As we consider general human psychology, we will acknowledge the "principle of inferability" speakers just uses sarcasm among people with whom he is sure that they can comprehend it. We found in reviews that sarcasm will probably be used between two persons, who know each other well than between the individuals who do not.

In sarcasm, some words are emphasized, we as a human can understand sarcasm in a conversation from the tone of that person. However, machine is unable to do so, to sarcasm in a sentence. Rajadesignam et al. approach suggests use of user current and past tweets, and we can extract their tweet patterns and generalized their knowledge level.

Generally individual's behavior on a social media platform also varies with their familiarity with that social platform as mention in Fig. 2. Survey regarding individual's familiarity with platform proves that people who are less familiar with any particular social platform tend to express opinions in a straightforward and simple way rather than making use of sarcasm in a sentence. As the fear of misunderstanding limits people from posting sarcastic tweets, but as a person gets familiar with platform, causes drastic change in the behavior. Thus we can check user's likelihood of being sarcastic or not on twitter. We have to consider this social media environment for analyzing user dynamic behavior. On social media, peoples follow other peoples and they also followed by others. As study of psychology conveys very crucial, individual mind gets influenced by whom they used to follow. "Our conduct is essentially get veered off as we ran over some solid energies around us" according to

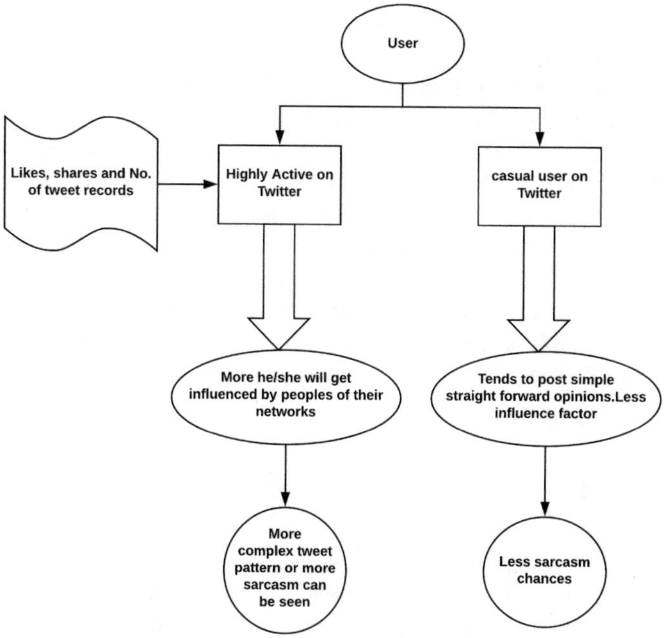

Fig. 2 User behavior trends

psychology. Therefore we have to consider this "Followers and Following" aspects while designing efficient sarcasm detection engine as no one acknowledge this.

4.2 Part B: Context Evaluation

Major difficulty in detecting sarcastic tweets is insufficient context knowledge. To understand sarcasm one must know the context to which it is used. Thus, context plays a crucial role in sentiment analysis and we can not ignore it while designing efficient sarcasm-detection engine. Very few researchers have worked using this method, which is why it is an ultimate aim of this paper, to shade light on this approach. Examples of contexts on Twitter are as follows:

1. NASA successfully launched Curiosity Rover to Mars.
2. Steven Smith found guilty in Ball-Tampering Scandal.
3. Stephan Hawking passed away.

Contexts can be manually examined by human experts and decide whether it is controversial, recreational, or grief-stricken topic. There is less probability of users using sarcasm in third context mentioned above than in the second context. In sad situations, use of sarcasm will be frowned upon as it is considered inappropriate, whereas in second context, people will use sarcasm to mock or insult.

Proper classification of contexts and how people reacted to it can be studied. We can then cluster users according to their reaction. Example, for a particular context, users will post their views or opinions. We will analyze their tweets and check, how many tweets are sarcastic (till this step, we are unaware about user profiles). For this, we can make use of current machine learning techniques. Then, we will find profiles of users who have reacted to it sarcastically. Consider a case where all users initially have score assigned as 0, i.e., a user is new to twitter and we have not yet determined his behavior, i.e., whether he uses sarcasm within his tweets or not. So now, if any user uses sarcasm in his current tweet, a score can be assign to him (for example, +1) and 0 will be assign to the tweets which do not contain sarcasm. By defining an appropriate threshold by statistical calculations, we can label a user to be sarcastic. Thus, this user will probably use sarcasm in his future tweets. Figure 3 shows diagramatic representation of context evalution.

5 Conclusion

There is still large amount of ongoing research in sentiment analysis. Sarcasm has been detected through linguistic patterns so far, but we need to consider the factors that causes the use of sarcasm. Researchers are trying to find various solutions, but we have to work toward a generalized approach which would be domain-independent.Therefore, we have introduced user behavioral approach which would

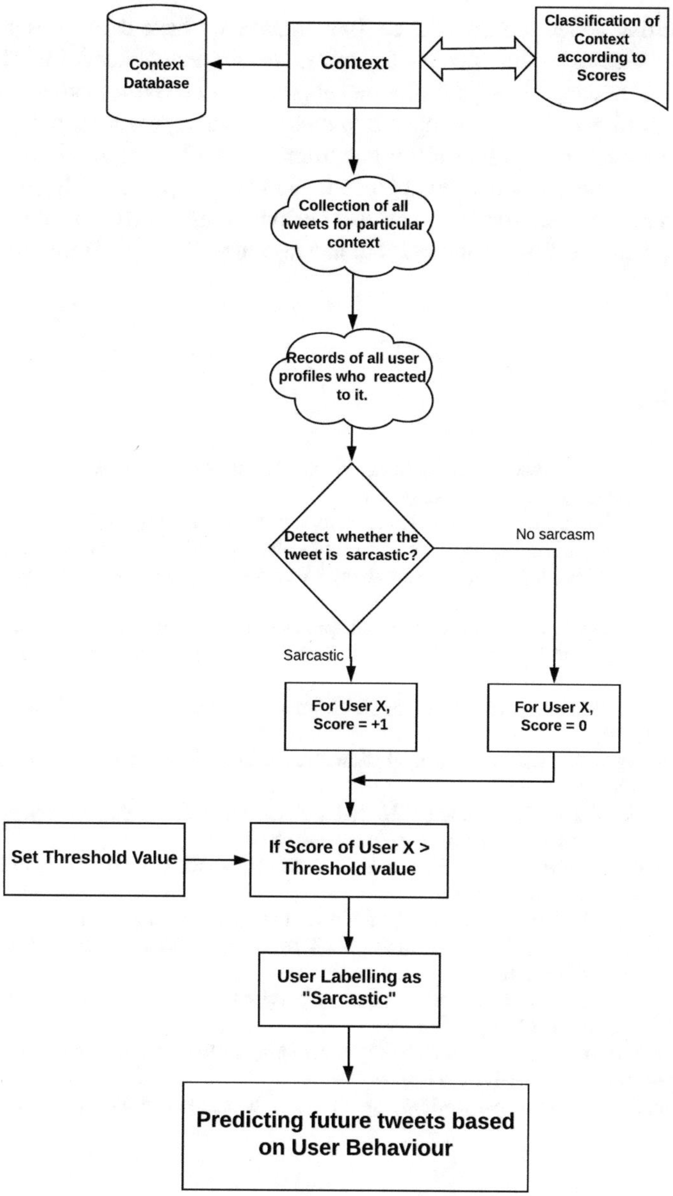

Fig. 3 Context evaluation

help in getting generalized results on Twitter.Once we have derived user behavior pattern and classes of their personality traits, we can predict about their future behaviors too. Research through diversified approach of behavior modeling approach which concern about determining user's mood and analyzes their past tweets for checking probability of sarcasm use in current tweet. Our proposed system would reduce the complexity of current approaches and also help in reducing workload in terms of domain-based frameworks. User behavioral approach can supplement existing sarcasm-detection technologies to increase the accuracy of sarcasm detection in tweets.

References

1. Carvalho, P., Sarmento, L., Silva, M.J., de Oliveira, E.: Clues for detecting irony in user-generated contents: oh...!! its so easy" ;-) (2009)
2. Kaushik, S., Barot, M.P.: Sarcasm detection in sentiment analysis (2016)
3. Yadav, P., Pandya, D.: SentiReview: sentiment analysis based on text and emoticons (2017)
4. Bharti, S.K., Babu, K.S., Jena, S.K.: Parsing-based sarcasm sentiment recognition in Twitter data (2015)
5. Bouazizi, M., Ohtsuki, T.: A pattern-based approach for sarcasm detection on Twitter (2016)
6. Khan, F.H., Qamar, U., Younus Javed, M.: SentiView: a visual sentiment analysis framework (2014)
7. Tayal, D.K., Yadav, S., Gupta, K., Rajput, B., Kumari, K.: Polarity detection of sarcastic political tweets (1999)
8. Rajadesingan, A., Zafarani, R., Liu, H.: Sarcasm detection on Twitter: a behavioral modeling approach (2015)
9. Bharti, S.K., Vachha, B., Pradhan, R.K., Babu, K.S., Jena, S.K.: Sarcastic sentiment detection in tweets streamed in real time: a big data approach (2016)
10. Reganti, A.N., Maheshwari, T., Kumar, U., Das, A., Bajpai, R.: Modeling Satire in English text for automatic detection (2016)
11. Thu, P.P. New, N.: Impact analysis of emotion in figurative language (2017)
12. Wicana, S.G., bisoglu, T.Y., Yavanoglu, U.: A review on sarcasm detection from machine-learning perspective (2017)
13. Davidov, D., Tsur, O., Rappoport, A.: Semi-supervised recognition of sarcastic sentences in Twitter and Amazon (2010)
14. Parveen, S., Surnar, A., Sonawane, S.: Opinion mining in Twitter: how to make use of sarcasm to enhance sentiment analysis: a review (2017)
15. Deshmukh, P., Solanke, S.: Review paper: sarcasm detection and observing user behavioral (2017)

Personalized Recommender Agent for E-Commerce Products Based on Data Mining Techniques

Veer Sain Dixit and Shalini Gupta

Abstract In this article, a recommender agent is designed to meet the increasing demand of consumers for the diversity of products offered by big e-commerce Web sites. Turning visitors of Web sites into customers and clicks made by them into purchases is a challenging task which is achieved through accurate product recommendation. The recommendation algorithm designed in the present work initially design a user-context click matrix that predicts each users' preference level based on deals offered by the company. The users are clustered on the basis of these preferences and neighborhood formation task is completed using collaborative filtering technique that is based on user-item category matrix. The matrix shows users' preference for the type of item user is interested in. After the neighborhood formation task, like-minded users are found using various similarity measures. Finally, the products that are clicked by similar users are marked to find association level among these products using association rule mining to generate user-product preference matrix. The proposed work is flexible and can be applied to Web sites that keep track of users' click-stream behavior. The experimental results clearly justify that the proposed work outperforms the conventional ones.

Keywords Recommender agent · E-commerce · Collaborative filtering · Association rule mining · Click-stream behavior

1 Introduction

The significance of recommender systems (RSs) [1, 2] is increasing in commercial market due to their ability to assists customers in making appropriate decisions. A

V. S. Dixit · S. Gupta (✉)
Department of Computer Science, Atma Ram Sanatan Dharma College,
University of Delhi, New Delhi, India
e-mail: sgupta@arsd.du.ac.in

V. S. Dixit
e-mail: vsdixit@arsd.du.ac.in

S. M. Thampi et al. (eds.), *Intelligent Systems, Technologies and Applications*,
Advances in Intelligent Systems and Computing 910,
https://doi.org/10.1007/978-981-13-6095-4_6

competitive effort is needed for an e-commerce business to survive in the market. This indicates the need of proper tool such as RS that boost the sales system. The main aim of RS is to endorse products to the customers which they might want to buy, based on their personal characteristics and purchase history. RSs offer a personalized shop to every customer that visits the Web site, thus increasing the relevance of every product shown on the Web site. This makes it possible to recommend products that a customer would have never heard of.

Traditional RS deals with explicit data [3–5] that is obtained from customers' ratings, queries, rankings and tags. The work done in the study is dealing with implicit behavior [6] of customers that is extracted while browsing e-commerce Web site. Strong attributes are identified from this click-stream data [6, 7] that clearly indicates customers' interest for a product. The attributes that can be identified from these sessions are number of visits to a particular product, time spent on reading the details of product, click context (users' interest such as promotional offer or regular offer), item category (such as branded product or regular product), cart placement status and purchase status. Like-minded users are identified on the basis of these attributes and association among the products is calculated.

The main contributions of this article can be summarized as below:

- The article proposes an efficient method to generate association rules between the products viewed by the user while browsing the e-commerce Web site.
- These associations are based on the products viewed by the target user. The like-minded users are identified by clustering them on the basis of similar context interests and then like users are recognized from these clusters using CF technique.
- On the basis of products viewed by the target user and products associated with these viewed products, top-*N* recommendations are generated.

As per our knowledge, this work is a first attempt to predict preferences for online products by finding association between them. Like-minded users are clustered on the basis of context clicks. The experimental study shows that the proposed work outperforms the existing ones. The rest of the paper is organized as follows. In Sect. 2, we discuss the related work which includes recommendations of online products based on implicit data. The proposed recommendation model and approach used are presented in Sect. 3. Section 4 justified our approach on the basis of performance evaluation based on experiments carried out. Finally, in Sect. 5, conclusions and suggestions for further developments are outlined.

2 Related Work

RS is a type of service that assists customers who are interested in browsing and making purchases online. The software analyzes the customer behavior that is helpful in making recommendations. One of the most frequent and traditional techniques is collaborative filtering (CF) [8, 9]. However, content-based filtering (CBF) [10] is used when item characteristics are taken into account. Knowledge [11, 12] and hybrid [13] techniques are also employed when multiple techniques are used to make recommendations. CF helps in discovering like-minded users based on their common interest and purchases made whereas CBF recommends products that share common characteristics. Hybrid recommenders combine two or more algorithms to enhance the accuracy of prediction level. This type of recommender removes the limitations faced by individual algorithms such as cold start [14], data sparsity [15], and over-specialization [16].

The work done in this study employs CF technique as it finds the similarity among the customers. Customers' previous history is used to rank products that is related to purchases made, rating given by them, queries made, comments, etc. Like-minded users form a neighborhood that helps to predict the preferences of customers for a particular product. However, the explicit data obtained is contextual in nature as it depends on demographic situations, time, budget and mood. Contextual recommenders [17, 18] deal with these types of conditions. Thus, to keep track of users' behavior, implicit data is taken into concern. This implicit data extracts users' browsing behavior from the click-stream path followed. Strong attributes such as a number of visits, reading time, types of product clicked, bookmarked, basket placement status and purchase made are identified to predict like-minded users. Several correlation measures are used to find similarity among customers. One of the traditional methods is Pearson correlation (PC) that is used to group similar users. A modified PC is named as constrained Pearson Correlation (CPC) in which average value of customers is replaced by median. Cosine vector (CV) is another important aspect to find user similarity.

The data that is mined from these click-through sessions is termed as implicit data. This data is traced when a user is browsing the Web site. It is a record of navigation path a customer follows. The data captures several attributes such as IP address, date, time, sequence, frequency, purchase status, etc. The data is pre-processed and strong attributes are identified to predict correct recommendations. In our work, users are clustered based on their interest in the type of item and similar users are identified. The associations among the products that are clicked by these similar users are identified using association rules [6, 19] to give top-*N* recommendations.

3 Recommendation Approach and Model

Figure 1 shows the overall working of proposed recommender system which integrates sequential pattern of click-stream data analyzing sequential behavior and absolute preference-based recommendation. The system consists of two phases: navigational behavior-based preferences and implicit data-based absolute preferences.

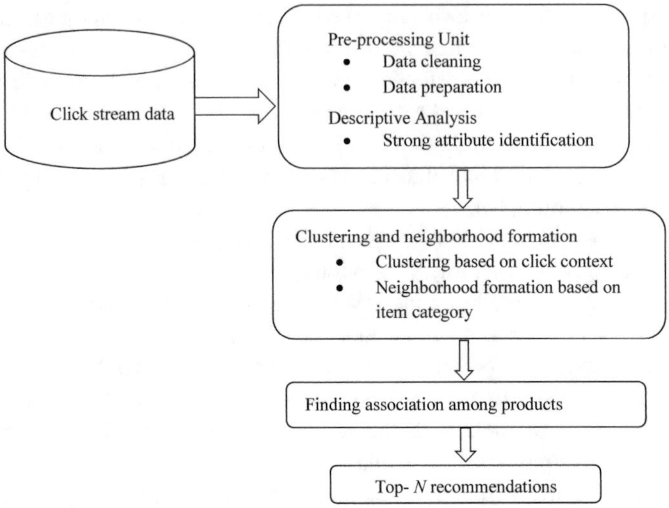

Fig. 1 Proposed methodology

3.1 *Phase I: Pre-processing Unit*

Dataset

Real dataset is inherited from click-through sessions of e-commerce Web site [20] that traces the navigation behavior of its customers. The Web site sells clothing, books, electronic gadgets, etc. The pre-processing phase identifies the strong attributes that indicates users' interest in a particular item. The data is collected for million users who visited the Web site in past one month. Due to enormous amount of click-stream sessions, only those users are filtered who have clicked more than 70 products. Data is pre-processed and experimented using RStudio 1.1.442. Only 57,540 users are taken into account to find preferences of 4300 items.

Descriptive statistics

Implicit data is extracted from these sessions to identify strong attributes. Some of the data variables extracted from the sessions include time stamp, user ID, product ID, purchase status, price, cart placement status, etc.

To identify the strong attributes from these variables, the probability of purchase for each variable is analyzed. As given in Table 1, the possibility of purchase of an item after its cart placement is 85.5. We can also see that a highly viewed item has higher chance of purchase.

Viewing time is also considered as strong attribute considering the fact that interested items' web page is likely to be read for more duration. The fact results are shown in Table 2. Finally, the attributes that are identified from the click-stream dataset are view count, visit duration, click context, item category, cart placement and purchase status.

Table 1 Statistics for cart placement and view count

	Purchase (1)	Purchase (0)	Total
Cart placement			
0	0	3643	3643
1	562	95	657
Total	562	3738	4300
View count			
1	305	3277	3582
2–4	153	298	451
5 or more	104	163	267
Total	562	3738	4300

Table 2 Duration of reading time (*t*-test at 5% significance level)

	N	Mean	Std. Dev.	Std. Err.	Pr > |*t*|
Duration of visit					
Purchase = 1	562	62.19	142.60	8.86	<0.0001
Purchase = 0	3738	28.31	68.88	2.04	

The click hierarchy of the sequence traced while navigating the Web site is shown in Fig. 2. The user starts from the home page of the Web site and finally reaches the product that is viewed, cart placed or purchased. Level 1 finds whether user is attracted toward promotional offer, regular offer or any other. Similarly at the next level, users' interest is depicted for regular or branded products.

3.2 Phase II: Clustering and Neighborhood Formation

The customers are grouped according to context clicks depending on whether they are interested in promotional offers, regular offers, discounted offers or any other offer given by the company. Customer-context preference matrix (C) is built (Fig. 3) by computing the frequency of clicks (C^1), basket placed items $\left(C^2\right)$ and purchased items (C^3). Here, c_{nk}^m represents nth user preference for kth context based on mth browsing behavior. K-means algorithm is applied on customer-context preference matrix to cluster users that share similar context preferences.

Weighted average Customer-context preference matrix is built for user i based on jth context by combining the three matrices as in Eq. 1.

$$C_{ij} = \frac{\propto_1 c_{ij}^1 + \beta_1 c_{ij}^2 + \gamma_1 c_{ij}^3}{\sum_{d=1}^{k} (\propto_1 c_{id}^1 + \beta_1 c_{id}^2 + \gamma_1 c_{id}^3)} \tag{1}$$

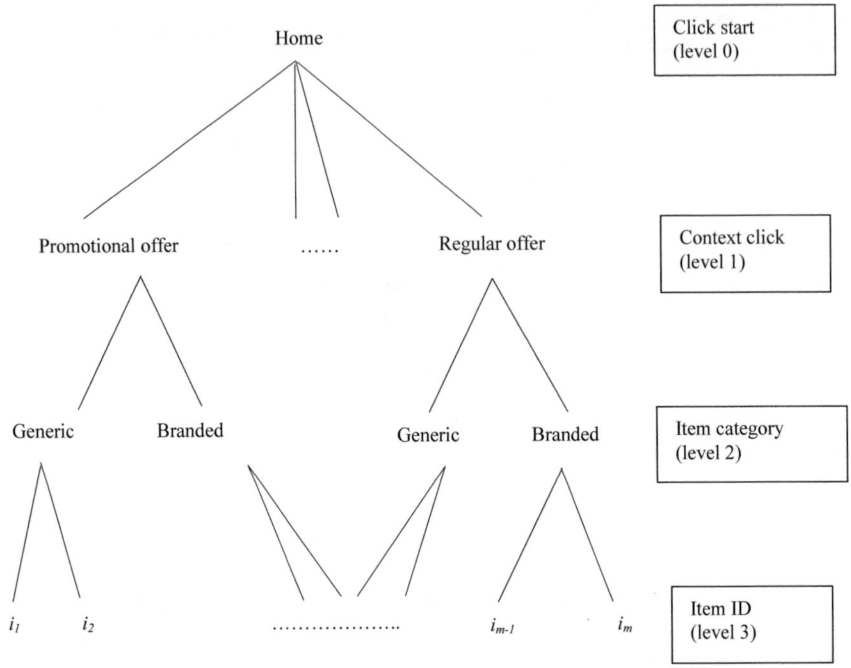

Fig. 2 Click sequence hierarchy

$$\begin{array}{ccc} context_1 \ldots context_k & context_1 \ldots context_k & context_1 \ldots context_k \\ \begin{pmatrix} c_{11}^1 & \cdots & c_{1k}^1 \\ \vdots & \ddots & \vdots \\ c_{n1}^1 & \cdots & c_{nk}^1 \end{pmatrix} & \begin{pmatrix} c_{11}^2 & \cdots & c_{1k}^2 \\ \vdots & \ddots & \vdots \\ c_{n1}^2 & \cdots & c_{nk}^2 \end{pmatrix} & \begin{pmatrix} c_{11}^3 & \cdots & c_{1k}^3 \\ \vdots & \ddots & \vdots \\ c_{n1}^3 & \cdots & c_{nk}^3 \end{pmatrix} \end{array}$$

(i) Viewed item (C^1) (ii) Cart placed item (C^2) (iii) Purchased item (C^3)

Fig. 3 Customer-context preference matrix based on browsing behavior

where $0 \leq \alpha_i, \beta_i, \gamma_i \leq 1$ and $\alpha_i + \beta_i + \gamma_i = 1$.

In the same fashion, user-item category matrix (Fig. 4) is built based on browsing behavior.

Here, T^1, T^2, T^3 represents the count of items viewed, items placed in basket and items purchased, respectively. Also t_{ij}^m represents ith user preference for jth item category (regular or branded) in matrix T^m. The average weighted preference of user i for category j is the combined results of T^1, T^2 and T^3. Customer-item-category preference matrix (T) is calculated by combining the result of three matrices as shown in Eq. 2. Here, $0 \leq \alpha_i, \beta_i, \gamma_i \leq 1$ and $\alpha_i + \beta_i + \gamma_i = 1$.

$$\begin{array}{ccc} category_1 \ldots category_k & category_1 \ldots category_k & category_1 \ldots category_k \\ \begin{pmatrix} t_{11}^1 & \cdots & t_{1k}^1 \\ \vdots & \ddots & \vdots \\ t_{n1}^1 & \cdots & t_{nk}^1 \end{pmatrix} & \begin{pmatrix} t_{11}^2 & \cdots & t_{1k}^2 \\ \vdots & \ddots & \vdots \\ t_{n1}^2 & \cdots & t_{nk}^2 \end{pmatrix} & \begin{pmatrix} t_{11}^3 & \cdots & t_{1k}^3 \\ \vdots & \ddots & \vdots \\ t_{n1}^3 & \cdots & t_{nk}^3 \end{pmatrix} \end{array}$$

(i) Viewed item (T^1) (ii) Cart placed item (T^2) (iii) Purchased item (T^3)

Fig. 4 Customer-item category matrices based on browsing behavior

$$T_{ij} = \frac{\propto_2 t_{ij}^1 + \beta_2 t_{ij}^2 + \gamma_2 t_{ij}^3}{\sum_{d=1}^{k} (\propto_2 t_{id}^1 + \beta_2 t_{id}^2 + \gamma_2 t_{id}^3)} \tag{2}$$

After constructing the user-context preference matrix (C) and user-item-category preference matrix (T), next task is to group the customers that share similar preference levels. Users that share similar context preferences are grouped on matrix (C) using k-means clustering. Like-minded users form a neighborhood within these clusters instead of complete dataset to reduce the time complexity of the algorithm. Similar users are traced using proximity measures as given by Eqs. (3–5), in which, CPC outperforms the results obtained by PC and CV.

$$\mathrm{PC}_{u^t,u^a} = \frac{\sum_{i_p \in \Sigma} \left(r\left(u^t, i_p\right) - \overline{u^t}\right)\left(r\left(u^a, i_p\right) - \overline{u^a}\right)}{\sqrt{\sum_{i_p \in \Sigma} \left(r\left(u^t, i_p\right) - \overline{u^t}\right)^2 \sum_{i_p \in \Sigma} \left(r\left(u^t, i_p\right) - \overline{u^a}\right)^2}} \tag{3}$$

$$\mathrm{CPC}_{u^t,u^a} = \frac{\sum_{i_p \in \Sigma} \left(r\left(u^t, i_p\right) - v\right)\left(r\left(u^a, i_p\right) - v\right)}{\sqrt{\sum_{i_p \in \Sigma} \left(r\left(u^t, i_p\right) - v\right)^2 \sum_{i_p \in \Sigma} \left(r\left(u^t, i_p\right) - v\right)^2}} \tag{4}$$

$$\mathrm{CV}_{u^t,u^a} = \frac{\sum_{i_p \in \Sigma} r\left(u^t, i_p\right) \times r\left(u^a, i_p\right)}{\sqrt{\sum_{i_p \in \Sigma} \left(r\left(u^t, i_p\right)\right)^2} \sqrt{\sum_{i_p \in \Sigma} \left(r\left(u^a, i_p\right)\right)^2}} \tag{5}$$

PC, CPC and CV find similarity among users u^t and u^a. The correlated value is based on the commonly clicked item i_p. The preference levels of users u^t and u^a for commonly clicked item i_p is given by $r\left(u^t, i_p\right)$ and $r\left(u^a, i_p\right)$. $\overline{u^t}$ and $\overline{u^a}$ represents the average values of user u^t and u^a preference levels, respectively, for all commonly clicked items. Assume v to be 0.5 since preferences lie between 0 and 1.

3.3 Phase III: Finding Association Among Products

Like-minded users obtained from above phase share similar interest among click contexts and category. In this phase, the click sequence of these similar users is taken into account. Based on this click sequence, the association among the products

Table 3 Converted strong attributes (sample)

Customer ID	Item ID	View count	Duration of visit	Click context	Item category	Cart placement	Purchase
C_1	i_1	Low	Low	Regular	Generic	0	0
C_1	i_3	High	Medium	Discounted	Generic	1	0
C_2	i_6	Low	Low	Regular	Branded	0	0
C_2	i_4	Medium	Low	Promotional	Branded	1	1
C_2	i_8	High	High	Promotional	Branded	1	0
C_2	i_1	Medium	High	Promotional	Branded	1	1

is derived using support, confidence and lift measures. The products that are used to build user-product preference matrix are only those which appear in the click sequence. To find association among the products, view count and duration of visit variables are converted to categorical attributes. The variables are divided according to the percentage levels. Less than 25% are categorized as low, more than 75% are categorized as high and rest are categorized as medium. A sample of the conversion is shown as in Table 3.

After converting discrete and continuous variables to categorical, the next step is to find association among the products that are clicked by clustered users. Inference rules denoted by the notation $i_a \Rightarrow i_b$ is expressed as 'if item i_a appears in the click sequence then item i_b also occurs.' Support, confidence and lift measures as shown in Eqs. (6–8) are used to access the usefulness of the rule.

$$\text{support} = P(i_a \cap i_b) \tag{6}$$

$$\text{confidence} = P(i_b|i_a) = \frac{P(i_a \cap i_b)}{P(i_a)} \tag{7}$$

$$\text{lift} = \frac{P(i_b|i_a)}{P(i_b)} = \frac{P(i_a \cap i_b)}{P(i_a)P(i_b)} \tag{8}$$

The corresponding support of all the products that are clicked in a cluster is calculated using Eq. (6). For all pair of products having support greater than or equal to 1%, corresponding lift values are calculated using Eqs. (6–8). If the lift value is greater than 1, then the pair of products is considered to generate inference rules.

To generate inference rules, antecedent part consists of clicked product, viewed time and number of visits. The consequent part consists of another clicked product. For example, items i_a and i_b form the following association rules. Confidence of each rule is calculated based on Eq. (7).

The rule having the highest confidence level is selected as the association level among the products. From the sample (Table 4), association among item i_a and i_b is considered as 0.7. Similarly, association among all the products belonging to similar cluster is found and products having higher confidence level are sorted for the target user.

Table 4 Association rules generated (sample)

Inference rules generated	Confidence
item(i_a) and view count(low) ⇒ item(i_b)	0.7
item(i_a) and view count(medium) ⇒ item(i_b)	0.3
item(i_a) and view count(high) ⇒ item(i_b)	0.6
item(i_a) and reading time(low) ⇒ item(i_b)	0.3
item(i_a) and reading time(medium) ⇒ item(i_b)	0.4
item(i_a) and reading time(high) ⇒ item(i_b)	0.2

4 Experiments and Results

4.1 Steps for Recommender Evaluation

Following assumptions are made to test the performance of proposed RS:

Step 1: 10% of the items that are purchased are considered as unseen items by the target user.
Step 2: Preference for that item is predicted using proposed method and top-N item list is generated for the target user.
Step 3: Recommended items are examined to see whether they are purchased or not.

4.2 Evaluation Measures

'Recall' and 'precision,' as mentioned in Eqs. (9–10), are used to test the accuracy of the RS.

$$\text{recall} = \frac{\sum_{i \in X} |\text{unseen}(i) \cap \text{Top\_}N(i)|}{\sum_{i \in X} \text{unseen}(i)} \tag{9}$$

$$\text{precision} = \frac{\sum_{i \in X} |\text{unseen}(i) \cap \text{Top\_}N(i)|}{N.|X|} \tag{10}$$

where

unseen(i) items that are not clicked by user i
N number of recommended items
Top_$N(i)$ Top-N items recommended to user i
$|X|$ a number of users who has unseen items

Combined measure $F1$ mentioned in Eq. 11 is used whose higher value indicates better accuracy of the RS.

$$F1 = \frac{2 \times \text{Recall} \times \text{Precision}}{\text{Recall} + \text{Precision}} \quad (11)$$

4.3 Evaluation of Proposed and Conventional RSs

The results presented in Figs. 5, 6 and 7 justify the performance of the proposed work as compared to conventional [2] one. The similarity measures such as PC_{u^t,u^a}, CPC_{u^t,u^a} and CV_{u^t,u^a} are used to find similarity among the users after the neighborhood formation process.

A number of neighbors selected varies from 5 to 10. Also a number of items recommended to each user varies from 5 to 20 in increments of 5. As we can see, CPC_{u^t,u^a} outperforms PC_{u^t,u^a} and CV_{u^t,u^a} in all measures of precision, recall and $F1$. The highest accuracy of the selected similarity measure is obtained when the number of recommended items is 20. Table 5 clearly shows that proposed method outperforms the conventional one. The comparison among the respective similarity measures with conventional approaches can be seen in Figs. 8, 9 and 10.

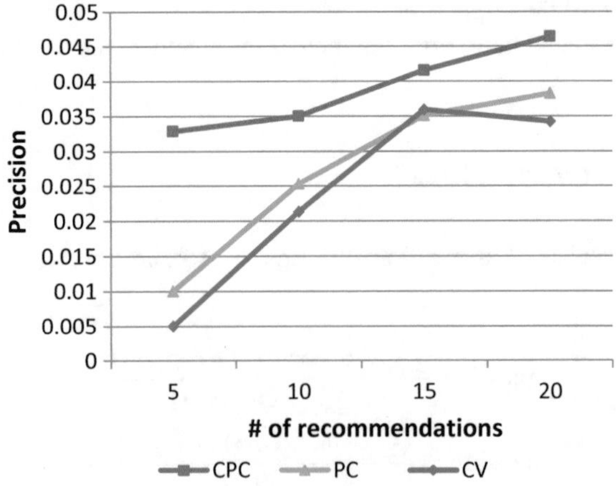

Fig. 5 Precision values among three similarity measures

Fig. 6 Recall values among three similarity measures

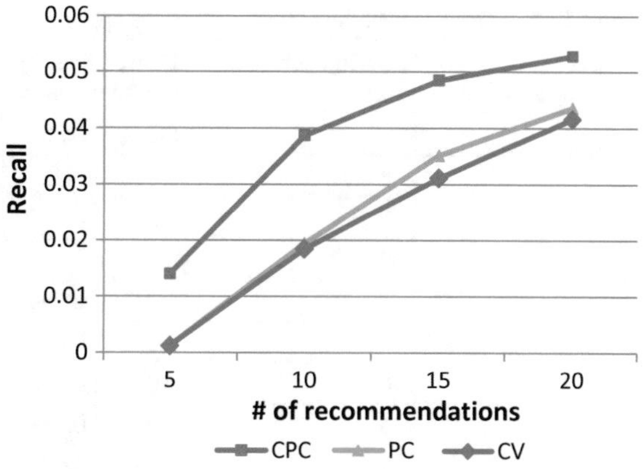

Fig. 7 F1 values among three similarity measures

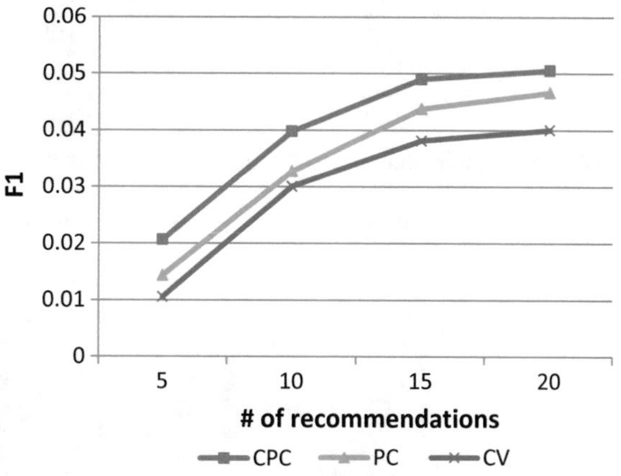

Fig. 8 Comparing similarity among users using PC measure

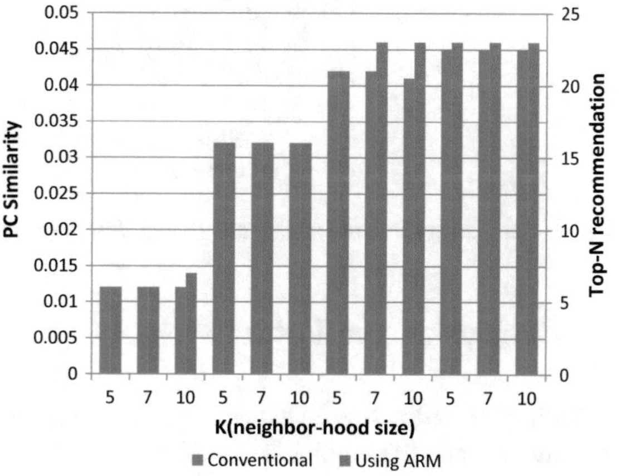

Table 5 $F1$ values: conventional versus proposed approach

Top-*N* recommendation	K (neighborhood size)	Conventional approach (clustered users)			Proposed approach (using ARM)		
		PC	CPC	CV	PC	CPC	CV
5	5	0.012	0.02	0.01	0.012	0.02	0.01
	7	0.012	0.02	0.01	0.012	0.02	0.01
	10	0.012	0.02	0.01	0.014	**0.022**	0.01
10	5	0.032	0.038	0.03	0.032	**0.040**	0.03
	7	0.032	0.038	0.03	0.032	**0.040**	0.03
	10	0.032	0.038	0.03	0.032	**0.040**	0.03
15	5	0.042	0.048	0.036	0.042	0.048	0.036
	7	0.042	0.048	0.036	0.046	0.048	0.038
	10	0.041	0.049	0.036	0.046	0.049	0.038
20	5	0.045	0.05	0.038	0.046	0.05	0.038
	7	0.045	0.05	0.038	0.046	**0.052**	0.038
	10	0.045	0.05	0.038	0.046	**0.052**	0.040

Bold digits justifies the work as it shows significant results than conventional approach.

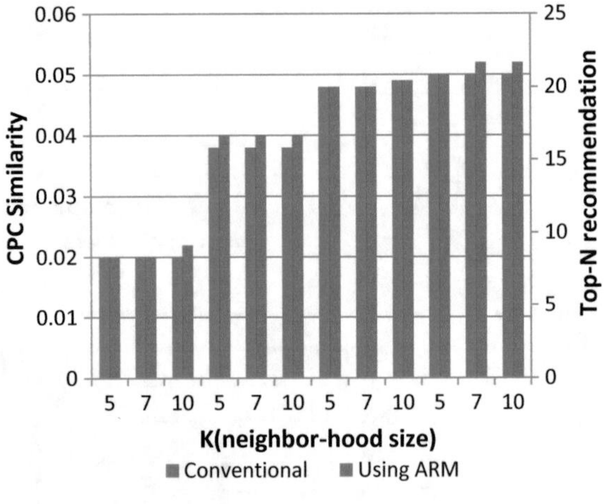

Fig. 9 Comparing similarity among users using CPC measure

5 Conclusion and Future Work

After implementing several experiments, we conclude the following outcomes. Taking into account the implicit behavior of users that are extracted from click-stream data, the strong attributes are identified. Users are clustered based on click contexts and neighbors are identified from the formed cluster. CPC gives the highest accuracy among the rest of the similarity measures. Experiments are carried out considering

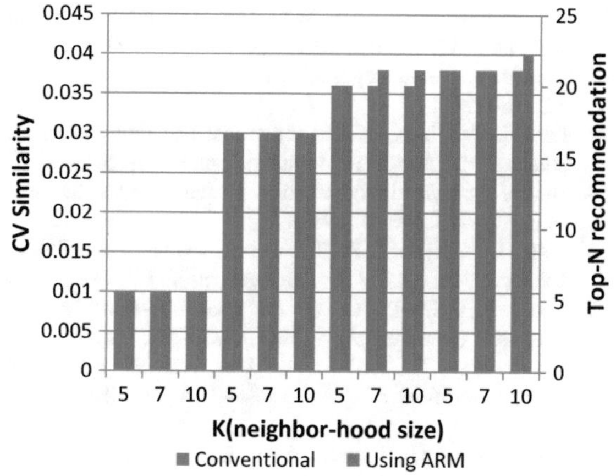

Fig. 10 Comparing similarity among users using CV measure

three values for *min-sup* as 1, 2 and 3%. The algorithm designed gives a better performance when *min-sup* is set to 1% as the number of products e-commerce Web site offers is very high. Finally, the proposed algorithm outperforms conventional approaches in all measures of precision, recall and $F1$. The accuracy of the system enhances if the implicit data captured is vast and accurate. Formation of clusters is a tedious task if number of users and context clicks is high. The above findings are based on relatively compact dataset. A fruitful area of future work may include applying proposed work on vast datasets and by finding the fuzzy associations levels among the products.

References

1. Ricci, F., Rokach, L., Shapira, B.: Introduction to recommender systems handbook. In: Recommender Systems Handbook, pp. 1–35. Springer US (2011)
2. Gupta, S., Dixit, V.S.: Scalable online product recommendation engine based on implicit feature extraction domain. J. Intell. Fuzzy Syst. **34**(3), 1503–1510 (2018)
3. Jeong, B., Lee, J., Cho, H.: An iterative semi-explicit rating method for building collaborative recommender systems. Expert Syst. Appl. **36**(3), 6181–6186 (2009)
4. Zhao, X., Niu, Z., Chen, W.: Interest before liking: two-step recommendation approaches. Knowl. Based Syst. **48**, 46–56 (2013)
5. Cleger-Tamayo, S., Fernández-Luna, J.M., Huete, J.F.: Top-N news recommendations in digital newspapers. Knowl. Based Syst. **27**, 180–189 (2012)
6. Kim, Y.S., Yum, B.-J.: Recommender system based on click stream data using association rule mining. Expert Syst. Appl. **38**(10), 13320–13327 (2011)
7. Li, Y., Tan, B.H.: Clustering algorithm of web click stream frequency pattern. J. Tianjin Univ. Sci. Technol. **3**, 018 (2011)
8. Kim, S.-C., Sung, K.-J., Park, C.-S., Kim, S.K.: Improvement of collaborative filtering using rating normalization. Multimedia Tools Appl. **75**(9), 4957–4968 (2016)
9. Sarwar, B., Karypis, G., Konstan, J., Riedl, J.: Item-based collaborative filtering recommendation algorithms. In: Proceedings of the 10th International Conference on World Wide Web, pp. 285–295. ACM (2001)

10. Belkin, N.J., Bruce Croft, W.: Information filtering and information retrieval: two sides of the same coin? Commun. ACM **35**(12), 29–38 (1992)
11. Trewin, S.: Knowledge-based recommender systems. Encycl. Libr. Inf. Sci. **69**(Supplement 32), 180 (2000)
12. Park, Y.-J., Chang, K.-N.: Individual and group behavior-based customer profile model for personalized product recommendation. Expert Syst. Appl. **36**(2), 1932–1939 (2009)
13. Burke, R.: Hybrid recommender systems: survey and experiments. User Model. User Adap. Inter. **12**(4), 331–370 (2002)
14. Kim, H.-N., Ji, A.-T., Ha, I., Jo, G.-S.: Collaborative filtering based on collaborative tagging for enhancing the quality of recommendation. Electron. Commer. Res. Appl. **9**(1), 73–83 (2010)
15. Lee, J.-S., Olafsson, S.: Two-way cooperative prediction for collaborative filtering recommendations. Expert Syst. Appl. **36**(3), 5353–5361 (2009)
16. Adomavicius, G., Tuzhilin, A.: Toward the next generation of recommender systems: a survey of the state-of-the-art and possible extensions. IEEE Trans. Knowl. Data Eng. **17**(6), 734–749 (2005)
17. Adomavicius, G., Tuzhilin, A.: Context-aware recommender systems. In: Recommender Systems Handbook, pp. 191–226. Springer, Boston, MA (2015)
18. Zheng, Y., Mobasher, B., Burke, R.: Similarity-based context-aware recommendation. In: International Conference on Web Information Systems Engineering, pp. 431–447. Springer, Cham (2015)
19. Forsati, R., Meybodi, M.R.: Effective page recommendation algorithms based on distributed learning automata and weighted association rules. Expert Syst. Appl. **37**(2), 1316–1330 (2010)
20. Ben-Shimon, D., Tsikinovsky, A., Friedmann, M., Shapira, B., Rokach, L., Hoerle, J.: Recsys challenge 2015 and the yoochoose dataset. In: Proceedings of the 9th ACM Conference on Recommender Systems, pp. 357–358. ACM (2015)

Pep—Personalized Educational Platform

Vivek M. Jude, A. Nayana, Reshma M. Pillai and Jisha John

Abstract There are many services that facilitate learning with ease. But effective learning is not limited to only one factor. As of now, there isn't a service that completely streamlines a students academics from his/her perspective without disregarding the relevance of the indulgence from the educational institution. With our project, we intend to bring about such a platform and thereby improve learning. A specific student using the application can manage his personal activities and set reminders as needed. A teacher can assign assignments and other tasks to a particular set of students or all of the students concerned. A shared database will be available to a group of students, and it is managed by the concerned teachers. Efficient assessment of the students as well as the teachers can be done, and also the personal and educational activities of the user can be managed. Educational institutions directly incorporating into the learning process can improve the results collectively. We will be using Ionic framework (built on Angular and Cordova) and Firebase as the backend.

Keywords Learning management system · E-learning · Activities · Firebase · Ionic · Real-time database

V. M. Jude · A. Nayana · R. M. Pillai · J. John (✉)
Department of Computer Science and Engineering, Mar Baselios College of Engineering and Technology, Trivandrum, Kerala, India
e-mail: jisha.json@gmail.com

V. M. Jude
e-mail: vivekmjude@gmail.com

A. Nayana
e-mail: nayana.7red@gmail.com

R. M. Pillai
e-mail: reshmampillai@gmail.com

S. M. Thampi et al. (eds.), *Intelligent Systems, Technologies and Applications*, Advances in Intelligent Systems and Computing 910,
https://doi.org/10.1007/978-981-13-6095-4_7

1 Introduction

The application allows a student to build a planner based on specific fields, and the information produced will be saved as the schedule. The student has the flexibility to edit personalized activities and can access documents from the shared database. The student will have the ability to edit or add activities. This platform will help the student to manage both their personal as well as their academic activities in an effective way. The most beneficiary of this work is the student as it will provide the required details for their studies on a digital device, and hence, the student can keep track of classes and concentrate more easily on their academic and non-academic purposes. The average student spends most of their time online, involved in social activities, and for this reason the need for an interactive student scheduler app will come to use for most of them.

The greater efficiency of an e-learning management system depends on whether or not the users are ready to embrace and receive the technology. Therefore, it is crucial for the makers of the application to understand the factors influencing the acceptance of Web-based learning systems in order to upgrade a student's learning experience. Researches have shown that e-learning is not simply a technological answer, but also a process of various factors such as social, individual, behavioral, cultural, and organizational factors [1]. Such vital elements play a significant part in how an online learning management technology is developed and used.

There are several educational platforms available of which the most common are Moodle and Blackboard Learn.

Moodle—An open-source learning management system developed to aid educational institutions manage their student's curriculum. Based on a modular design, it allows teachers to build their own curriculum using plug-ins for various workflows and activities. The users can either install a Moodle account on their servers or in the cloud [2].

Blackboard Learn—A Web-based learning management system that can be used in academic or business environments to help students and employees enhance their learning experience. The main features of Blackboard Learn include testing, assessments, discussions, and a dedicated user profile [2].

Pep is much more user-friendly and an easily accessible product as compared to other apps. It has excellent documentation, strong support for security, and administration. The Personalized Educational Platform is a student–teacher learning management system designed for both the student and teacher to reduce each other's workload. This app can be used all over the world by universities, schools, companies, independent teachers, and the students.

This application will assist the student to manage their personal activities in an effective way. If the user doesn't want to be a part of any shared database, he can use the app to schedule and manage his personal activities. Otherwise if the user is a part of a shared database, he gets to manage his personal activities along with the activities from the shared database in such a way that no clash occurs between his activities.

2 Methodology

The Personalized Educational Platform includes two applications- the "Pep" app for the students and the "Pep-Teacher's" app for the teachers.

2.1 Modules

The two modules involved are the student and teacher modules.
Student Module Anybody can register into the app as a student/user. The student can add, edit and delete their own personalized activity logs and reminders. They get notified about their pending activities and works from time to time (Fig. 1).
Teacher Module A teacher can add activities and set their deadlines. They can assign those activities to the concerned students. When a student marks an activity as done, the teacher gets notified (Fig. 2).

2.2 Features

Student App-Pep The student logs into the application to a homepage, displaying the tasks pending this month and this week. It also displays a card with an inspirational quote followed by a link to a plagiarism checker and finally the logout button. The right sliding menu has the Schedule option which navigates the student to the Schedule Page where the Pending Tasks for the day and the Done Tasks are listed as shown in Fig. 3.

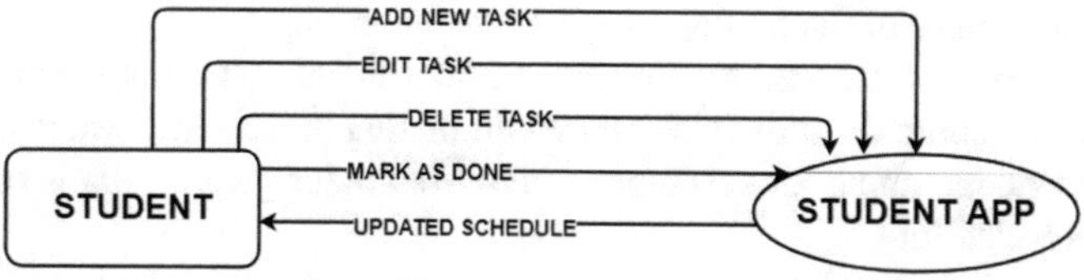

Fig. 1 Student data flow diagram

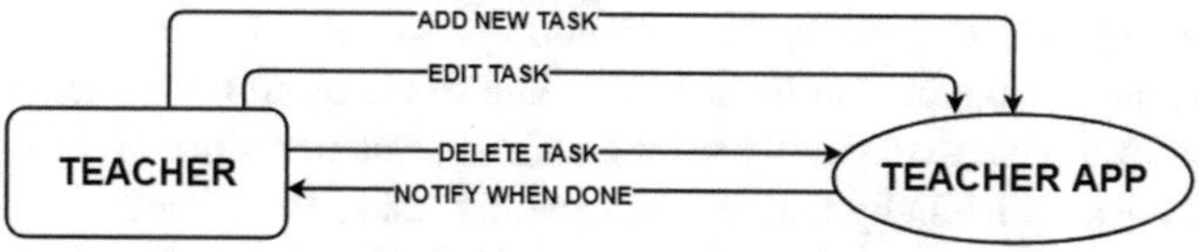

Fig. 2 Teacher data flow diagram

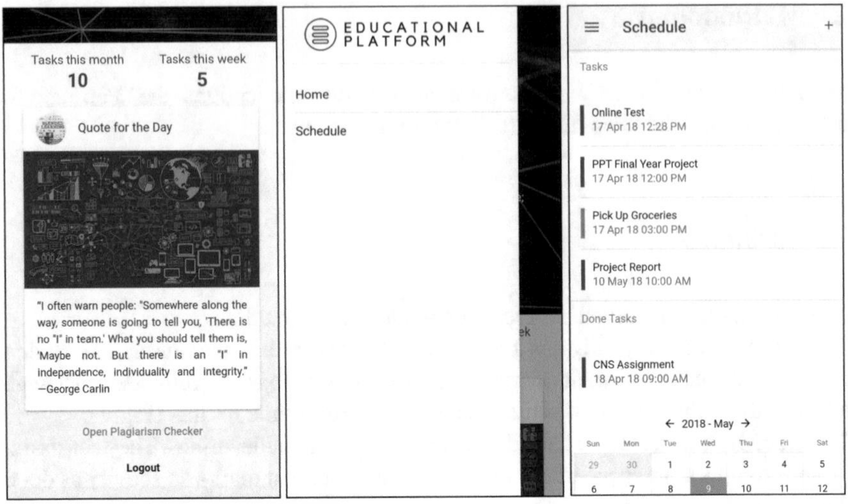

Fig. 3 Homepage, sliding menu, schedule page

The '+' button on the navbar navigates the user to an Add New Task page where the student can add personal tasks as well as academic tasks (by marking the Academic box) that he wishes to do. The personal tasks may include gym, movie, shopping etc., while an academic task may include record completion, project report, etc. He can also set whether a task is fixed or not, i.e., whether it has to be completed during a fixed time or not.

The tasks on the schedule page can be edited by selecting them. The Done box on the edit page is to be marked when the task is done thus sending that activity into the Done Task List on the Schedule Page. If the task marked done is one which was given by the teacher, then the teacher also gets notified about it. The Add New Task and Edit Task pages of the Student App are shown in Fig. 4.

When a new task is added by the teacher, the student gets a notification on their mobile devices about the added task hence eliminating the scenario where the student has no knowledge about an academic activity/reminder given. This is enabled by Firebase Notifications.

Teacher App-Pep-Teacher's The teacher logs into a page displaying all the tasks that she has given. The page has another section that appears on selecting the Done button next to All button (as shown in Fig. 5). The Done Tasks section shows the tasks which are marked done by the students.

The floating '+' button can be used to add a new task to the students. Here the teacher does not have to explicitly set the task as academic since all tasks given by the teacher are automatically marked as academic tasks. The teacher can also assign fixed tasks like online tests. Selecting a task from the Task list in the Homepage would navigate to a page where the teacher can edit the task in cases where he/she

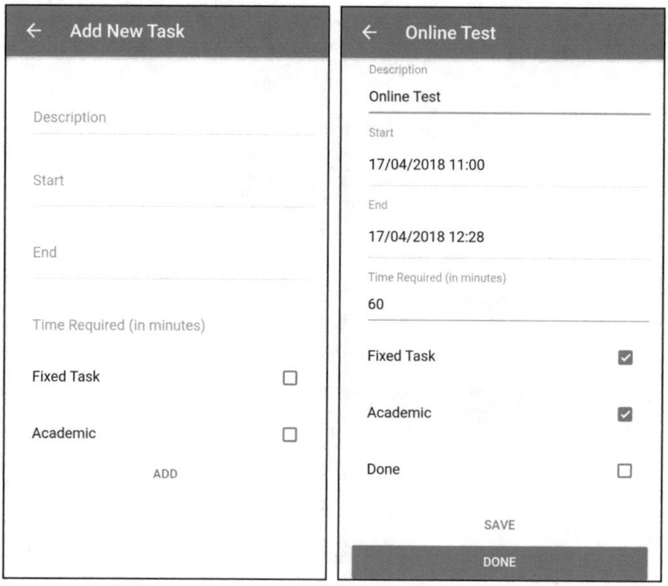

Fig. 4 Student-Add new task and edit task

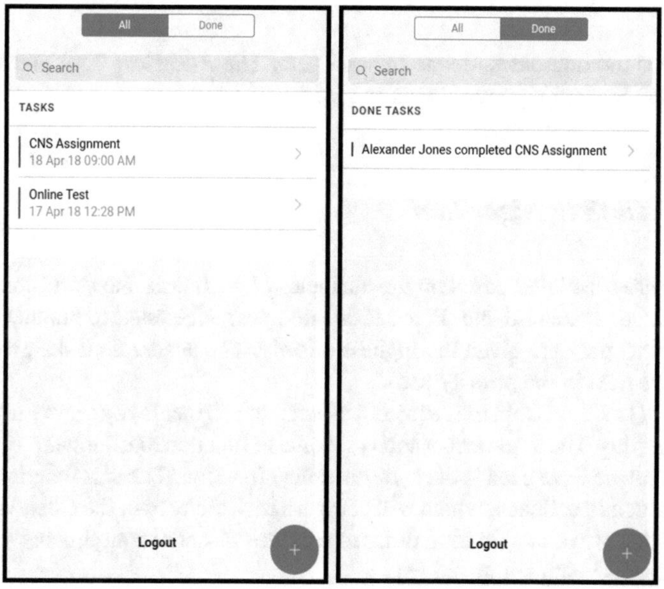

Fig. 5 All tasks and done tasks

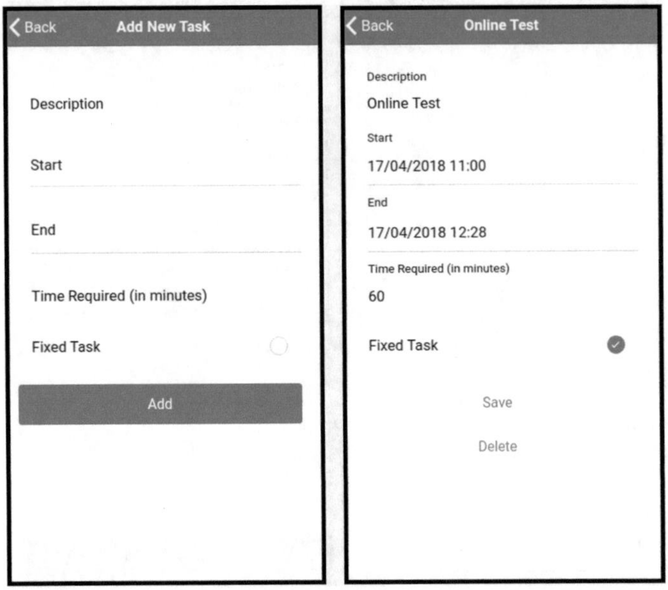

Fig. 6 Teacher-Add new task and edit task

needs to extend due dates, delete the task, etc. The Add New Task and Edit Task pages of the Student App are shown in Fig. 6.

2.3 *Scheduling Algorithm*

The tasks are scheduled based on the attributes of each task like start time, whether they are fixed, academic, etc. Fixed tasks are never scheduled to another time and to enable this they are given the highest priority. The academic tasks given by the teachers are next in the priority list.

The sort() method from the Array object sorts the elements of an array in place and returns the array. The argument of sort () specifies a function that defines the sort order. The function we have used is the sortSchedule() function. The sort function receives a comparator as a callback, which will compare each element of the Observable array with the others (two at a time) in order to sort the dataset, the arguments a and b are the two objects being compared [3].

Algorithm—**sortSchedule(a,b)**

Step 1: If the start time of task a is greater than the start time of task b, return 1.
Step 2: If the start time of task b is greater than the start time of task a, return −1.
Step 3: If both the a and b have the same start time and a is a task provided by the teacher and b is not, return −1.

Step 4: If both the a and b have the same start time and b is a task provided by the teacher and a is not, return 1.
Step 5: Otherwise, return 0.

3 Technologies Used

One of the crucial problems encountered while building a smartphone application is the multiplicative cost of developing a native application across various platforms. While front-end developers have started into the development of various hybrid platforms that would help in resolving this issue, the Ionic Framework and Firebase are a dynamic duo that together provide us amazing flexibility in building real-time smartphone applications using JavaScript and HTML5 (Fig. 7).

Ionic is a hybrid application development framework built using the AngularJS library. AngularJS is a JavaScript MV* library used for developing single page applications for Web and mobile. It is a product of Drifty, a Silicon Valley start-up. Ionic does not fail to provide any the functionalities available in the native mobile development SDKs. The users can build their applications, customize them for Android or iOS, and deploy through Cordova. Ionic is an npm module and requires Node.js [5].

Main components of the Ionic Framework are:

- A SASS-based UI framework designed and optimized for mobile UIs.
- An AngularJS front-end framework which is to develop scalable and swift applications.
- A compiler (Cordova or PhoneGap) for native mobile applications with HTML, CSS, and JavaScript.

The Ionic Framework also contains a lot of CSS components [6].

Firebase is a cloud-based platform which can be integrated into real-time applications on any Web or mobile platforms. It enables applications to write data into the database and get updates about changes immediately. It also supports offline capability and all the data is synchronized instantly as soon as the app is connected to the network. Firebase stores your data as JSON documents so that you are free to choose the schema of your data [5].

Fig. 7 Ionic and Firebase [4]

Firebase is a backend-as-a-service schema-free data system which provides real-time data syncing without requiring any custom code to be written. The multi-platform development time is greatly reduced as a result of the obsolete backend development. Firebase Storage offers secure uploads and downloads of files for Firebase apps, whatever the network quality may be and it is backed by Google Cloud Storage.

Main features of Firebase are:

- The data changes without any change in the code. All changes in the data are published to clients at once, without the need of any backend code modification.
- A wide number of adapters are available. Adapters with excellent support and documentation are present for all popular JavaScript frameworks and mobile platform SDKs.
- Greater ease of authentication. Authentication in Firebase is very simple as a single method call, regardless of the method in which the authentication is done. Firebase supports simple email and password, Google, Facebook, Twitter, or GitHub-based logins. Firebase Auth is a service that can authenticate users using only the code at the client-side.
- Enables offline mode activities. All of Firebase's data is offline-enabled, and therefore, an application can be fully (or close to fully) functional even in the disconnected mode. When the network connection is regained, the application is automatically synchronized.
- Availability of the configuration dashboard. Much of the Firebase rules can be easily configured through Firebases intuitive dashboard interface.
- JSON-centric database. In Firebase, all data is stored and retrieved in the form of JSON objects [6].

Firebase also offers cloud services for hosting the front-end code which can save a notable amount time in the deployment and maintenance of the application. It also has an ML Kit that would let us bring powerful machine learning features into the app by which we could study the daily routines of the student like his wake-up hours, lunch breaks, sleep hours etc., and schedule tasks accordingly.

4 Implementation

A small-scale version of the Pep app was deployed in our department with support from the faculty. The results collected from the students were as foreseen, as the students were 73% more likely to complete their daily curriculum work. This is also related to the fact that students tend to pay more attention to real-time notifications and thereby increase throughput. The faculty reported that they were more updated on their students curricular activity by the detail that they received immediate updates. The implementation was successful, gave more ideas and inspired for further development.

The Pep app also intended to demonstrate how the right blend of the technological set that is available at this point of time could bring about increased levels of efficiency in a field which is as crucial as education and also how such a system has not been introduced before. There is still plenty scope for incessant innovation and addition of more features to Pep as it has capabilities to be introduced as a product into the market.

5 Conclusion

The evolution of the computer systems in this digital era leads to a scenario where mobile applications are more preferred than Web applications. It would be simpler if the student can connect to the system from anywhere around the world. They will be prompted for due dates making it difficult for them to forget deadlines.

The handling of information is very easy and secured as everything is managed by Firebase Storage. A student can only view the information that concerns him. This is achieved by using Firebase Authentication where each user gets a unique id which is used to identify the user and what parts of the database they are authorized to access [7].

This app provides an optimized way of e-learning by efficiently optimizing the workflow while studying. It is a storehouse of education, information, training, knowledge, and performance management and a service that brings all of the necessary utilities under one hood which is absent at present. Not always can an educational institution actively participate in a students learning post-classes. The Personalized Educational Platform or Pep provides a streamlined approach to learning by providing an interactive space between the students and faculty even after getting home from their educational institutions.

5.1 Future Perspectives

The future perspectives of the application can include the integration of a machine learning model into it to study the daily routine of the student and to help the scheduling of tasks accordingly. This can be done with Firebases ML Kit [8].

The application can also have a shared database where the teacher can provide study materials to the students and where the students can also exchange materials among them. This would reduce the case where a student will have to search throughout their inboxes for the study materials that they received long back.

Such an educational platform can influence how we learn in many different ways and also involve institutions to bring forth more positive results.

References

1. Kattoua, T., Al-Lozi, M., Alrowwad, A.: A review of literature on e-learning systems in higher education. Int. J. Bus. Manage. Econ. Res. (IJBMER) **7**(5), 754–762 (2016)
2. https://www.betterbuys.com/lms/blackboard-vs-moodle/
3. http://4dev.tech/2016/09/angular2-how-to-sort-a-json-dataset-by-field/
4. https://medium.com/@dalenguyen/migrate-ionic-auth-to-firebase-auth-ac9491c52ddd
5. https://www.airpair.com/angularjs/posts/build-a-real-time-hybrid-app-with-ionic-firebase
6. https://www.toptal.com/front-end/building-multi-platform-real-time-mobile-applications-using-ionic-framework-and-firebase
7. https://firebase.google.com/products/auth/
8. https://firebase.google.com/products/ml-kit/
9. Machado, M., Tao, E.: Blackboard vs. Moodle: comparing user experience of learning management systems. In: 37th ASEE/IEEE Frontiers in Education Conference (2017)
10. Chourishi, D., Buttan, C.K., Chaurasia, A., Soni, A.: Effective e-learning through moodle. Int. J. Adv. Technol. Eng. Res. (IJATER) (2011)
11. Yarboi, T., Tetteh, N.: Interactive student planner application. CSUSB ScholarWorks (2014)
12. Sanchez, H., Enrique Agudo, J., Ricob, M.: Adaptive learning using moodle and handheld devices. In: Frontiers in Artificial Intelligence and Applications (2009)
13. https://firebase.google.com/docs/storage/security/user-security
14. https://www.seguetech.com/desktop-vs-web-applications/
15. https://cloud.google.com/solutions/mobile/mobile-app-backend-services

Maneuvering Black-Hole Attack Using Different Traffic Generators in MANETs

Fahmina Taranum and Khaleel Ur Rahman Khan

Abstract Malicious attacks in the network should be detected and examined by exploring its type, intension, nature, cause, and effects to handle security threats. The attack could be just for an acquaintance or with determination to harm a network. The type of attack implemented is black hole in which the traffic is attracted by node and destination is kept deprived of the transmission. The analysis is done using FTP, VBR, and CBR generators at the application layer. VBR is supported by IPv4 only. The scenario is designed using IPv4 over DYMO routing protocol as it is an advanced version of AODV for the traffic generators that supports IPv4. The traffic monitoring protocols are applied to check the packet delivery rate at the sink and in case of any possibility of leakage, the traffic is directed through some other shortest path with the assumption of availability of malicious node which needs to be eliminated from transmission as a preventive measure. Black hole usually assimilates traffic of the network around it and tries to harm the network by reducing the performance metrics of the system. The aim is to enhance different characteristics for diverting the traffics toward a black-hole node in the initial scenario, later on detecting a malicious node, preventive steps are applied, viz the node is taken in a loop of suspension and is deactivated from transmission by using a phenomenon called Human-in-the-loop. To measure performance metrics like utilization, average transmission delay, end-to-end delay, traffic analysis, and packet delivery rate are used.

Keywords DYMO-DYnamic On-demand routing protocol · CBR · FTP+ · VBR · Black hole · IPv4 · IPv6

F. Taranum (✉)
Computer Science and Engineering, Osmania University, Hyderabad, India
e-mail: ftaranum@mjcollege.ac.in

F. Taranum
Muffakham Jah College of Engineering and Technology, Hyderabad, India

K. U. R. Khan
Computer Science and Engineering, J.N.T.U.H., Hyderabad, India
e-mail: khaleelrkhan@gmail.com

K. U. R. Khan
ACE Engineering College, Hyderabad, India

S. M. Thampi et al. (eds.), *Intelligent Systems, Technologies and Applications*, Advances in Intelligent Systems and Computing 910,
https://doi.org/10.1007/978-981-13-6095-4_8

1 Introduction

The present era is an evolving telecommunication epoch, where the contrivance of emerging ideas and proposal were practiced to integrate a plethora of different inventions for producing incredible results. To meet standard escalating assail as a requirement for enhancing security ultimatum—various factors are examined as a measure to provide protection such as type of attack, intension, cause and impact of attack. The fast-changing revolution in the itinerant platform and wireless technologies has become the reason of interest for the researchers to explore and apply different adaptable strategies for preventing a system. With the enhancement in technologies, there are still many pitfalls to work on, viz limited performance, loss of data, re-routing of packets, and threat to network. The reasons for loss of packet include—no route, forcefully drop, forcefully due to aging, buffer overflow, expired TTL, queue overflow, retransmission limit, and other reasons. The delays at the transport layer for unicast transmission includes average unicast delay, average delivery delay, average jitter and average delivery jitter; the aggregate of all these are taken to calculate the delay at transport layer in the designed scenario. DYMO is reactive, on-demand, multi-hop, unicast routing protocol that do not update route information periodically and is compatible with IPv4 and IPv6 protocols. Link breakage on active routes is monitored using Hello messages. Gateway will use IP forwarding table to route packets to external network. The designed scenario includes deployment of black hole by enhancing additional privileges like increasing the antenna height and power supply for transmission of data to a node. The detection of a malicious node and its elimination from data transmissions are incorporated with DYMO. For detection, the packet-drop rate or detrimental behaved nodes are observed for final confirmation of malicious node and for prevention, the node is eliminated from transmission. The malicious node could be a hidden node or be a part of network, further it may be active or passive. The type of attack implemented is just to mesmerize the traffic and drop packets without modifying the content of packet. A node with the maximum number of packet drops and captured rate is selected as a black-hole node. Upon detection of hacker, the command of .hitl file is used to deactivate node by interacting with simulator over the socket using iterative interface during scenario execution. Traffic generators are configured and analyzed using .config, .app, .trace, and .stat files. Some protocols are used directly at application layer such as FTP and Telnet, while others are used to simulate large number of real network applications by mimicking traffic patterns, viz CBR, VBR, and dynamic CBR which are UDP-based application while FTP, FTP generic, and HTTP are TCP-based application. The following types of traffic generators are used for analysis.

- Constant Bit Rate (CBR): is a UDP-based client–server application which sends data from client to server at a constant bit rate. Parameters: <source, destination, items-to-send, item-size, interval, start-time, end-time>
- File Transfer Protocol (FTP): This tcplib application generates TCP traffic based on historical trace data and is secured in nature.

- Generic FTP: It is similar to the FTP model, which helps user for controlling traffic properties and uses FTP to transfer a user-specified amount of data in more configurable form.

The rest of the paper is organized as follows: Sect. 2 provides a brief description of related work; Sect. 3 describes about methodology; Sect. 4 discusses the result analysis; Sect. 5 portrays conclusion and future research scope.

2 Related Work

The technique proposed was to implement a wormhole attack detection and prevention using NS-2 for AODV multipath routing algorithm in [1]. The wormhole attack attracts the traffic and modifies the content of the packet before forwarding it to the neighboring nodes till it reaches the destination. The proposed approach uses some of the parameters for measuring performance metrics as is defined in this paper.

Author has used Qualnet simulator for the analyses of different wireless routing protocols using multiple parameters. The simulation was conducted using CBR and FTP traffic. The packet delivery rates are used along with other components to measure performance metrics in [2]. The other parameters used for comparison includes end-to-end delay, average jitter, packet delivery rate and throughput.

Black-hole detection and prevention scheme using DYMO is proposed to attract traffic by Nitnaware et al. [3]. The parameters are set to provide extra privileges to hacking node, viz transmission power and antenna height are used for deploying a malicious node in the network. For detection and prevention, a threshold is set to check the suspicious value for the neighboring node. Suspicious node is detected if the threshold exceeds its set value and is eliminated from transmission. The malicious node may or may not belong to the network. It may also be a hidden node receiving transmission of node in vicinity. Suspicious node advertises itself in the network by promising with the shortest path so as to attract traffic and drop packets, thereby preventing it from forwarding packets to the destination. Parameter used for result analysis includes number of nodes, terrain size, speed, and pause time. This approach is adapted in designing the scenario of our paper.

Ranjan et al. discusses different security issues of black-hole attacks in MANETs [4]. The paper focuses on a cooperative or standalone black-hole attack and finally concluded that the identification of malicious node is more difficult in case of cooperative attack. Author did a survey on types, issues, and solutions of different network attacks and concluded that security is a major concern to be provided for the network to behave in a normal manner.

Prevention of black-hole on AODV is highlighted in [5] by analyzing the sequence number of the nodes, if the destination sequence number is greater than source sequence number, then the node is declared as malicious and is removed from the route request table generated in case of proactive protocols. Performance evaluation is done using packet delivery ratio and end-to-end delay. Author has also analyzed the behavior and challenges for securing network to detect and handle threats in MANET. The formulas used to measure the performance metric have been adopted in the proposed approach from [5].

$$\text{Packet delivery ratio} = \frac{\text{Number of data packets received}}{\text{Number of data packets sent}} \times 100 \quad (1)$$

$$\text{End to End Delay} = D = \sum_{k=1}^{n} \frac{x^k}{n} \quad (2)$$

where x represents types of delays like route discovery queuing, processing at intermediate nodes propagation time, data acquisition, and n represents the number of delays considered in transmission. Formulas 1 and 2 have been adapted in the proposed scenarios along with other performance metrics.

Jaiswal et al. [6] The Author has classified nodes as of type 1, 2, and 3 based on their behavior. The analysis was conducted using AODV and DYMO for the packet-drop rate under the presence of misbehaved node under varying degrees of types. Type 1 nodes drop some or all data packets which have source- or destination-node address, but it participates in route discovery and route maintenance phases. Type 2 uses energy for its own communication; it does not participate in route discovery and route maintenance phases but acts as passive eavesdropper snooping the information in the network. Type 3 is similar to type 1 and behaves in a selective manner, i.e., it starts dropping the data packets when the energy is below the set threshold to conserve the energy with an attempt to disrupt the communication.

Agarwal et al. [7] Author has examined the performance of the network by deploying a black hole using AODV and finally deduced that its deployment reduces the packet deliver rate to a greater extent.

Tamilselvan and Sankaranarayana [8] Author has proposed an elucidation for an enhancement of the basic AODV routing protocol for avoiding black hole by monitoring the replies from all the neighboring nodes to find a safe route. After receiving the first request, it sets timer in the Timer Expired Table and stores the sequence number with the time at which the packet arrives in "Collect Route Reply Table (CRRT)". The waiting time is directly proportional to the distance of node from the source. It calculates the "timeout" value based on arriving time of the first route request, then it checks for repeated next hop node in CRRT. If any repeated next hop node is available in the reply path, then chance of malicious path is limited, thereby reducing and eliminating the possibility of malicious node.

Taranum and Khan [9] The proposal is to measure the performance analyses of the routing protocols of network-layer- and application-layer-based protocols used for IPv4 and IPv6 standards to decide an appropriate protocol for IPs. The performance metrics is analyzed using unicast-offered load, average delay, average jitter overhead, and throughput. The analysis proves that RIPng outperforms network layer protocols.

3 Methodology

Black hole disseminates transmission with the shortest path to destination which is spurious and attracts traffic toward itself with the intension to maltreat the network, thereby decreasing the performance of the network by reducing packet delivery rate, increasing transmission delay and reducing efficiency of the network.

Architecture

The scenario is designed using the hardware configuration listed in Table 1.

1. The network is created using subnets, mobile devices connected via a wireless link (blue-dotted line) with source and destination node connected with green solid line representing either CBR or FTP traffic. The traffic used is CBR, Dynamic CBR, VBR, FTP, and generic FTP. The CBR application connects source-node-1 to destination-node-16 for sending 100 data packets of 512 bytes. The scenario initially works with a black-hole node capturing traffic and on its detection; the node is eliminated from transmission in prevention phase. The routing table entries of DYMO consists of destination address, sequence number, HopCount, NextHopInterface, NextHop Address, Is Gateway, Valid Timeout, and Delete Timeout.

The algorithm to deploy, detect and prevent a network from malicious attack is shown in Fig. 2, while Fig. 1 briefs about the scenario demonstrating layerwise hardware changes incorporated. Node 3, node 17, and node 8 are gateway nodes. Node 1 does not have a direct route to target, hence it generates RREQ routing message and broadcast message with the TTL value of 1 to all the reachable neighbour till

Table 1 Parameters in the designed scenario

Parameter	Value
Qualnet version	7.4
Terrain size	1500 * 1500 m
Physical layer	802.11b Radio
Mobility model	Random waypoint
Routing protocol	DYMO
Application	CBR, FTP, VBR
Network layer	IPv4
Max speed	6 and 10 m/s
Simulation time	600 s
Pause time	0, 30 and 60 s
Frequencies	2.4 GHz
No. of channels	3
Packet size	100
Seed	1

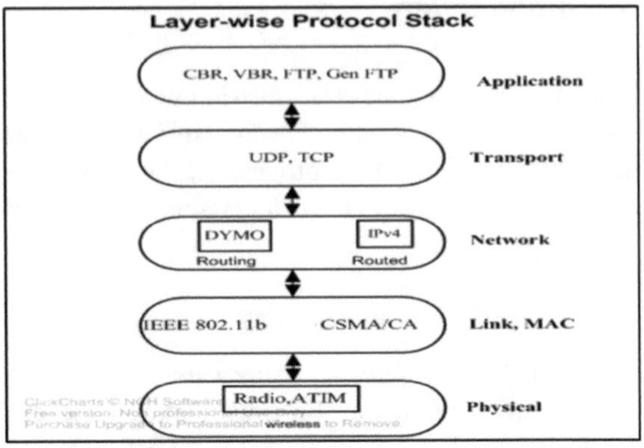

Fig. 1 Layer-wise Configuration Setting

the destination node is discovered. Node 17 is deployed as a malicious node by enhancing the hardware level features like reception transitivity at 2 Mbs is set as −69; antenna height is 30 m; maximum speed is 6 mps; noise factor is 20; buffer size for storing maximum packets to 100; maximum transmission retries to 10; and maximum hop limit to 20. The packets sent through other routes are forwarded to the destination using shortest-distance algorithm by eliminating the route containing a black-hole node. The transmission acceptance power becomes more for malicious node as the antenna height is intentionally increased to capture more packets in the designed scenario. The propagation-limit parameter sets the limit of transmission. The limit should be lower for highly sensitive low-power radio signal acceptance. Signals with power above propagation limit of −110 dBm are delivered. Therefore, to deliver a signal, the value of propagation limit is set to −110 dBm. In routing protocol properties-append own address is enabled to add the address of the node in the relayed request/reply message during transmission for active communication. Source node will continuously send data packets to the target node till the end of simulation time. The fundamental access method of IEEE 802.11 MAC used is distributed coordinated function based on carrier sense multiple access with collision avoidance for adhoc mode to enable control frames, viz request to send (RTS) and clear to send (CTS) with power-save mode set as on and the management frames selected as adhoc traffic indication map (ATIM). Each node remains awake for a short interval at the start of the beacon interval. This short interval is called the ATIM window used to save power during transmission for reducing the power consumption at the station nodes using Awake and Doze states at the station.

The detection and prevention strategies are employed for the malicious nodes to improve the performance metrics. For detection mechanism, the traffic at the destination is supervised to check the packet delivery rate. If the packets are not delivered after the set threshold, then there is some malicious behavioral node which is stopping traffic from reaching the destination. The node which attracts and stops traffic from reaching the destination may also be referred as a malicious. The aim of the prevention mechanism is to eliminate this malicious node from data transmission.

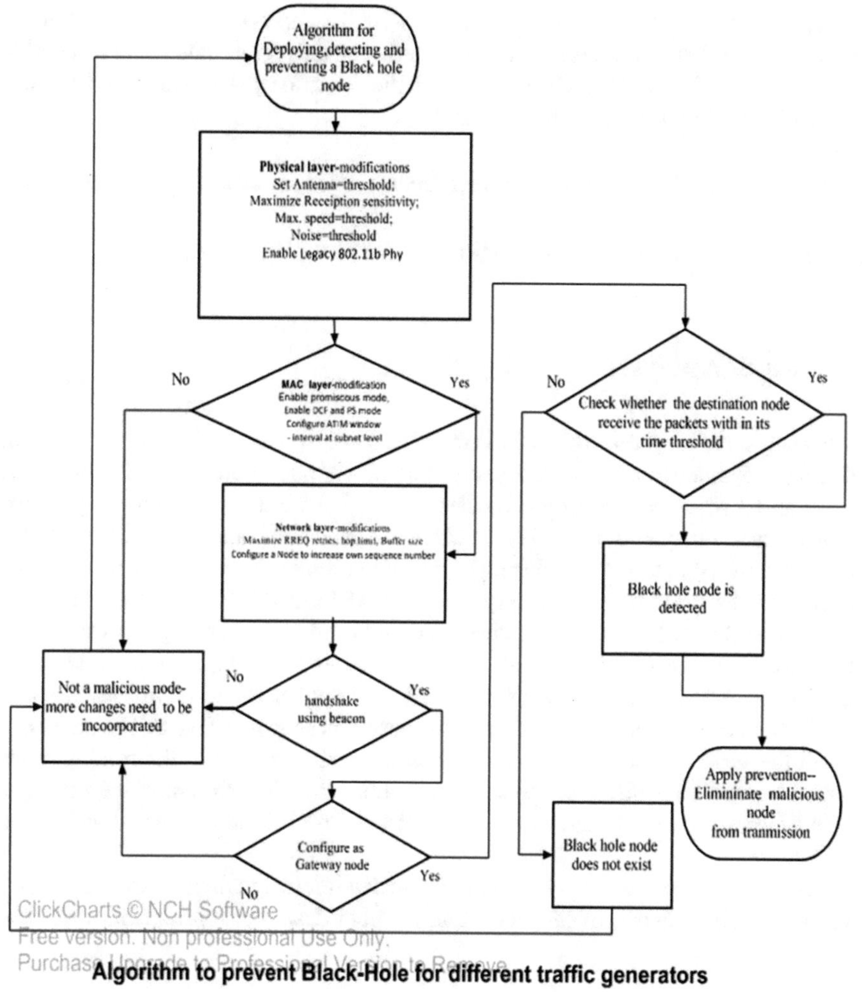

Fig. 2 Algorithm for preventing black hole

Once the behavior is monitored and a malicious node is detected, the next step is toward its elimination. A file .hitl is applied as a supplementary file for deactivating a node during scenario execution in an interactive mode. Human-in-the-loop files are used to activate deactive nodes and change traffic priorities along with the rate of CBR sessions represented by L parameter in the .HITL file.

To use this command, application layer dynamic parameters must be enabled for the scenario.

```
HITL file: 1000S D 17; 4000S A 17; 400S T 100; 450S L
.2
```

This command deactivates node 17 from 1000 s and activate it at 4000 s if the scenario last till this time, else it remains deactivated only till the end of the simulation and L changes the rate of all CBR sessions in the scenario by changing the CBR inter-packet interval to 0.02 using Eqs. 3 and 4.

$$\text{New\_inter\_packet\_interval} = \text{current\_inter\_packet\_interval} * (\text{rate\_factor}) \quad (3)$$

$$\text{New\_inter\_packet\_interval} = 0.1 * (0.2) = 0.02 \quad (4)$$

4 Result Analysis

The analysis of preventing black hole in DYMO is based on the file statistic (.stat) of qualnet 7.4 obtained after executing the scenario configuration and application files.

Figure 3 shows the red-color directed link of VBR traffic for connecting source with destination, blue-dotted links for wireless connectivity and green color links for route discovery with gray circles interpreting the propagation limit of reachability. Data transmission after the discovery process of RREQ and RREP is depicted in Fig. 4 with bidirectional links and gray circles around each node showing its propagation limit for transmission. The initial scenario of deploying a black-hole node in the network with the source, destination, CBR, and subnet is shown in Fig. 5, and a malicious node prevented from data transmission is shown in Fig. 6. The addresses of all the visited intermediate nodes are stored and appended in the routing table, which is used after discovery to select the shortest path for running the scenario. Hello messages are enabled to do the handshake for all the connected and active nodes.

As soon as the network is detected with active internal malicious attack, the prevention mechanism is applied and the traffic is not transmitted through this node. The

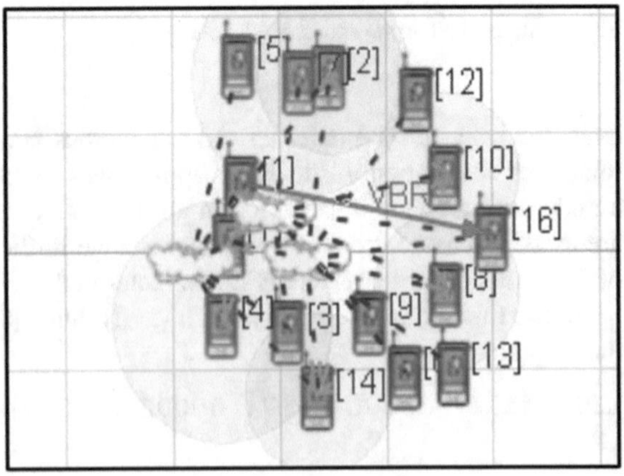

Fig. 3 Scenario with VBR traffic

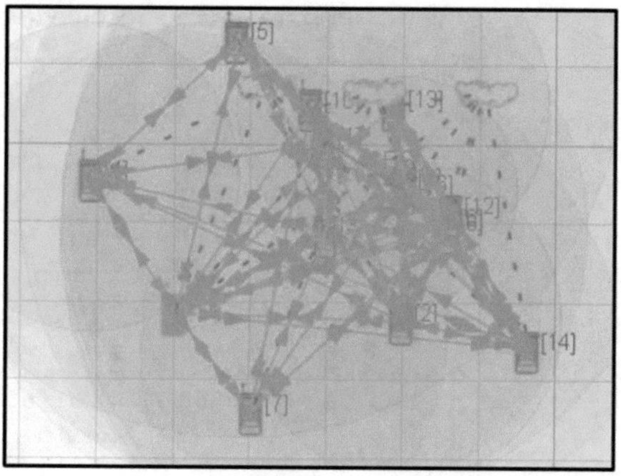

Fig. 4 Data transmission using DYMO

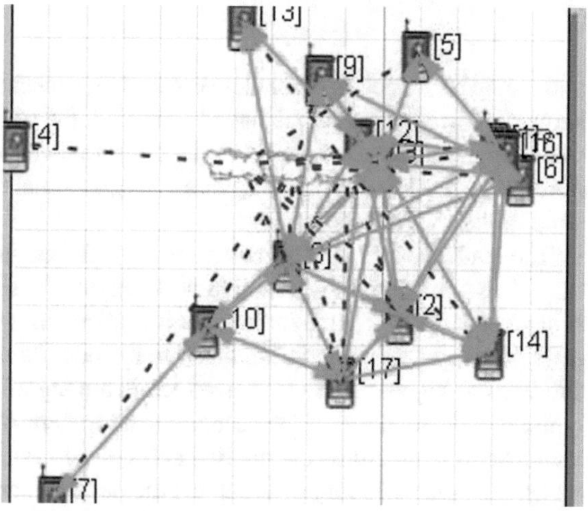

Fig. 5 Network with mobile nodes and subnet

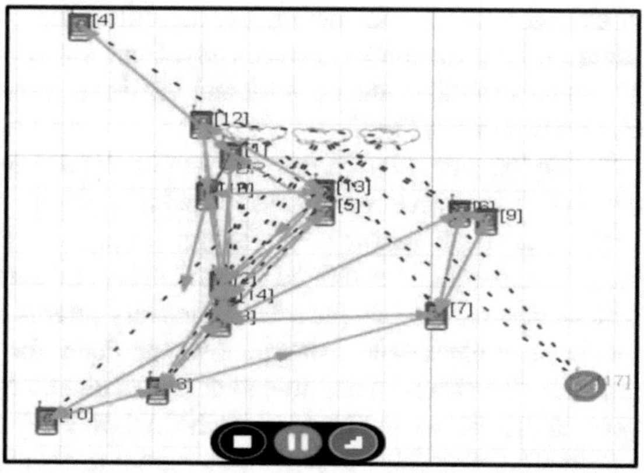

Fig. 6 Malicious node is taken into a loop

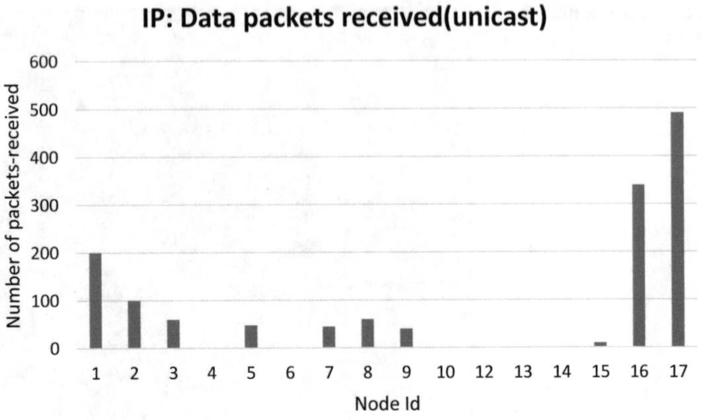

Fig. 7 Unicast data packets

main logic for detecting black hole is based on packet dropped rate at any node in the network. Once the malicious node is detected, the search for other shortest-distance route begins and selected path is used for data transmission. The phenomenon human-in-the-loop used to deactivate a node acting as a malicious node is represented with a red no-parking symbol in Fig. 6.

The statistics in Fig. 7 shows node 17 with maximum unicast packets captured rate when compared to destination node. Figure 12 shows that the packet-drop rate is highest at node 17, concluding that it is acting as a malicious node because of the extra privileges assigned to the black-hole node. Its reception probability increases and has the highest RREQ received value among all the nodes connected to IPv4 in the initial execution of the scenario used to deploy a malicious node. Therefore, the preventive measures are applied and node 17 is made deprived of transmission by deactivating it in Fig. 8. It depicts the deactivation period of node 17 as 1000 s after start of simulation time, whereas Fig. 7 with 150 s of deactivation. The rest of execution is used for deploying, detecting, and preventing the malicious nodes using Figs. 9, 10, and 11, respectively. Hence, it is proved that a black-hole node-17 imbibes traffic of the network and the destination-node-16 is being deprived of transmission from Fig. 7.

It may be noted that in the presence of attacker node, the packet received from network is less when compared to its elimination, i.e,. 1:6 in its proportion. The analysis at the transport layer for a non-black-hole scenario is compared with a black-hole scenario for jittered delivery delay and average delay as shown in Fig. 12 with an observation that the average delivery jitter is less for pause time 0 and more for 60 for a black-hole scenario. Average delay for black-hole is more for pause time 60 s. The average end-to-end delay for unicast transmission of different traffic generators is shown in Fig. 13, with the highest delay for FTP generic and lowest for CBR traffic. The delay in network terminology is used to specify the time it takes for a bit to travel from source to destination. The analysis carried out at the transport

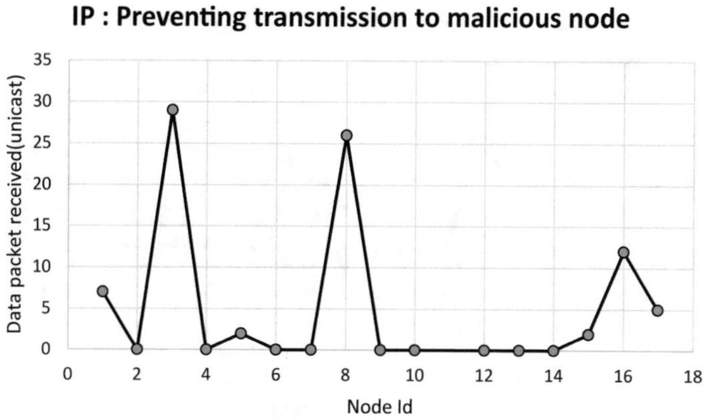

Fig. 8 Preventing transmission to malicious node

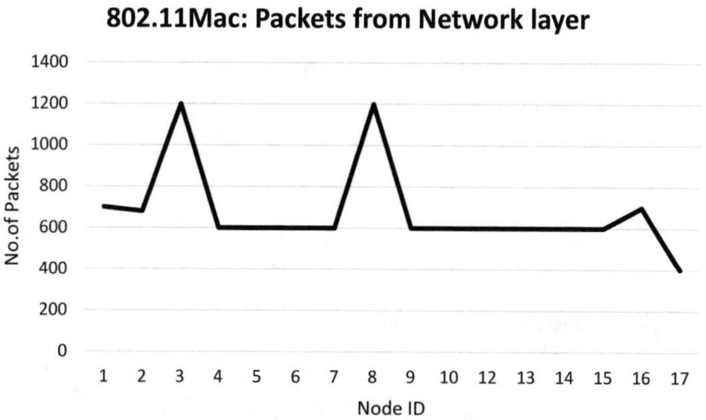

Fig. 9 Packets—network-to-physical layer

layer for unicast transmission for the metrics of black hole versus non-black hole is shown in Fig. 11, with three parameters to generate the tri-axis graph, viz black hole (BH) versus non-black hole (NBH) on X-axis; Jitters on Y-axis; and delay on Z-axis. The three colored lines in Fig. 12 are for pause times 0, 30, and 60 s.

The utilization refers to effective use of different parameters that helps in transmission to improve performance metrics. The analysis of utilization factor at the physical layer is shown in Fig. 14 testament that the utilization is maximized using prevention approach, X-axis represents Node-id and Y-axis represents the utilization factor. The standard IEEE 802.11 is used along with IPv4 protocol. Other standards of IEEE can also use to do the analysis at different layers be used to do the analysis. The transmission delays at the physical layers using IEEE 802.11b architecture for all the nodes of the network is represented in Fig. 15. The distributed coordinated function analysis at the physical layer is depicted in Fig. 16 showing that after

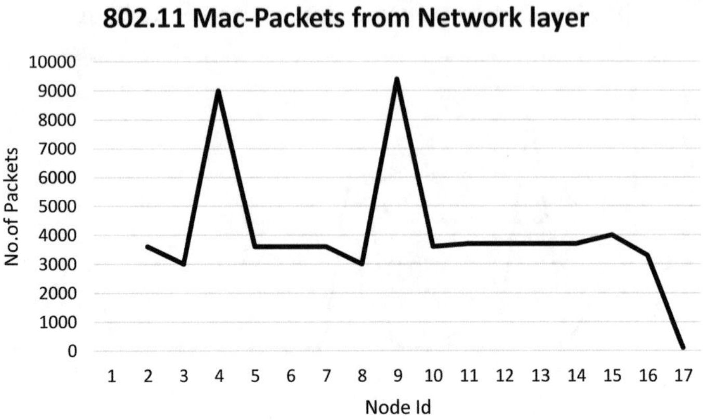

Fig. 10 Packets—physical layer after prevention

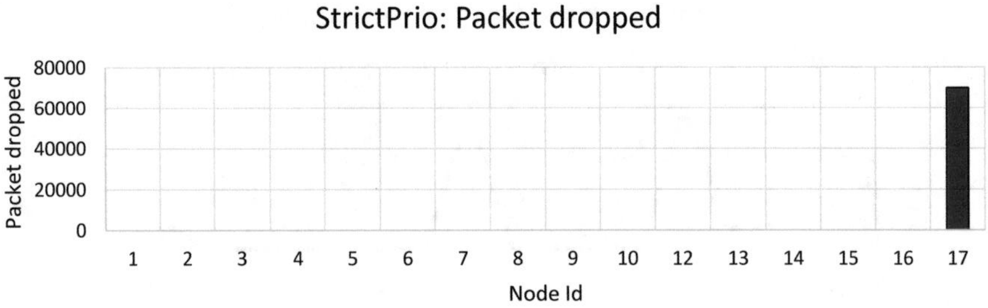

Fig. 11 Packet dropped rate at network layer

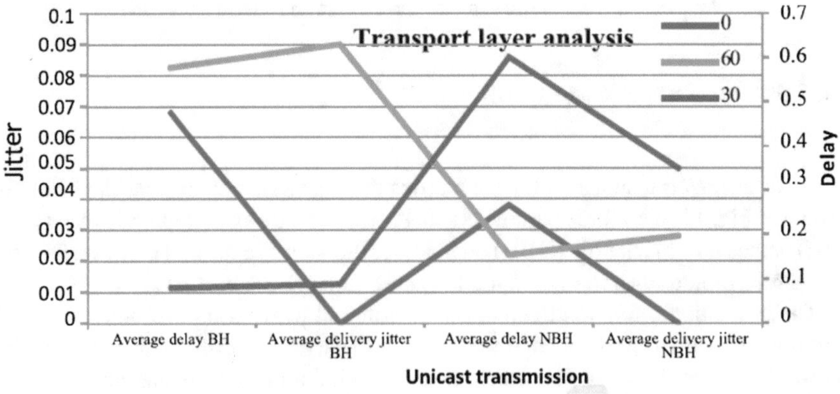

Fig. 12 Comparison with different pause times

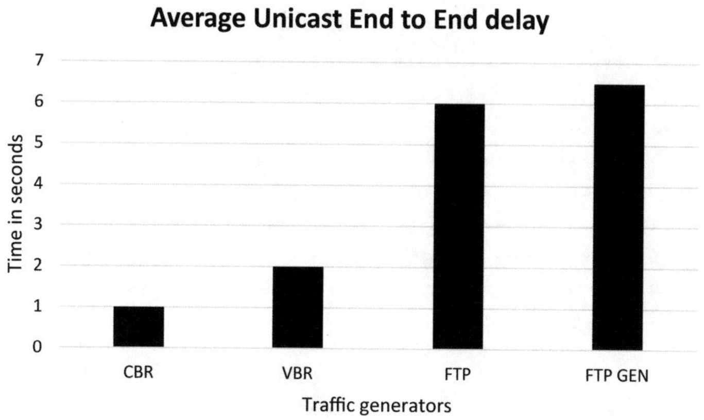

Fig. 13 End-to-end traffic generator

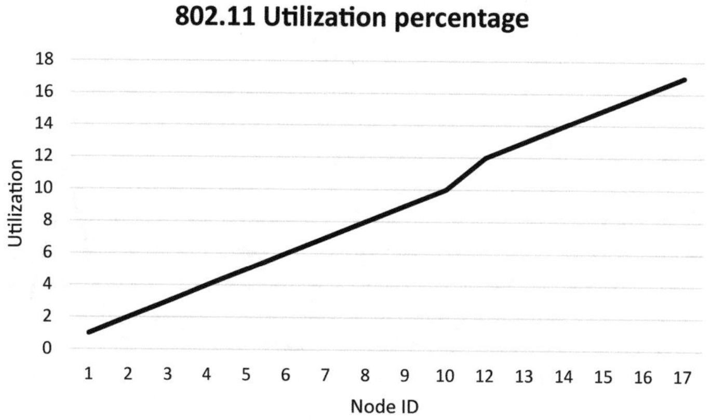

Fig. 14 Utilization at physical layer

prevention the unicast received rate at node 17 is quite less when compared to destination. The option of using point coordinated or distributed coordinated function with inter-framing space is available in this simulator at the physical layer. Distributed coordinated function (DCF) is used in adhoc mode and is being applied in designed architecture. The option to set inter-framing space and guards are available to avoid cross talking. The transmission delays at the physical layers using IEEE 802.11b architecture for all the nodes of the network is represented in Fig. 15. The distributed coordinated function analysis at the physical layer is depicted in Fig. 16 showing that after prevention, the unicast received rate at node 17 is quite less when compared to destination. The figures are generated as result of packet-tracer file. The simulation consists of all mobile nodes without any backbone to connect with different type of traffic to analyze the performances for both normal and a malicious network.

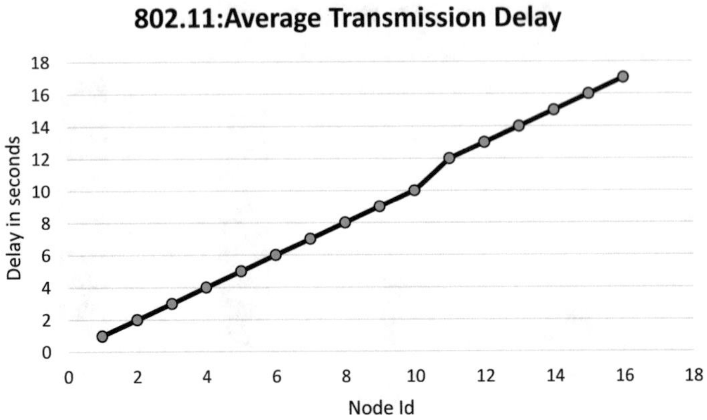

Fig. 15 Average transmission delay

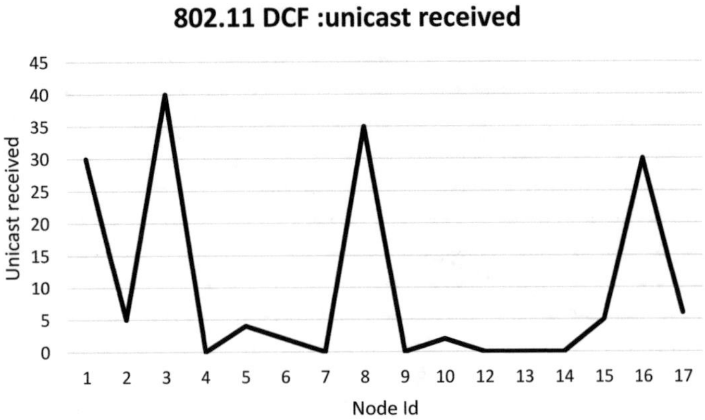

Fig. 16 DCF analysis at physical layer

The traffic sends through FTP and FTP generic in Fig. 13 is almost same. The utilization of data transmission at the destination and an attacker node is almost same in both cases at the physical layer.

Two different color lights—blue and ink blue—are used for two different interfaces used in communication.

5 Conclusions

A power aware routing algorithm—DYMO—is used to deploy, detect, and prevent a malicious attack. Different types of traffic generators are used to check the end-to-end delay in transmission. The approach used for deployment is of type 1 in which the attacker just captivates traffic and then drops the packets without any modification to the data being transmitted. Upon detection of malicious node in the network, the node is abolished from transmission. With the parameters used to analyze performance of the scenarios, it may be concluded that by applying prevention mechanism; the efficiency of the system shows an improvement to a greater extent when compared to malicious network. The analysis is done using different traffic, pause time, simulation time, speed and comparison of a non-malicious network are used.

6 Enhancement

The main challenge in MANET after deployment of black hole is to design a robust security solution to detect and protect the network from various types of routing attacks, viz wormhole and grayhole. The solution to handle hidden and cooperative attacks can also be explored for other types of attack using different approaches.

References

1. Amish, P., Vaghela, V.B.: Detection and prevention of wormhole attack in wireless sensor network using AOMDV protocol. In: 7th International Conference on Communication, Computing and Virtualization, Science Direct Procedia (2016)
2. Kumar, J., Singh, A.: Study and performance analysis of routing protocol based on CBR. Science Direct, Procedia (2016)
3. Nitnaware, D., Thakur, A.: Black hole attack detection and prevention strategy in DYMO for MANETs. In: 3rd International Conference on Signal Processing and Integrated Networks (SPIN) (2016)
4. Ranjan, R., Singh, N.K., Singh, A.: Security issues of black hole attacks in MANET. In: International Conference on Computing, Communication and Automation, IEEE (2015)
5. Ghonge, M., Nimbhorkar, S.U.: Simulation of AODV under blackhole attack in MANET. Int. J. Adv. Res. Comput. Sci. Softw. Eng. **2**(2) (2012)
6. Jaiswal, P., Kumar, R.: Prevention of black hole attack in MANET. Int. J. Comput. Netw. Wirel. Commun. **2**(5), 599–606 (2012)
7. Agarwal, S., Sanjeev, S., et al.: Mobility based performance analysis of AODV and DYMO under varying degree of node misbehavior. Int. J. Appl. **30**(7), 36–41 (2011)
8. Tamilselvan, L., Sankaranarayana, V.: Prevention of blackhole attack in MANET. J. Netw. **3**(5), 13–20 (2008)
9. Taranum, F., Khan, K.U.R.: Performance analysis of routing protocol based on IPv4 and IPv6. In: International Conference on Innovative Technologies in Engineering, April (2018)

Variant of Nearest Neighborhood Fingerprint Storage System by Reducing Redundancies

K. Anjana, K. Praveen, P. P. Amritha and M. Sethumadhavan

Abstract Biometric security is really important when it is the case of proving a individual's identity. Fingerprint, iris, face, and gesture are the main biometric technologies. Fingerprint is the most convenient biometric which is used for proving an individual's identity. Minutiae are said to be the unique representation of a fingerprint. There are different schemes in the literature for efficient storage of minutiae. Recently, a binary tree-based approach for efficient minutiae storage was proposed in the literature by removing the redundancies. We found out that the existence of redundancy in nearest neighborhood method reduces the efficiency. In this paper, we propose nearest neighborhood method by reducing redundancies for better efficiency. Comparative study of these proposed systems with existing scheme is done. As a result, we found out that, even though the complexity of algorithm is high, storage will be efficient.

1 Introduction

The uniqueness of a fingerprint is being extracted mainly using two features called ridge ending and bifurcation. A ridge ending is abrupt end of a ridge, and bifurcation is a single ridge that divides into two ridges. We can find out the minutiae points by analyzing the fingerprint patterns. Each fingerprint pattern will contain different data such as position, distance, direction, and type, and from these data, we will generate

K. Anjana · K. Praveen (✉) · P. P. Amritha · M. Sethumadhavan
TIFAC-CORE in Cyber Security, Amrita School of Engineering,
Amrita Vishwa Vidyapeetham, Coimbatore, India
e-mail: k_praveen@cb.amrita.edu

K. Anjana
e-mail: anjanakodoth@gmail.com

P. P. Amritha
e-mail: pp_amritha@cb.amrita.edu

M. Sethumadhavan
e-mail: m_sethu@cb.amrita.edu

S. M. Thampi et al. (eds.), *Intelligent Systems, Technologies and Applications*,
Advances in Intelligent Systems and Computing 910,
https://doi.org/10.1007/978-981-13-6095-4_9

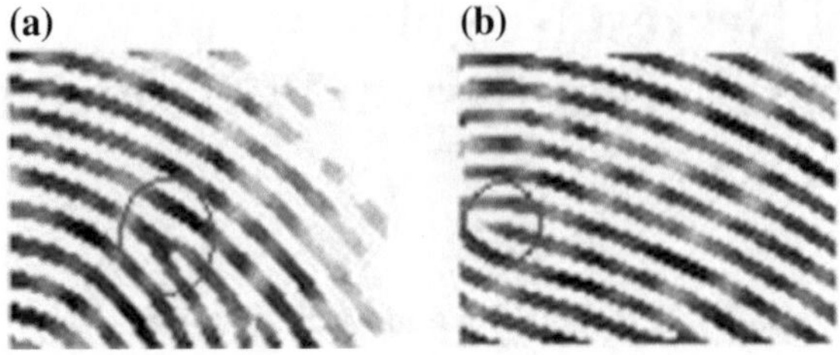

Fig. 1 Minutiae features **a** Ridge ending. **b** Bifurcation [22]

a unique template for each fingerprint. Apart from the features of a fingerprint, the uniqueness depends upon the way, how they are being stored. There are different minutiae representations like pattern based, pixel based, ridge based, geographical structures, and skeleton structures. Apart from ridge ending and bifurcation, there are some more features which can be included are mentioned in future work. Ridge ending and bifurcation are shown in Fig. 1 [22].

Galton [7] introduced the study of classification of fingerprint for proving individual identity by inventing roller machine. Arch, loops, and whorls are the three main classifications of fingerprint patterns according to Galton's study. Using the roller machine, people can take their fingerprint along with full palm. The very first method of taking fingerprint for storage was using a glass sheet and little oil on the finger. Fingerprint impression will be taken into glass sheet after applying a little oil on fingertip. Recognition of this fingerprint is done by looking at the inclined glass sheet according to the sunlight. Henry added some extension to Galton's classification mechanism. In Henry's mechanism [9], each finger is being identified by a number. If a finger is identified with a whorl pattern, then an additional score will be added to number which corresponds to that finger. Another way of uniqueness extraction is by analyzing the relationship of local ridges in a fingerprint [13, 16]. Kawagoe and Tojo have identified more local ridge patterns in fingerprint for feature extraction [16]. But the problem is, most of the features they identified were depending upon the clarity of the image. These features can only be extracted if the image having a high quality which is not commonly available in fingerprint images. From position of each minutia point, some other features like distance and orientation can be calculated. These derived attributes can also be used to represent the uniqueness of fingerprint [6, 23]. Level 1, Level 2, and Level 3 are the main three classifications of the fingerprint in distinctive levels [23]. Level 1 and Level 2 are different in characteristics, but Level 3 is the combination of Level 1 and Level 2 (Fig. 2).

1.1 Structural Method

Extracting the features from the fingerprint is an important part of fingerprint recognition system. Accurate extraction and good storage mechanism will give a better

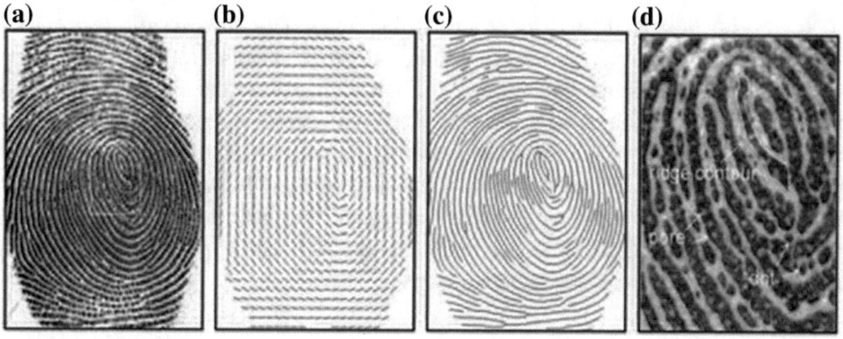

Fig. 2 Three levels of features in fingerprint **a** Binary fingerprint image. **b** Level 1—orientation field. **c** Level 2—ridge skeleton. **d** Level 3—ridge contour, pore, and dot [6, 23]

fingerprint recognition system. Moayer and Fu [19] introduced method for the structural approach. First, a fingerprint image will be taken, and then preprocessing will be done. These preprocessed images will be divided into different blocks. The pattern is in image format, and corresponding literal in alphabet is given in a predefined table. Examples are as follows: Left curve in left side will be represented using literal *A*, and right curve in right side will be represented using *B*. Maio and Maltoni [17] introduced a new structural method which is combination of distinct grid method [5] and dynamic clustering algorithm [17]. Firstly, computation of directional image using distinct grid is done. Patterns are divided into macro patterns. Then, the development of the relational graph will be done. For the maximum matching of macro patterns with the patterns in the image, dynamic clustering algorithm will be used. Because then only we will get the maximum matching in macro patterns in a fingerprint image, matching a pattern in an image and predefined pattern will be tough without a machine learning algorithm. In Wahab's method [21], feature extraction is done in the way that original image of size 320 × 240 is divided into piece of 40 × 30. We have eight directional codes to represent the direction of the ridges in the small area. Calculation of the direction with the use of directional code is less time consuming. Each 40 × 30 pixel area which have any type of feature is compared with directional images. From each comparison, a fluctuation value will be calculated. Fluctuation value is the mean value of the grey level of the pixels in directional image and each piece of original image. Degree value corresponds to the largest fluctuation value which is taken as the direction for that feature.

1.2 Ridge Orientation Method

In ridge orientation approach, ridge structures are taken as the feature. Problem with this method are creation of spurious minutiae and loss of actual minutiae. This is happening because when the image having very poor quality, blurred image, and image of damaged finger is given as input to feature extraction. Even clear image of fingerprint will lead to an addition of the ridge. This also results in a wrong localization of the pixels, so ridge orientation approach needs a image enhancement step for the proper feature extraction [20]. Binary images are the images where ridge pixel in a fingerprint image is represented using white color and non-ridge pixel using black. We can convert a grey-level fingerprint image to a binary image using a ridge extraction algorithm. The clarity of the ridge structure in a binary image will depend upon the efficiency of the ridge extraction algorithm, but in some cases original data can also be lost. Fast fingerprint enhancement algorithm [10] was introduced to improve the properties of fingerprint image.

1.3 Pixel-Level Method

Abutaleb and Kamel [1] used grayscale fingerprint image for feature extraction. In pixel-level approach, the parameter used for feature extraction is the width of edge and non-edges. A genetic algorithm was introduced for generating global features [4]. Kaur [15] introduced a new method of feature extraction by dividing the image into different blocks. In this method, blocks are taken separately and processed. Gammasi [8] and Alibeigi [2] proposed a square base method, where a pixel is chosen and a square around that pixel is considered. After that, average of the pixel value in square is calculated. Using this average value, we will be extracting feature. Using this square method, orientation direction of the ridge ending is also calculated.

1.4 Geometric Method

Min and Thein [18] proposed geometric approach, which is the combination of geometry and statical approaches. In each fingerprint, core point will located. Segmentation will be done to the fingerprint image, so that foreground and background can be separated. After that, core point of the fingerprint will be detected. In the matching process, extracted feature that is stored in the database is compared with the extracted feature of the current fingerprint image. The next idea of storing the fingerprint is storing the topological relationship between minutia point. In 5-nearest neighbor (NN) method [25], type of each minutia in template will be taken, and then this point will be connected with its 5 nearest neighbors. Distance between this parent and corresponding five nearest points and the type of minutia are the two properties used to indicate each neighbor. Jain [12] introduced a method of minutiae storage. In this method, nearest neighbor vector (NNV) for each minutiae points is being stored. NNV for each center minutiae will be stored as unique representation like (x, y, θ) and $(D, \alpha, \beta, \gamma)$. Kannavara and Bourbakis [14] introduced another method of minutiae storage as centroid global graph. Centroid of the fingerprint will be connected with the mean value of starting point and ending point of each ridge. The Euclidean distance between the centroid and each mean point will be stored as the unique representation.

1.5 Multichannel Method

Anil Jain and Salil Prabhakar introduced a new method [11] for minutiae storage. In this method, central point of fingerprint is detected and a circular region around this point is taken as interested area. This circular area is then divided into sectors. Mean and variance of each sector calculated using Gabor filter are stored as feature. The fingerprint template is called finger code. $I(x, y)$ is the grey level of a pixel (x, y) where (x_c, y_c) is the center of fingerprint. Center point detection algorithms are done based on the detection of a point which have maximum curvatures. Normalization is done to the region of interest so that pressure different in fingerprint images and noise because of dirt can remove. Then, Gabor filter is used for filtering the normalized image in different orientation. Filtered sectors will combine together to form a 192 dimensional finger code. In filtering, noise will be removed and some enhancement will be done to ridges and furrows. Here, we have five classes of fingerprint patterns. Zhao and Tang introduced a new method [24] where fingerprint valley is being used for the feature extraction. Rutovitz crossing number is used to extract minutiae features from this thinned image. Rutovitzs crossing number CN can be defined as. $\mathrm{CN} = \frac{1}{2}\sum_{i=1}^{k} |p_i - p_{i+1}|\ where\ P_i{-}Binarypixelvalue.$

1.6 Tree Structure Method

A method of minutiae storage using tree representation is proposed in the paper [3]. In this method, center point of the fingerprint is taken as the root of a binary tree. Then, two nearest neighbors of root will be found out. Then, two nearest neighbors which are not inserted into the tree will be selected for each parent and inserted as the child. This process will continue to the end of minutiae points. Nearest neighbor is selected based on the Euclidean method. *POS* is a data structure used to store position and type of minutiae. *DIST* is data structure which stores distance between each point in the minutiae set. *S_DIST* is the sorted *DIST*, and it is sorted based on the distance and all other detailed like position and type of minutiae points stored accordingly. TREE is the data structure which holds output, first entry will be center point and all other minutiae points are arranged as in binary tree. Algorithm for insertion into binary tree is given below [3].

Input: Minutiae points with features *POS[1, …, n*], Total number of minutiae points, *n*.

Output: Array of tree structure, *TREE*.

1. Start.
2. Add central point to minutiae points to *POS*.
3. Find distance between each point in the *POS* in *DIST*.
    a. $\sqrt{(x_1 - x_2)^2 + (y_1 - y_2)^2)}$ where (x_1, y_1) and (x_2, y_2) are any two points (Euclidean method).
4. Sort the distance matrix *DIST* to *S_DIST*.

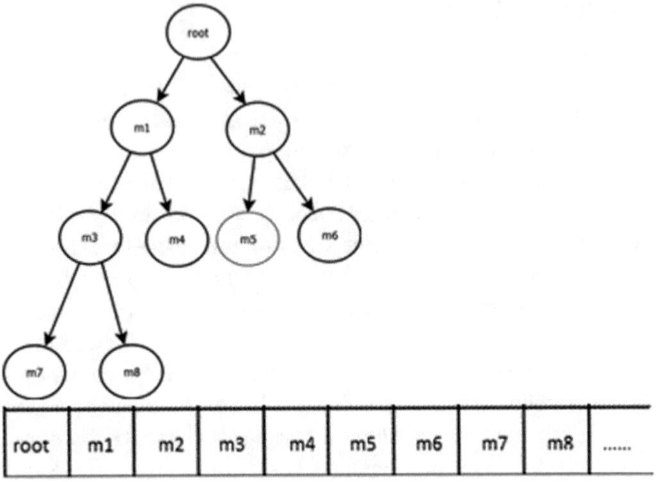

Fig. 3 Tree representation of minutiae and its corresponding array representation

5. Insert central point to TREE.
6. For $i = 1$ to $\frac{n}{2}$
   a. Find two children which are in minimum distance from ith node
   b. If the position corresponding to children is already in TREE, search for next minimum child.
   c. Insert children into $(2 \times i)$th and $(2 \times i + 1)$th position in TREE.
7. Stop.

The output of this algorithm will be in a tree structure. Details of minutiae will be in a sequence. The tree will be represented using its corresponding array as shown in Fig. 3. If parent is in ith position of the array, then child will be on 2 × i and 2 × (i+1). If $m1$, $m2$, $m3$, …are the children, then they are represented using tree structure and the corresponding array is used store data in database.

1.7 5 Nearest Neighbor Method

Five NNs is a geometrical way of storing minutiae [25]. In this method, minutiae points are stored in a graphical structure. $X = \{\alpha, (\beta, \gamma)_i$, where $i = 1, 2, 3, 4, 5\}$ is the representation of 5 NNs using set notation. Here, α is the type of minutiae points in fingerprint. β is type of 5 nearest neighbor, and γ is the distance between minutia and its five nearest neighbors. An array representation of 5 NNs in shown in Fig. 4. Here, distance between minutia and all five nearest neighbors and its type are stored regardless of duplication check.

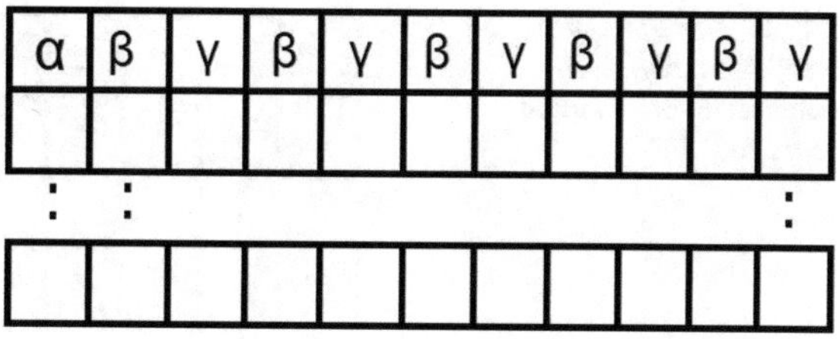

Fig. 4 Five nearest neighbor array representations

2 Proposed System—Reducing redundancy in *K*-Nearest Neighbor method

The proposed system is a method to store the minutiae points. Here, we use type of minutiae points and distance as unique features. A standard algorithm is used for minutiae points extraction. Then type of minutia, type of its K (=5)-nearest neighbor minutia point and distance between minutia point and its K (=5)-nearest neighbor are stored by reducing duplicates. Time complexity of K (=5) nearest neighbor method is linear. But memory requirement is very large. Memory requirement is large because K (=5) NN stores type of each minutia along with type and distance of its K (=5)-nearest neighbors without duplication check. Binary tree along with feature bit [3] is free from duplication. So applying this duplication elimination in K (=5) NN will give K (=5) NN with less memory requirement similar to binary tree. Graphical representation of this idea is shown in Fig. 5. Depending upon the distance between minutia and its neighbor, number of neighbors may vary from zero to five. For the first minutia in the fingerprint, the number of nearest neighbor will be five. It may reduce to four, three, two, one, or zero. There is no connection between minutia point which is not in set of its 5 nearest neighbors. Minutia point with no child (isolated minutia) is shown in Fig. 5 as an array with only entry, α. So the data structure created will be a forest. So in order to represent each minutia point with at least one nearest neighbor, we can add type of nearest neighbor and distance between nearest neighbor and minutia point as the single child of minutia point as given in Fig. 6. The minutiae representation graph is better than forest because graph will have connec-

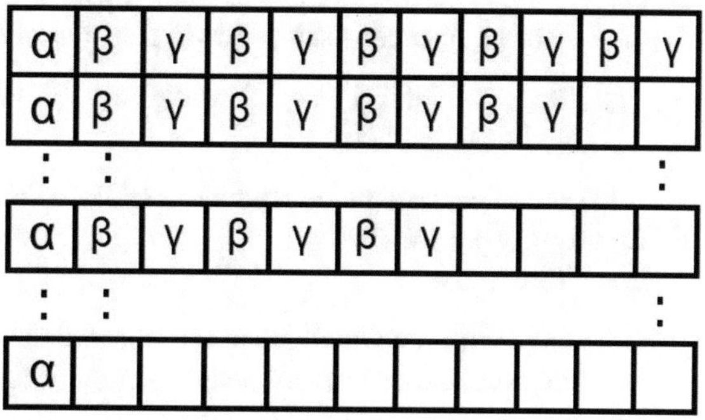

Fig. 5 K (=5)-nearest neighbor array representation—Forest (with isolated point)

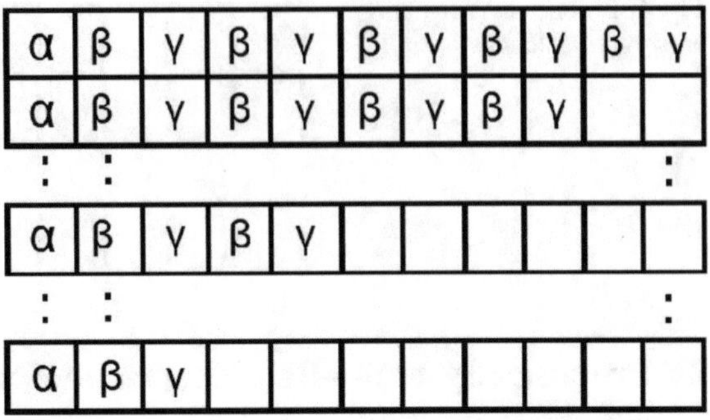

Fig. 6 *K* (=5)-nearest neighbor array representation—without isolated point

tion between minutia point which implies that for every minutia point, neighborhood information is stored unlike forest. So we designed a algorithm which will insert at least one child for each minutia point to ensure the connectivity between minutia point. *K*(=5)-nearest neighbor will be inserted as the neighbor for first minutia point. For the second minutia point, we will insert most nearest neighbor as the first neighbor. Then for second, third, fourth, and fifth neighbors, duplication check will be done. After duplication check, the number of neighbor may change. If one neighbor of second minutia is already inserted as the one of the neighbors of first minutia, then check that neighbor is nearest or not. If it is nearest, then insert that neighbor as first neighbor else that neighbor would not be inserted as neighbor for second minutia point. *POS* is the data structure with position and type of each minutia point. Even though we are storing distance and type as feature in proposed system, we need an extra data structure *INSPOS* for duplication check. *INSPOS* stores position of minutiae points which are inserted in to *TEMPLATE*. *TEMPLATE* is the output template where all distance and type are stored.

Inputs: Minutiae points with features (*POS[1, ..., n]*), Total number of minutiae points (n) and $K > 1$.
Output : Array of unique representation, *TEMPLATE*.

1. Start.
2. Set *K* (=5). (This algorithm is explained with *K* as 5.)
3. Find distance between each points in the *POS* in *DIST*.
    a. $\sqrt{(x_1 - x_2)^2 + (y_1 - y_2)^2)}$ where (x_1, y_1) and (x_2, y_2) are any two points (Euclidean method).
4. Add type of each minutiae points to *TEMPLATE*.
5. Create an array *INSPOS*.
6. For i from 1 to n
    a. Find 5 children which are in minimum distance from ith node.
    b. Add type and distance of nearest neighbor to *TEMPLATE*.

   c. Check whether position corresponds to second, third, fourth, and fifth nearest neighbors are there in *INSPOS*
      i. If yes, do not insert type and distance in *TEMPLATE*.
      ii. If no, insert type and distance in *TEMPLATE* and position in *INSPOS*.

7. Stop.

3 Comparison

In $K(=5)$ NN [12], the details like type of minutiae points and distance between minutiae points are being stored. So, complexity of the algorithm will be in the order of n only. But the problems with this method are entry duplication, because of entry duplication which lead to huge memory storage. Also, this method is variant to rotation. Any change in image translation will reflect the uniqueness of fingerprint. Next method is parse tree method [19]. In this method, fingerprint image is divided into block, and then these blocks are compared with predefined pattern—literal table and literal sequence will be generated. Based on this literal sequence, a grammar will be generated and this grammar will be stored as unique representation. For storing a grammar, only less memory is required. But grammar may be ambiguous, so storing as grammar is not good solution. NNV method is rotation invariant because orientation degree is stored as a feature. The disadvantage includes difficulty in location identification and high memory requirement. In central method, core point is calculated, and then distance between core point and average point is stored. The average point calculated would not be unique, but memory required will be less. General equations used to calculate memory used for different minutiae storage mechanism are given below. If the number of minutia point in fingerprint image is n, then for ***K*(=5) NN method**: $2 \times 5 \times n + n = 11 \times n$, $11 \times n$ for storing type of parent along with type and distance between parent and child of $K(=5)$-nearest neighbors for all n minutia point. **4 NNV method**: $3 \times n + 4 \times n = 7 \times n$, $3 \times n$ for storing position and orientation and $4 \times n$ for Nearest Neighbor Vector. ***K*(=5) NN method extended**: $2 \times n + n + 2 \times n = 3 \times n$, n for storing type of minutiae and $2 \times n$ for storing type and distance of first neighbor and $2 \times n$ for storing type and distance of second, third, fourth, and fifth neighbors. **Binary tree with feature bit**: $2 \times n + n = 3 \times n$, n for storing feature bit and $2 \times n$ for position. The 5-nearest neighbor system will store type and distance between parent and each five nearest neighbor in minutiae set [12]. $K(=5)$-nearest neighbor will have an $\mathcal{O}(n)$ complexity because we can directly store type and distance. The disadvantage of this method is, this method needs a large memory for storage. If we have 100 minutiae in an image, 100×11 memory unit will be needed for unique representation. This is because $5 \times 2 + 1$ (each nearest neighbor will need two memory unit, one for distance and one for type + one for storing the type of parent). Extending this idea by removal of the redundant entry in minutiae set is done. Reducing redundant entry is done by selecting 5 nearest neighbors for each minutiae points, and then delete the duplicate entry from the nearest neighbor

Table 1 Comparison of minutiae storage

Method	Advantages	Disadvantages
5 nearest neighbors	Complexity is in order of n	Entry duplication and rotational variant, high memory
Parse tree	Comparatively less memory needed	Ambiguous grammar may generated
NNV-4 neighbors	Rotation invariant	Comparatively high memory, difficulty in location identification
Central	Less storage needed	Average point may not be unique

set. We will keep nearest neighbor of each minutia as the first neighbor. Only for second, third, fourth, and fifth neighbors, duplication checking is done. So that the connectivity between minutia point persist. The complexity of the extended version will be in order of $\mathcal{O}(n^2)$ because we will be deleting duplication of children for each parent. This method will need 500 memory unit, if the total number of minutia in an image is 100, $100 + 2 \times 100 + 2 \times 100$ (100 storing the type of each parent and 200 for storing the distance and type 100 nearest neighbor corresponds to each minutia + 200 for storing second, third, fourth, and fifth nearest neighbors after duplication check). But we need an extra data structure for checking the duplication because we store type and distance only. So we have to keep a data structure which has the position of minutiae points which corresponds to type and distance. Using this data structure, duplication of entries is found out. Another method of storing minutiae is using NNV. Here, each minutiae points are center point and 4 nearest neighbors of each minutiae points are calculated. Center point is represented using position and direction, and four neighbors are represented using a vector which includes four features. So total 7 memory units are needed for each minutiae points. A centroid global graph is another method used to represent minutiae. In this method, center of each ridge and ridge arc is calculated. Then, the Euclidean distance between the center of ridges and core point of the fingerprint is calculated. This distance is stored as the unique representation. The center of ridge and ridge arc may not be correct always, and in case of bifurcation, there will be two centers instead of one are the disadvantage of this method. The complexity of our proposed method is the order of $\mathcal{O}(n^2)$ because we will be checking duplication of entry for each parent. Here unique feature chosen for minutiae creation are position and type of minutiae points. So no need of an extra data structure for position storage. Table 2 shows the memory needed to store minutiae by different methods with different number of minutia point (Table 1).

Table 2 Comparison of minutiae storage schemes: Let us assume number of isolated point in $K(=5)$ with isolated point (shown in Fig. 5) as r

Minutia point	K(=5) NN	4-NNV	K(=5) with isolated point	K(=5) NN with out isolated point	Binary tree with feature bit
90	990	630	270+2r	450	270
100	1100	700	300+2r	500	300
110	1210	770	330+2r	550	330
120	1320	840	360+2r	600	360
130	1430	910	390+2r	650	390
140	1540	980	420+2r	700	420
150	1650	1050	450+2r	750	450
160	1760	1120	480+2r	800	480
170	1870	1190	510+2r	850	510
180	1980	1260	540+2r	900	540

4 Conclusion

Comparison of the minutiae storage mechanism based on its memory requirement is done. In parse tree method [19], grammar corresponding to the fingerprint pattern is saved as unique representation, and memory required for this grammar storage is less, but the grammar may be ambiguous. In 4-NNV method, orientation is also stored as the feature, but location identification of minutiae may be different. In binary tree representation [3], position and feature of minutiae are stored. This method is free from redundancy and need less memory. In K(=5) NN method, memory required is huge because same entries are stored multiple times, but complexity of the algorithm is of order $\mathcal{O}(n)$. So extending the 5 NN methods by reducing the redundancy is proposed. Comparison of proposed method with binary tree representation is also done. Memory required for proposed system is more than memory required for binary tree method [3]. But in this proposed method, more optimized information about neighbors are stored than binary tree method [3] of order $\mathcal{O}(n^2)$. In proposed version of K(=5) NN, an extra data structure is needed to store position of the feature inserted for duplication reduction which makes the algorithm complexity as $\mathcal{O}(n^2)$.

References

1. Abutaleb, A.S., Kamel, M.: A genetic algorithm for the estimation of ridges in fingerprints. IEEE Trans. Image Process. **8**(8), 1134–1139 (1999)
2. Alibeigi, E., Rizi, M.T., Behnamfar, P.: Pipelined minutiae extraction from fingerprint images. In: Canadian Conference on Electrical and Computer Engineering, 2009. CCECE'09, pp. 239–242. IEEE, May 2009

3. Anjana, K., Praveen, K., Amritha, P.P.: Binary tree based fingerprint representation along with feature bit. IJPAM **118**(20), 3751–3760 (2018)
4. Ceguerra, A.V., Koprinska, I.: Integrating local and global features in automatic fingerprint verification. In: 2002 Proceedings of 16th International Conference on Pattern Recognition, vol. 3, pp. 347–350. IEEE (2002)
5. Donahue, M.J., Rokhlin, S.I.: On the use of level curves in image analysis. CVGIP: Image Underst. **57**(2), 185–203 (1993)
6. Feng, J., Jain, A.K.: Fingerprint reconstruction: from minutiae to phase. IEEE Trans. Pattern Analy. Mach. Intell. **33**(2), 209–223 (2011)
7. Galton, F.: Finger Prints. Macmillan and Company, New York (1892)
8. Gamassi, M., Piuri, V., Scotti, F.: Fingerprint local analysis for high-performance minutiae extraction. In: 2005 International Conference on Image Processing, ICIP 2005, vol. 3, pp. III-265. IEEE, Sept 2005
9. Henry, E.: Classification and Uses of Finger Prints. [Sl], George Routledge and Sons (1900)
10. Hong, L., Wan, Y., Jain, A.: Fingerprint image enhancement: algorithm and performance evaluation. IEEE Trans. Pattern Anal. Mach. Intell. **20**(8), 777–789 (1998)
11. Jain, A.K., Prabhakar, S., Hong, L.: A multichannel approach to fingerprint classification. IEEE Trans. Pattern Anal. Mach. Intell. **21**(4), 348–359 (1999)
12. Jain, M.D., Pradeep, S.N., Prakash, C., Raman, B.: Binary tree based linear time fingerprint matching. In: 2006 IEEE International Conference on Image Processing, pp. 309–312. IEEE, Oct 2006
13. Kamijo, M.: Classifying fingerprint images using neural network: Deriving the classification state. In: 1993 IEEE International Conference on Neural Networks, pp. 1932–937. IEEE (1993)
14. Kannavara, R., Bourbakis, N.G.: Fingerprint biometric authentication based on local global graphs. In: Proceedings of the IEEE 2009 National Aerospace & Electronics Conference (NAECON), pp. 200–204. IEEE, July 2009
15. Kaur, R., Sandhu, P.S., Kamra, A.: A novel method for fingerprint feature extraction. In: 2010 International Conference on Networking and Information Technology (ICNIT), pp. 1–5. IEEE, June 2010
16. Kawagoe, M., Tojo, A.: Fingerprint pattern classification. Pattern Recogn. **17**(3), 295–303 (1984)
17. Maio, D., Maltoni, D., Rizzi, S.: Dynamic clustering of maps in autonomous agents. IEEE Trans. Pattern Anal. Mach. Intell. **18**(11), 1080–1091 (1996)
18. Min, M.M. and Thein, Y.: Intelligent fingerprint recognition system by using geometry approach. In: 2009 International Conference on the Current Trends in Information Technology (CTIT), pp. 1–5. IEEE, Dec 2009
19. Moayer, B., Fu, K.S.: A tree system approach for fingerprint pattern recognition. IEEE Trans. Pattern Anal. Mach. Intell. **3**, 376–387 (1986)
20. Vaikole, S., Sawarkar, S.D., Hivrale, S., Sharma, T.: Minutiae feature extraction from fingerprint images. In: 2009 IEEE International Advance Computing Conference, IACC 2009, pp. 691–696. IEEE, Mar 2009
21. Wahab, A., Chin, S.H., Tan, E.C.: Novel approach to automated fingerprint recognition. IEE Proc. Vis. Image Sign. Process. **145**(3), 160–166 (1998)
22. Zafar, W., Ahmad, T., Hassan, M.: Minutiae based fingerprint matching techniques. In: 2014 IEEE 17th International Multi-Topic Conference (INMIC), pp. 411–416. IEEE, Dec 2014
23. Zhang, P., Hu, J., Li, C., Bennamoun, M., Bhagavatula, V.: A pitfall in fingerprint bio-cryptographic key generation. Comput. Secur. **30**(5), 311–319 (2011)
24. Zhao, F., Tang, X.: Preprocessing and postprocessing for skeleton-based fingerprint minutiae extraction. Pattern Recogn. **40**(4), 1270–1281 (2007)
25. Zhong, W.B., Ning, X.B., Wei, C.J.: A fingerprint matching algorithm based on relative topological relationship among minutiae. In: 2008 International Conference on Neural Networks and Signal Processing, pp. 225–228. IEEE, June 2008

Evaluation of Water Body Extraction from Satellite Images Using Open-Source Tools

K. Rithin Paul Reddy, Suda Sai Srija, R. Karthi and P. Geetha

Abstract Chennai is a metropolitan city in India. It has many lakes and reservoirs which are changing due to urbanization. Remote sensing and GIS techniques are widely used for water body extraction and water body change detection. This study evaluates water body extraction from satellite images of Chennai city using machine learning methods. Many classifiers are trained to extract water bodies from satellite images. Landsat 5 images of Chennai were taken from USGS Earth Explorer for the year 2009. The study aims to compare the classification results of different machine learning algorithms such as J48 decision tree, naive Bayes, multilayer perceptron, k-nearest neighbor, iso-cluster and random forest in extracting the water bodies. The tools used are ArcGIS for geospatial analysis, Weka tool for classification and R for the visual interpretation of the results. The results illustrate that naive Bayes classifier is able to identify lake regions better when compared to all other classifiers.

Keywords Remote sensing · Water body extraction · Open-source tools · Classification

---

K. Rithin Paul Reddy · S. S. Srija · R. Karthi (✉)
Department of Computer Science and Engineering, Amrita School of Engineering, Amrita Vishwa Vidyapeetham, Coimbatore, India
e-mail: r_karthi@ch.amrita.edu

K. Rithin Paul Reddy
e-mail: rithin.bannu@gmail.com

S. S. Srija
e-mail: srijasuda05@gmail.com

P. Geetha
Center for Computational Engineering and Networking (CEN), Amrita School of Engineering, Amrita Vishwa Vidyapeetham, Coimbatore, India
e-mail: p_geetha@cb.amrita.edu

S. M. Thampi et al. (eds.), *Intelligent Systems, Technologies and Applications*, Advances in Intelligent Systems and Computing 910,
https://doi.org/10.1007/978-981-13-6095-4_10

1 Introduction

Water from lakes and rivers is very vital for survival and agriculture. So it is very important to prevent shrinkage of these water resources. Sustainable development and usage of land and water resources are necessary to conserve water and natural habitat [1]. Many lakes and water bodies provide water for irrigation and domestic use. Finding out the water surface of these water bodies is essential to observe the changes in their surface area. Data from the past plays an important role in determining the extent of change in water area. Remote sensing techniques play an important role in monitoring earth surface changes in a consistent way. There are many satellite images of earth available in United States Geological Survey (USGS) Earth Explorer for use which can be accessed for free by all users. These images are used for many applications: Land cover and land use mapping, geological mapping, air pollution monitoring, sea surface temperature mapping and deforestation [2]. A geographic information system (GIS) is a framework which is intended to store, examine and analyze spatial or geographic information [3].

There are several remote sensing, GIS and machine learning techniques to identify water body's area from satellite images using standard proprietary softwares. The propriety softwares ENVI, ERDAS IMAGE and ArcGIS supports image analysis, mapping and visualization. These softwares are costly and require expertise to work and analyze the various modules available in them. Many open-source machine learning and data mining tools are extensively developed by the open-source community [4]. They are used in a number of classification and clustering applications and can be adapted and applied for GIS and remote sensing applications [5].

This paper aims to develop machine learning methods for identifying water bodies from satellite images using open-source tools. Witten [6] developed Waikato Environment for Knowledge Analysis (Weka) and different classifiers from Weka are used to classify satellite images for identification of water bodies. R is used as visualization tool to display satellite images. The satellite images required for this project were obtained from USGS Earth Explorer. Manual and automated techniques are adapted for validating the results obtained by the classifiers.

Satellites with different sensors provide enormous amount of data with different spatial, spectral and temporal resolution. Many techniques and water indices are proposed by researchers for detecting water surfaces. A brief review of the methods is given below. Ronki [7] studied spatiotemporal changes that occurred in Lake Urmia using Landsat 5, 7 and 8 data. Different water indices were tested, and a new method based on principal component was proposed to measure change in lake water surface. Du [8] analyzed Qingjiang River basin for water surface changes using Multispectral Scanner System, Thematic Mapper and Enhanced Thematic Mapper (ETM) Landsat data. Surface water was estimated using water indices and image processing methods. The results indicate a growing trend in surface water due to construction of dams in the region. Elsahabi [9] evaluated the performance of supervised, unsupervised learning algorithms and water indices to extract water bodies from Landsat ETM data. From the study, the authors concluded that unsupervised image classification gave the maximum accuracy and errors occurred at the edge pixels of the lake. Acharya [10] used J48 a decision tree algorithm to identify water and non-water bodies using Landsat 8 images with its reflectance bands. The authors concluded that J48 decision tree performs better than other methods such as normalized difference water index (NDWI), modified NDWI, SVM classifiers.

Yang [11] proposed a new water index based on fuzzy clustering method (FCM) and tested the accuracy based on Landsat 8 data. The results were compared with SVM classifier and other water indices. The accuracy of FCM-based water index in classification was not affected by sessional changes that occur in water throughout the year. Zhai [12] analyzed the performance of water indices in identifying water bodies in Landsat 8 images. Four water indices are considered for water body extraction from Landsat data and compared the results with thematic mapper data. The results indicate that automatic water extraction index and normalized water indices perform better and can be used for detection of changes in water bodies over a long period of time.

The main objective of this study is to compare and contrast the performances of water surface area extraction techniques using open-source tool. Weka and R are used in this study for implementation.

2 Study Area and Data Collection

Puzhal lake and other water bodies located in Chennai city around Red Hills, Tiruvallur district, Tamil Nadu were selected as the study area. The latitude and longitude of Puzhal lake are 13.158203 N, 80.175824 E. The topographic map details of the study area were collected from Survey of India Open Series Map with No: D4404 (66C/3) in 1:50,000 scale. Figure 1 shows the area of topographic map under study.

Landsat data of the area is collected from USGS Earth Explorer. The image data used in this study were acquired on 15 Feb 2009 by Landsat 5 image (Path/Row = 142/51). The image data that is acquired is already geo-referenced by USGS to the WGS84 ellipsoid with Universal Transverse Mercator (UTM) projection, zone 44P. The features of the Landsat 5 bands have been summarized in Table 1.

Figure 2 shows the original satellite image acquired and Fig. 3 shows the clipped image, the region of interest (ROI) image used for classification. The original Landsat image contains 7091 $\times$ 7021 pixels with seven band values. The image is cropped to the region of interest which has lake regions. The size of the clipped image is 333 $\times$ 300 pixels with seven band values.

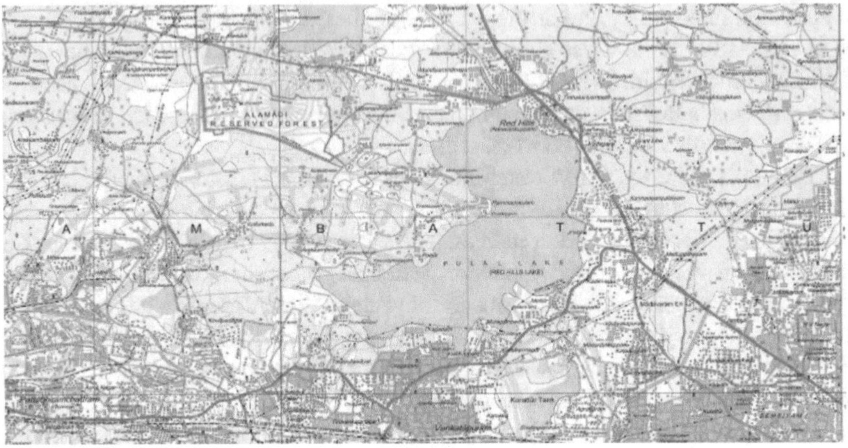

Fig. 1 Location of study area in Red Hills, Chennai city, Tiruvallur District

Table 1 Landsat 5 bands

Band	Name	Wavelength(λ)	Spatial resolution (m)
1	Blue	0.45–0.52	30
2	Green	0.52–0.60	30
3	Red	0.63–0.69	30
4	Near Infrared (NIR)	0.76–0.90	30
5	Short wave IR 1 (SWIR1)	1.55–1.75	30
6	Thermal IR	10.40–12.50	120
7	Short wave IR 2 (SWIR2)	2.08–2.35	30

3 Methodology

There are several techniques for water body extraction from Landsat imagery. Figure 4 shows the methodology used for identification of water bodies in this study. Multispectral satellite images of the study area which are of medium resolution were downloaded from USGS. From the downloaded image, subregions are cropped and used to generate the feature vector for the classifier. Each pixel has seven features which represent the spectral feature of the point. Sampled water subregions and land cover are identified from topographic maps of the area. All the sampled pixel points are assigned to any one class: water and non-water pixels. The feature vector with class labels is used for training and testing the classifier.

Weka tool is used for building the different classifier model. Classification algorithm J48, naive Bayes (NB), multilayer perceptron (MLP), k-nearest neighbor clas-

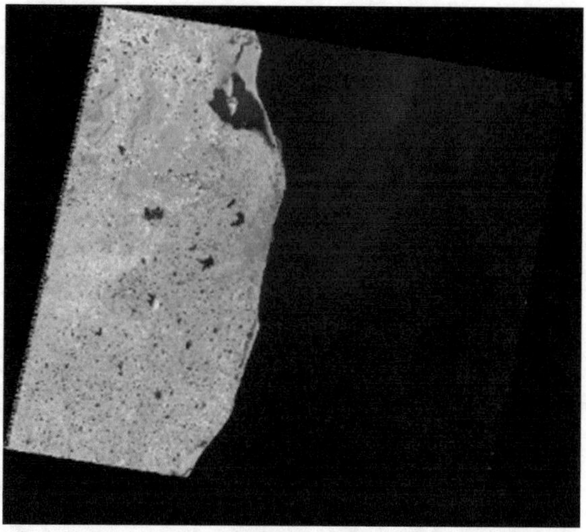

Fig. 2 Original image from Landsat 5 (Band4)

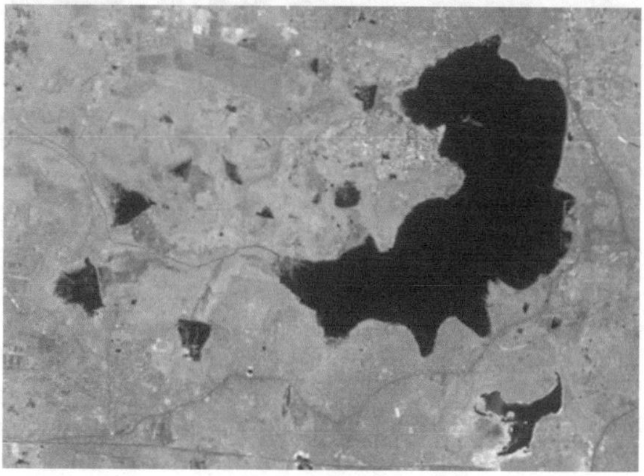

Fig. 3 Location of study area (region of interest)

sifier (KNN), random forest (RF) are used in this study. Each classifier model built from Weka is saved, and the model is validated using new test data. Images are clustered using iso-data in ArcGIS tool.

3.1 Algorithm

Figure 4 shows the flowchart adopted to extract the water surface area from satellite images. These phases are:

- *Image classification process using unsupervised classification technique (iso-cluster).*

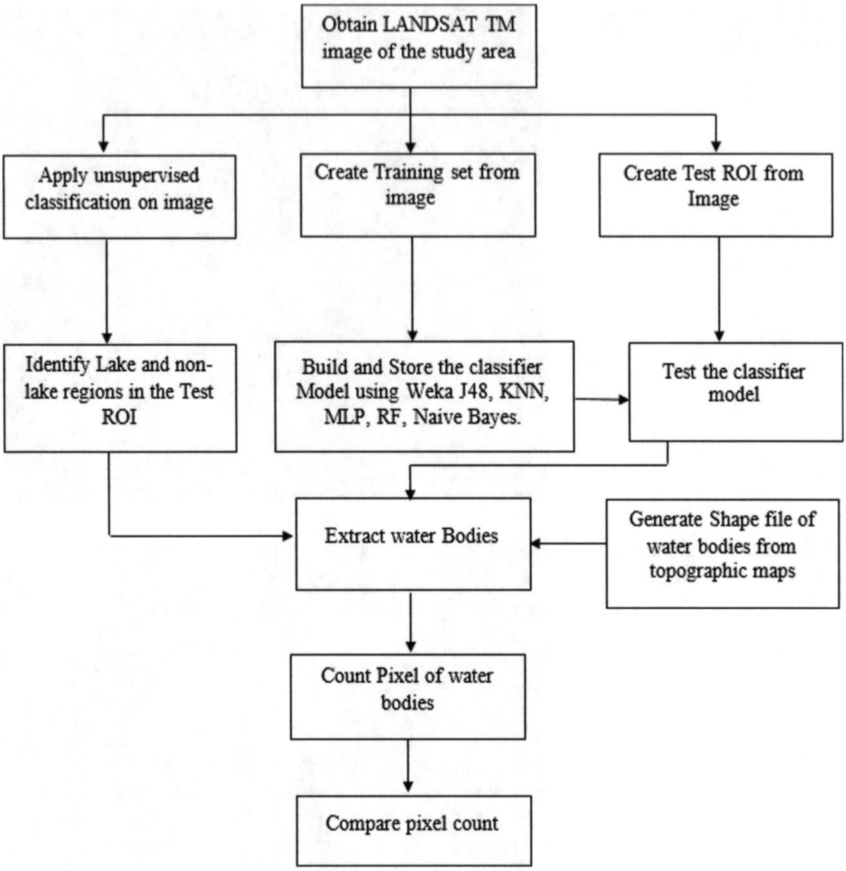

Fig. 4 Extraction of water bodies from satellite images

This classification is done in ArcGIS. The user specifies the number of classes, i.e., maximum class size. The image is segmented to the specified number of classes. By visual inspection, the user differentiates the classes as lake and non-lake regions and merge nearby regions into blocks.

- *Creating shape file from a topographic map:*

Shape files are created from topographic maps for all lakes within the region. The collected toposheets were scanned and uploaded in GIS platform and geo-referenced. Shape files of lake area and water bodies are extracted and stored in ArcGIS platform.

- *Creating training samples from satellite image for classifier.*

Topographic maps are used as reference to select portions of lake and non-lake part from the image. Each selected area has pixels to represent lake and non-lake region. Each pixel has seven bands which indicate the feature vector for the pixel.

The class labels are assigned based on the region to which the pixel belongs. The training set is created from these pixels with the class labels.

- *Build, Store and Test the classifier model*

Build the classifier using the training pixel set in Weka. Each of these classifiers is stored and tested using test image. The test image is a ROI image generated from original downloaded image. The accuracy is evaluated by comparing pixel count of water bodies and by overlaying on topographic sheet and iso-cluster results.

- *Validation of results*

Three methods are used for validation of the classifiers:

1. Using Toposheets data.
2. Using clustering techniques.
3. Using Overlaying techniques.

Using Toposheets data
The results are validated by comparing with the topographic map of the area. Shape files of lake area and water bodies are extracted from ArcGIS platform. The pixel count is calculated for all the water bodies and stored as a reference value which is represented as reference pixel count (RPC).

The test image is passed to each classifier, and the classifier identifies water bodies in the image. The extracted count of water pixels from the test image by the classifier is represented as Classifier Pixel Count (CPC). Comparison between CPC and RPC is done to evaluate the accuracy of the classifier.

Using clustering techniques
Clustering is an unsupervised classification technique where class labels are unknown for the data. ArcGIS supports iso-clustering which separates pixels into distinct groups based on the spectral feature value of each pixel. The distinct groups represent different regions in the image such as land cover, water body, vegetation and urban areas. Water bodies are identified manually in the test image using toposheets. The pixel count of these water regions is measured for the test image.

Using Overlaying techniques
The water bodies extracted by the classification algorithm on test image are overlaid with the shape files of the region. The overlay shows the difference in area of pixels identified by the classifier with respect to water regions in the image.

4 Results and Discussion

4.1 Data Set

To evaluate the performance of different classification algorithms, we have constructed a training data set and test data set. Images are downloaded from the standard USGS repositories. Training data set is generated by sampling regions in the image which contain water regions and non-water regions. The regions are manually cropped, and pixel values are extracted to construct the training set. Total of 10,000 pixels are used to train and model the classifier.

A single clipped image with Puzhal lake and neighbor water bodies are used for testing. This clipped image contains water and non-water regions. The clipped image contains 99,900 pixels with seven bands which is given as input to the classifier.

4.2 Experimental Setup

The algorithms in this study were implemented using R Studio 3.4.3, Weka 3.8.1 and ArcGIS 10.3.1 and ran over a PC with Intel i-7 2.5 GHz processor and 4 GB RAM. R offers many packages for preprocessing and visualization of multispectral images. Weka is used for building the classifiers models. ArcGIS is used for creation of shape files and for validation of the results.

4.3 Results

Different techniques for classification such as J48, naive Bayes, multilayer perceptron, k-nearest neighbor classifier, random forest are used to extract the water surface area from the Landsat TM data. Figure 5 shows the test ROI image generated from Landsat image. The ROI image is generated in R, and the spectral features are obtained for each pixel (seven band values)

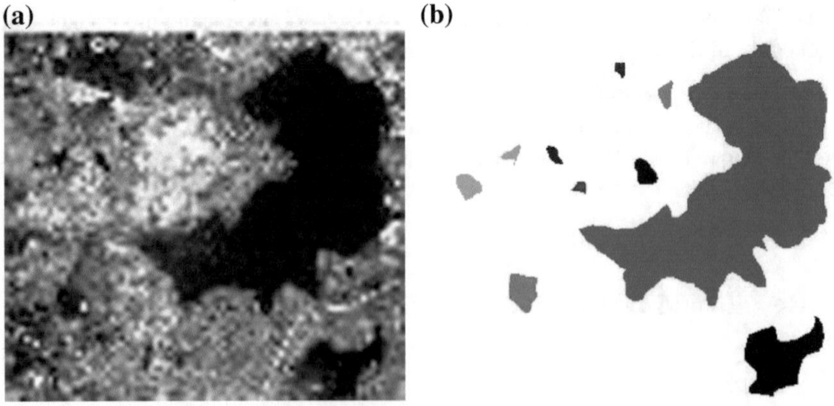

Fig. 5 **a** ROI of test image **b** shape files obtained from topographic maps

Shape files are created from topographic map. Figure 5b shows the shape files of water bodies extracted from topographic maps using ArcGIS tool. The shape file served as a ground truth information and is a reference for comparison of the results obtained from each algorithm.

Visual Interpretation of the water surface extracted using different algorithms.

Figure 6 shows the extracted water surface area using topographic map, iso-cluster, KNN, J48, naive Bayes, MLP and random forest. The image pixels are extracted using R and classified using Weka tool. Visual inspection of images is done using R which show that all algorithms are able to find water bodies in the test image.

Figure 6a represents the shape file and 6b represents classification of image using iso-cluster algorithm. These are obtained using ArcGIS tool. Figure 6c–g represents the results of classifiers obtained using Weka tool. The results are displayed using R tool after classification in Weka. Tree-based classifiers such as J48, random forest, probabilistic-based classifiers such as naive Bayes, neural network-based classifiers MLP and k-nearest neighbor classifier are used. All algorithms are able to identify water and non-water bodies in the test image. From the images, we identify that J48 and random forest algorithm are not able to identify smaller lake regions in the test image. Naïve Bayes and iso-cluster are able to identify many water bodies in the image.

Comparison of extracted water surface area's pixel count with reference topographic map.

The pixel count of water bodies in the image is compared to evaluate the accuracy of classifier. The pixel counts are compared with reference value taken from topographic maps. The total count of pixels in the shape files is taken as a reference measure. The pixel count of the extracted water surface areas by different classification algorithms is presented in Table 2.

From Table 2, we can infer that naive Bayes is able to extract and identify more water pixels compared to other algorithms. Naive Bayes classifier uses conditional probability and provides results, i.e., correct classification of pixels that was closest to the topographic map. Tree-based classifier J48 and random forest have identified the minimum number of water pixels in the image which infer that the algorithm has classified some pixels erroneously.

The difference in pixels compared with topographic maps is shown in Table 3. From the table, we infer that Naïve Bayes classifier is able to identify many water pixels present in the image.

Comparison of extracted water surface area's by overlaying with reference topographic map.

The extracted water surface area results of different algorithms are overlaid with the reference shape file obtained from topographic map for comparison. Each pixel corresponds to 30*30 m^2. When the visual results were compared, the boundaries of the extracted water surfaces match the actual boundaries of the water bodies in the

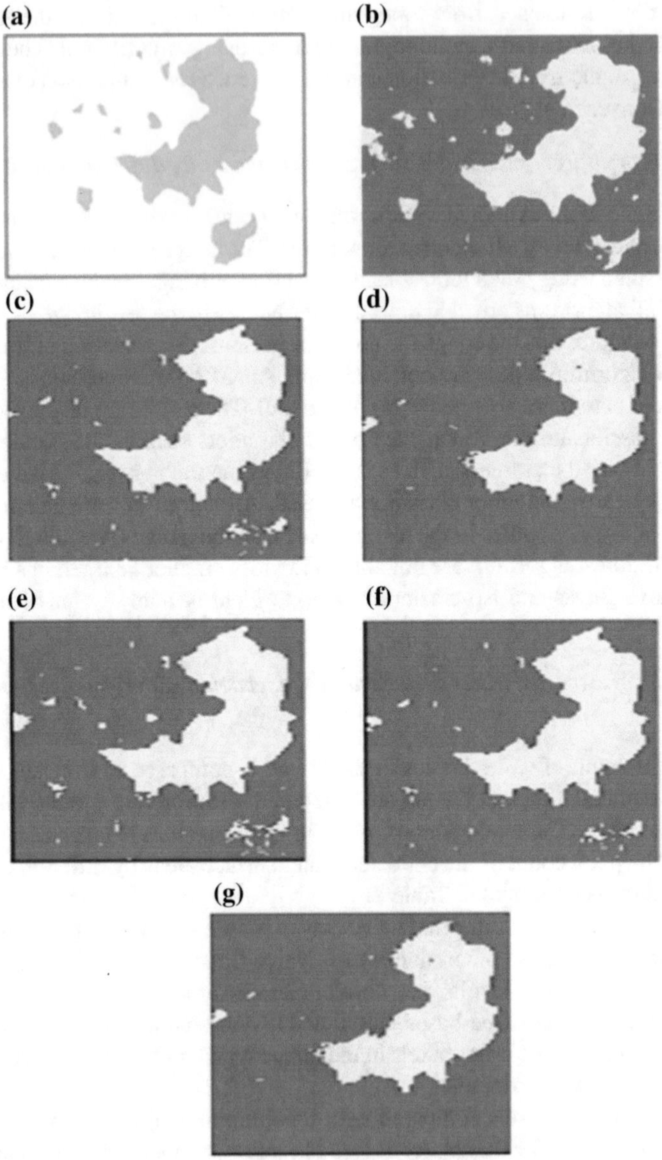

Fig. 6 Extracted water surface area **a** topographic maps **b** iso-cluster **c** KNN **d** J48 **e** Naïve Bayes **f** MLP **g** random forest

Table 2 Pixel count of the extracted water surfaces

Topographic map	Iso-cluster	KNN	J48	Naive Bayes	MLP	Random forest
25,732	24,901	23,525	21,087	25,051	23,735	21,080

Table 3 Difference in pixel count compared to topographic Map

Iso-cluster	KNN	J48	Naive Bayes	MLP	Random forest
831	2207	4645	681	1997	4652

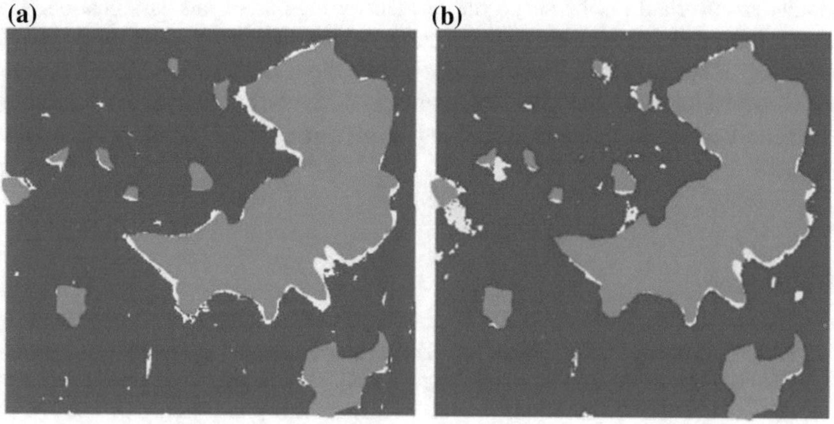

Fig. 7 **a** Shape files overlaid on output of Naïve Bayes classifier **b** shape files overlaid on output of iso-cluster algorithm

geo-referenced shape file very closely. Figure 7a shows the results of shape files and naive Bayes algorithm when overlaid using ArcGIS tool. The green patches represent boundary of shape file that is used as ground truth information. The white pixels in the image represent misclassified pixels in the image. Figure 7b shows the results of shape files and iso-cluster algorithm when overlaid using ArcGIS tool. From the results we infer that using iso-cluster, the misclassification rate is higher compared to Naïve Bayes algorithm.

The results show that classification of satellite images using Weka open-source tool can be implemented with good accuracy. Classifiers from Weka tool such as KNN, J48, Naïve Bayes, MLP and random forest can be used for satellite image classification.

5 Conclusion

The paper discusses the classification results of different machine learning algorithms for the extraction of water surfaces using Landsat 5 images. Classification of images into water and non-water parts is done using open-source tools Weka and R. Different types of classifier such as iso-cluster, k-nearest neighbor, J48, naive Bayes, MLP and random forest are used for classification.

To validate the estimated results of the water body classification, we compared it with the topographic map from Survey of India. Water bodies are extracted as shape file from ArcGIS using the topographic map. The pixel count of water bodies identified by the algorithm is used estimate to accuracy of the classifier. The classified results are overlaid on the shape file to identify regions of misclassification. The results indicated that our methodology extracted the different types of water body's efficiency. The results indicate that open-source tools like Weka and *R* can be used for classification of images. The error or misclassification of pixels by the algorithms might be due to the inability of identifying the pixels across the boundary of the water bodies.

References

1. Paul, A., Chowdary, V.M., Dutta, D., Sharma, J.R.: Standalone open-source GIS-based tools for land and water resource development plan generation. In: Environment and Earth Observation, pp. 23–34 Springer, Cham (2017)
2. Balaji, S.A., Geetha, P., Soman, K.P.: Change detection of forest vegetation using remote sensing and GIS techniques in Kalakkad Mundanthurai tiger reserve-(a case study). Indian J. Sci. Technol. **9**(30) (2016)
3. Lillesand, T., Kiefer, R.W., Chipman, J.: Remote Sensing and Image Interpretation. Wiley (2014)
4. Kanevski, M., Pozdnukhov, A., Timonin, V.: Machine learning algorithms for geospatial data. Appl. Softw. Tools (2008)
5. Karthi, R., Rajendran, C., Rameshkumar, K.: Neighborhood search assisted particle swarm optimization (npso) algorithm for partitional data clustering problems. In: International Conference on Advances in Computing and Communications, pp. 552–561. Springer, Berlin, Heidelberg. (2011, July)
6. Witten, I.H., Frank, E., Hall, M.A., Pal, C.J.: Data Mining: Practical Machine Learning Tools and Techniques. Morgan Kaufmann (2016)
7. Rokni, K., Ahmad, A., Selamat, A., Hazini, S.: Water feature extraction and change detection using multitemporal Landsat imagery. Remote Sens. **6**(5), 4173–4189 (2014)
8. Du, Z., Bin, L., Ling, F., Li, W., Tian, W., Wang, H., Zhang, X.: Estimating surface water area changes using time-series Landsat data in the Qingjiang River Basin, China. J. Appl. Remote. Sens. **6**(1), 063609 (2012)
9. Elsahabi, M., Negm, A., El Tahan, A.H.M.: Performances evaluation of surface water areas extraction techniques using Landsat ETM + data: case study Aswan High Dam Lake (AHDL). Procedia Technol. **22**, 1205–1212 (2016)
10. Acharya, T.D., Lee, D.H., Yang, I.T., Lee, J.K.: Identification of water bodies in a Landsat 8 OLI image using a j48 decision tree. Sensors **16**(7), 1075 (2016)
11. Yang, Y., Liu, Y., Zhou, M., Zhang, S., Zhan, W., Sun, C., Duan, Y.: Landsat 8 OLI image based terrestrial water extraction from heterogeneous backgrounds using a reflectance homogenization approach. Remote Sens. Environ. **171**, 14–32 (2015)
12. Zhai, K., Wu, X., Qin, Y., Du, P.: Comparison of surface water extraction performances of different classic water indices using OLI and TM imageries in different situations. Geo-Spat. Inf. Sci. **18**(1), 32–42 (2015)

A Modification to the Nguyen–Widrow Weight Initialization Method

Apeksha Mittal, Amit Prakash Singh and Pravin Chandra

Abstract Weight initialization is important factor for determining the speed of training in feedforward neural networks. In this paper, a variable parameter α is identified through statistical analysis in Nguyen–Widrow weight initialization method. The value of α is varied from 1 to 10 and is tested on nine function approximation tasks. The results are compared for each value of α using single-tail t-test. An optimal value of α is derived, and a new weight initialization technique is hence proposed.

Keywords Feedforward networks · Weight initialization · Function approximation

1 Introduction

Feedforward artificial neural networks are universal approximators, i.e., given a sigmoidal nonlinearity at the hidden layer (minimum one hidden layer is required for the network to possess Universal Approximation Property given sufficient number of nodes in the layer), a one hidden layer feedforward neural network can approximate almost any continuous function [7, 9, 12, 14] . A single hidden layer feedforward artificial neural network is given in Fig. 1.

$x_1, x_2 \dots x_I$ are the inputs to the network where I is the total number of input. $w_{11}, w_{12} \dots w_{ij}$ are input to hidden layer weights where w_{ij} denotes the weight from ith input layer node to jth hidden layer node. $\theta_1, \theta_2 \dots \theta_H$ are bias to hidden layer nodes where H is the total number of hidden layer nodes. $\beta_1, \beta_2 \dots \beta_H$ are hidden to output layer weights, and γ denotes bias at the output layer node.

A. Mittal (✉) · A. P. Singh · P. Chandra
University School of Information, Communication & Technology, Guru Gobind Singh Indraprastha University, Sector-16C, Dwarka, Delhi 110078, India
e-mail: apekshamittal3@gmail.com

A. P. Singh
e-mail: amit@ipu.ac.in

P. Chandra
e-mail: chandra.pravin@gmail.com

S. M. Thampi et al. (eds.), *Intelligent Systems, Technologies and Applications*, Advances in Intelligent Systems and Computing 910,
https://doi.org/10.1007/978-981-13-6095-4_11

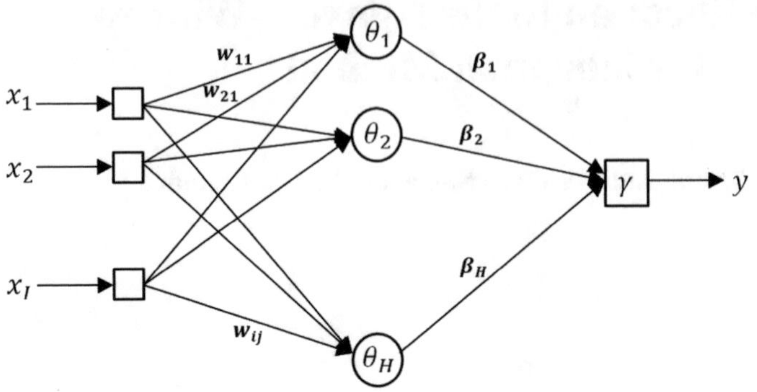

Fig. 1 Schematic diagram of single hidden layer (one input layer, one hidden layer and one output layer) feedforward neural network

The net input to a hidden node j is given by

$$n_j = \sum_{i=1}^{I} w_{ij}x_i + \theta_j \tag{1}$$

To provide nonlinearity at hidden layer, sigmoidal activation function is used. A sigmoidal activation function is defined as [3] a bounded, monotone increasing, continuous and differentiable mapping $\sigma : R \to R$, if for $x \in R$ it satisfy the following condition:

$$\lim_{x \to \infty} \sigma(x) = a \tag{2}$$

$$\lim_{x \to -\infty} \sigma(x) = b \tag{3}$$

where $a, b \in R$ and $a > b$

Both logistic and hyperbolic tangent functions defined in Eqs. (4) and (5) belong to this class; that is, they are sigmoidal functions. For logistic function, $a = 1$ and $b = 0$; whereas for hyperbolic tangent function, $a = 1$ and $b = -1$.

$$\sigma_1(x) = \frac{1}{e^x + e^{-x}} \tag{4}$$

$$\sigma_2(x) = \tanh(x) = \frac{e^x - e^{-x}}{e^x + e^{-x}} \tag{5}$$

The output of a hidden layer is given by

$$h_j = \sigma_p(n_j) \tag{6}$$

where value of p can be 1 or 2.

The net output of the network is given as

$$y = \sum_{j=1}^{H} \beta_j h_j + \gamma \tag{7}$$

Activation function at the output node is linear (though sigmoidal nonlinearity can also be used at the output layer).

Weight Initialization affects the speed of training in feedforward neural networks [1, 4, 8, 15, 17, 18, 20–22]. A weight initialization technique was proposed by Bottou in [1] such that weights are distributed uniformly in the interval $(-p/\sqrt{I}, p/\sqrt{I})$ where I is the total fan-in to the node. p is the maximal curvature of threshold function.

Another technique was proposed by Nguyen and Widrow in [18] where weights are initialized such that the magnitude of weights to a layer L is $0.7N_L^{1/N_{L-1}}$ where N_L is the number of nodes in the current layer, and N_{L-1} is the number of nodes in the previous layer. The current layer biases are initialized in the range $(-|W|, |W|)$ where $W = 0.7N_L^{1/N_{L-1}}$.

Drago and Ridella in [8] proposed a maximum bound of weights as a function of PNP (paralized neuron percentage). Kim and Ra in [15] initialized weights using decision boundaries in input space and proposed minimum bound on weights to be $\sqrt{\eta/I}$ where η is learning rate parameter, and I is the total fan-in to a node.

Sodhi and Chandra in [21] initialized weights in separate regions (without overlaps) of maximal curvature of threshold function. Mittal and Chandra in [17] initialized the weights in the useful range of activation function also ensuring that weights to each hidden nodes are initialized in statistically resilient range.

The weight initialization method proposed by Nguyen and Widrow in [18] is widely used as one of the better methods and is default weight initialization method for one of the generally used platform, that is, MATLAB. In this paper, a variable parameter α is identified in the weight initialization technique proposed by Nguyen and Widrow in [18] through statistical analysis and its value is varied from 1 to 10. The performance of obtained methods are compared to each other using single-tail t-test to check whether each method is statistically different from other methods. An optimal value of α is obtained, and a new weight initialization method is hence proposed.

This paper is organized as follows. In Sect. 2, the framework for proposition of weight initialization method is discussed. In Sect. 3, the network architecture for performing function approximation task is discussed. In Sect. 4, experiment design is discussed in detail. The Proposed Weight Initialization Method is discussed in Sect. 5. Results are presented in Sect. 6, and the conclusions are presented in Sect. 7.

2 Framework for Proposition of Weight Initialization Algorithm

Nguyen–Widrow proposed a new weight initialization method in [18] in which the norm of weights w_{ij} from ith node of layer $L-1$ to a jth node of layer L is given by

$$\|w_{ij}\| = 0.7 * N_L^{\frac{1}{N_{L-1}}} \tag{8}$$

Here, value of i ranges from 1 to N_{L-1}, N_{L-1} denoting the number of nodes in layer $L-1$ and N_L denoting the number of nodes in the layer L. All the bias values to a layer L are initialized in the range $(-\|w_{ij}\|, \|w_{ij}\|)$.

The net input to a node j is given by

$$n_j = \sum_{i=1}^{N_{L-1}} w_{ij}x_i + \theta_j \tag{9}$$

where w_{ij} is the weight from ith node in layer $L-1$ to jth node in layer L. θ_j denotes the bias value of jth node in layer L. Analyzing the effect input to hidden layer weights w_{ij}, thus replacing n_i with $\tilde{n}_i$.

$$\tilde{n}_j = \sum_{i=1}^{N_{L-1}} w_{ij}x_i \tag{10}$$

Squaring both sides of (10).

$$(\tilde{n}_j)^2 = \left(\sum_{i=1}^{N_{L-1}} w_{ij}x_i\right)^2 \tag{11}$$

By chebyshev's sum inequality

$$\tilde{n}_j^2 \leq \sum_{i=1}^{N_{L-1}} (w_{ij})^2 \sum_{i=1}^{N_{L-1}} (x_i)^2 \tag{12}$$

Taking expected value on both sides

$$\langle \tilde{n}_j^2 \rangle \leq \left\langle \sum_{i=1}^{N_{L-1}} (w_{ij})^2 \sum_{i=1}^{N_{L-1}} (x_i)^2 \right\rangle \tag{13}$$

$$\langle \tilde{n}_j^2 \rangle \leq \left\langle \sum_{i=1}^{N_{L-1}} (w_{ij})^2 \sum_{i=1}^{N_{L-1}} (x_i)^2 \right\rangle \tag{14}$$

Using Cauchy Schwartz inequality

$$\langle \tilde{n}_j^2 \rangle \le \left\langle \sum_{i=1}^{N_{L-1}} (w_{ij})^2 \right\rangle \left\langle \sum_{i=1}^{N_{L-1}} (x_i)^2 \right\rangle \tag{15}$$

$$\langle \tilde{n}_j^2 \rangle \le \sum_{i=1}^{N_{L-1}} \langle w_{ij}^2 \rangle \sum_{i=1}^{N_{L-1}} \langle x_i^2 \rangle \tag{16}$$

$$\langle \tilde{n}_j^2 \rangle \le \sum_{i=1}^{N_{L-1}} w_{ij}^2 \sum_{i=1}^{N_{L-1}} \langle x_i^2 \rangle \tag{17}$$

Rewriting (17) using (8),

$$\langle \tilde{n}_j^2 \rangle \le 0.7 * N_L^{\frac{2}{N_{L-1}}} * \frac{1}{3} \tag{18}$$

$$\sqrt{\langle \tilde{n}_j^2 \rangle} \le 0.483 * N_L^{\frac{1}{N_{L-1}}} \tag{19}$$

This implies that the net input to a node in layer L is dependent upon number of nodes in layer L. For large number nodes in the layer L, the net input to jth node $\tilde{n}_j$ becomes large and thus network enters saturated state. There is thus a need to limit the value $N_L^{\frac{1}{N_{L-1}}}$.

Let $N_L^{\frac{1}{N_{L-1}}}$ in (19) be denoted by α. Thus, square root of expected value of net input to a hidden node is given by

$$\sqrt{\langle \tilde{n}_j^2 \rangle} \le 0.483 * \alpha \tag{20}$$

Thus, limiting the value of α limits the net input $\tilde{n}_j$ to a certain range. The value of α is assigned different values to find the optimal value. Replacing $N_L^{\frac{1}{N_{L-1}}}$ in (8)

$$\|W\| = 0.7 * \alpha \tag{21}$$

In this paper, α is assigned values $1, 2, 3, \ldots 10$. Thus, 10 different weight initialization methods are obtained. Each of these 10 methods are tested on nine function approximation tasks. Left tailed t-test is used to compare the performance of each of these methods to others and to test that each method is statistically different from other methods. The value of α for which performance is best for maximum number of cases and worse for least number of cases is considered the optimal value. A new weight initialization method is derived using this optimal value of α.

3 Network Architecture

There are five choices to be made before training of a network: number of hidden layers, number of nodes in hidden layer, training algorithm, activation function and weight initialization.

1. *Number of hidden layers* Given nonlinearity at hidden layer node (refer Eqs. (4) and (5)), one hidden layer is sufficient to solve function approximation problems [7, 9, 12, 13].
2. *Number of nodes in Hidden Layer* The number of hidden layer nodes is determined by conducting exploratory experiments. The number of hidden layer nodes is varied from 1 to 20, and the number of nodes for which the performance was satisfactory was taken to be appropriate number of hidden nodes.
3. *Training Algorithm* In this work, resilient backpropagation is used for training the network as the performance is comparative to other high-order training methods [19].
4. *Activation function* For nonlinearity at hidden layer, sigmoidal activation function is used as it is continuous and differentiable. Hyperbolic tangent function (as in Eq. (5) is used at the hidden layer, and linear activation function is used at the output layer [10, 11].
5. *Weight initialization* A number of weight initialization are proposed in [1, 8, 15, 17, 18, 21] as discussed in Sect. 1. In this work, the weights are initialized using above methods discussed in Sect. 2 with $\alpha = 1, 2, 3, \ldots 10$.

Thus, for each of nine function approximation problems, the first four factors are kept as a constant and weight initialization techniques are varied and the results are compared using single-tail t-test. The best value of α is derived, and new weight initialization method is proposed.

4 Experiment Design

Nine function approximation problems from [2, 5, 6, 16] are solved using above weight initialization methods with $\alpha = 1, 2, 3, \ldots 10$ (Refer Table 1). A total of 750 input and output data sets are created of which 500 data sets are used for training the network, and 250 data sets are used for testing the network.

For each of the nine function approximation task, 30 three-layered feedforward networks (one input layer, one hidden layer and one output layer) with sigmoidal activation function at hidden layer and linear activation function at output layer are trained using resilient backpropagation training algorithm varying the weight initialization methods. Thus, in total $30 \times 10 \times 9 = 2700$ networks are trained for testing the weight initialization methods.

These experiments are conducted using MATLAB R2015a version on 64 bit Intel i5 processor, Microsoft Windows 10 system with 4 gb RAM.

Table 1 List of functions used to generate Input–Output dataset for function approximation task

	Functions used to generate data set	No. of inputs i	Range of x_i
F1	$y = \sin(x_1 * x_2)$	2	$[-2, 2]$
F2	$y = e^{x_1 * \sin(\pi * x_2)}$	2	$[-1, 1]$
F3	$a = 40 * e^{8*((x_1-0.5)^2+(x_2-0.5)^2)}$	2	$[0, 1]$
	$b = e^{8*((x_1-0.2)^2+(x_2-0.7)^2)}$		
	$c = e^{8*((x_1-0.7)^2+(x_2-0.2)^2)}$		
	$y = a/(b+c)$		
F4	$y = (1 + \sin(2x_1 + 3x_2))/(3.5 + \sin(x_1 - x_2))$	2	$[-2, 2]$
F5	$y = 42.659(0.1 + x_1(0.05 + x_1^4 - 10x_1^2x_2^2 + 5x_2^4))$	2	$[-0.5, 0.5]$
F6	$y = 1.3356[1.5(1 - x_1)$	2	$[0, 1]$
	$+e^{2x_1-1}\sin(3\pi(x1 - 0.6)^2)$		
	$+e^{3(x_2-0.5)}\sin(4\pi(x_2 - 0.9)^2)]$		
F7	$y = 1.9(1.35 + e^{x_1}\sin(13(x_1 - 0.6)^2)e^{-x_2}\sin(7x_2))$	2	$[0, 1]$
F8	$y = 10\sin(\pi x_1 x_2) + 20(x_3 - 0.5)^2 + 10x_4 + 5x_5 + 0x_6$	6	$[-1, 1]$
F9	$y = e^{2x_1 \sin(\pi x_4)} + \sin(x_2 x_3)$	4	$[-0.25, 0.25]$

Table 2 Network size summary

	No. of input	No. of hidden nodes	No. of output
F1	2	15	1
F2	2	15	1
F3	2	15	1
F4	2	15	1
F5	2	15	1
F6	2	11	1
F7	2	13	1
F8	6	19	1
F9	4	11	1

The network architecture for each of the nine function is given in Table 2. For each function, the number of input and output nodes is determined from the function definition and the number of hidden nodes is determined through exploratory experiments. The number of nodes for which the performance of network was satisfactory was taken to be appropriate number of hidden nodes for that function approximation task.

5 Results

The Train Data Result and Test Data Result for each of 9 function approximation problems, mean, median, standard deviation, maximum and minimum among the 30 trained networks for $\alpha = 1, 2, \ldots 10$ is given in Tables 3 and 4, respectively. Sym-

Table 3 Train data results ($\times 10^{-3}$). Mean, median, standard deviation, maximum and minimum is across 30 networks for all nine function approximation problems F1, F2,...F9 for $\alpha_1, \alpha_2, \ldots \alpha_{10}$ denoting weight initialization method corresponding to $\alpha = 1, 2, \ldots 10$, respectively.

		α_1	α_2	α_3	α_4	α_5	α_6	α_7	α_8	α_9	α_{10}
F1	Mean	9.376	8.893	7.205	5.997	5.512	6.492	6.924	7.397	9.839	13.408
	Median	9.556	8.701	6.308	4.754	4.208	5.254	5.106	5.794	9.26	11.155
	St. Dev.	1.562	1.858	3.68	3.45	3.892	4.251	4.368	4.298	4.203	7.875
	Max	13.097	14.597	17.799	15.216	18.339	15.517	18.375	18.485	17.258	29.207
	Min	6.193	5.488	1.536	1.43	1.858	1.349	2.235	2.027	3.24	2.496
F2	Mean	13.466	11.385	10.907	12.037	10.24	11.91	13.146	14.993	14.955	16.988
	Median	13.006	11.748	10.825	10.818	10.064	10.515	11.83	13.615	13.926	16.14
	St. Dev.	3.107	1.78	1.782	3.483	2.728	4.273	5.352	6.662	5.764	6.031
	Max	25.267	16.094	15.312	19.756	19.892	23.677	29.45	33.652	30.795	34.382
	Min	9.28	6.853	7.818	7.239	6.401	6.084	5.155	7.141	7.118	9.036
F3	Mean	0.185	0.0917	0.143	0.244	0.352	0.495	0.628	1.202	1.031	1.488
	Median	0.145	0.085	0.114	0.239	0.268	0.375	0.519	0.944	0.83	1.273
	St. Dev.	0.128	0.0317	0.103	0.168	0.212	0.438	0.423	0.778	0.716	1.101
	Max	0.652	0.153	0.495	0.825	0.906	2.329	2.184	2.916	3.722	5.718
	Min	0.0679	0.0432	0.0374	0.0549	0.108	0.134	0.109	0.187	0.302	0.296
F4	Mean	10.932	23.638	6.991	6.467	6.118	6.587	7.903	16.152	16.216	16.453
	Median	10.871	9.007	6.449	5.649	5.588	5.087	6.159	8.348	7.578	10.369
	St. Dev.	3.236	55.122	2.72	2.752	3.502	5.405	4.828	26.942	37.072	16.474
	Max	16.902	236.578	16.358	17.397	16.963	33.308	21.406	132.232	209.644	86.465
	Min	4.707	4.71	2.707	3.442	2.122	3.303	3.239	3.429	4.095	5.527
F5	Mean	5.994	4.653	4.05	3.043	2.853	3.545	3.795	4.864	6.761	7.489
	Median	5.884	4.341	3.395	2.635	2.409	3.165	3.542	4.448	6.447	7.104
	St. Dev.	1.53	1.231	2.028	1.335	1.234	1.495	1.303	2.201	2.599	2.909
	Max	10.007	7.93	10.21	7.381	6.05	6.632	6.889	14.138	13.969	15.373
	Min	4.107	2.952	1.866	1.697	1.389	1.107	0.959	2.158	2.926	2.801
F6	Mean	3.446	2.382	2.408	2.185	2.289	3.229	3.328	5.383	4.656	8.241
	Median	3.68	1.331	2.165	1.399	1.772	2.933	2.755	3.785	4.629	6.128
	St. Dev.	1.775	1.76	1.568	1.795	1.71	2.186	2.238	5.782	2.591	8.103
	Max	6.428	6.658	5.932	5.346	7.021	9.006	10.046	28.732	11.132	36.757
	Min	1.106	0.636	0.536	0.254	0.206	0.578	0.233	1.095	0.993	1.181
F7	Mean	8.435	3.524	2.82	2.516	3.153	3.492	4.082	6.635	9.743	8.488
	Median	5.648	2.725	2.362	2.177	2.946	2.969	3.458	5.537	9.065	7.097
	St. Dev.	6.397	2.543	1.546	1.356	1.491	1.776	1.923	3.452	5.546	4.56
	Max	23.42	12.29	9.211	8.22	9.169	7.327	10.102	18.109	27.919	26.137
	Min	2.254	1.419	1.55	0.937	1.491	1.187	1.731	2.737	3.056	3.388
F8	Mean	0.286	0.319	0.413	0.639	1.043	1.253	1.522	1.934	2.062	2.832
	Median	0.285	0.314	0.353	0.623	0.989	0.996	1.253	1.828	1.679	2.635
	St. Dev.	0.0526	0.0717	0.134	0.214	0.473	0.586	0.742	0.843	1.12	1.206
	Max	0.461	0.48	0.735	1.204	2.189	2.506	3.622	3.651	4.81	5.777
	Min	0.201	0.225	0.234	0.26	0.35	0.458	0.465	0.695	0.645	0.999
F9	Mean	0.207	0.276	0.299	0.415	0.718	0.842	1.474	1.598	2.981	4.015
	Median	0.175	0.176	0.227	0.341	0.573	0.658	1.315	1.249	1.541	2.314
	St. Dev.	0.149	0.282	0.203	0.253	0.593	0.659	0.888	1.008	3.36	3.815
	Max	0.791	1.128	1.045	1.109	3.264	3.225	5.102	5.058	15.441	15.168
	Min	0.0909	0.0846	0.108	0.112	0.211	0.281	0.588	0.31	0.577	0.92

Table 4 Test Data Results($\times 10^{-3}$). Mean, Median, Standard Deviation, Maximum and Minimum is across 30 networks for all 9 function approximation problems F1, F2,...F9 for $\alpha_1, \alpha_2, \ldots \alpha_{10}$ denoting weight initialization method corresponding to $\alpha = 1, 2, \ldots 10$, respectively.

		α_1	α_2	α_3	α_4	α_5	α_6	α_7	α_8	α_9	α_{10}
F1	Mean	21.692	19.572	15.565	12.951	12.141	13.264	14.804	14.921	20.286	28.652
	Median	21.841	18.669	15.813	11.093	10.602	12.492	12.199	13.348	19.781	27.106
	St. Dev.	3.131	3.173	5.271	5.824	6.44	7.565	7.03	5.737	7.382	12.367
	Max	29.802	27.82	28.914	25.926	33.811	33.799	28.367	27.678	35.931	52.634
	Min	15.172	13.771	4.812	4.713	5.057	2.342	5.93	5.246	8.988	8.485
F2	Mean	29.03	24.723	24.968	27.765	26.836	30.876	34.637	40.847	40.046	43.643
	Median	28.538	25.203	25.252	26.657	25.703	30.165	30.397	38.587	37.071	43.558
	St. Dev.	8.217	4.174	4.31	7.064	5.943	8.194	14.105	16.829	15.607	14.828
	Max	66.035	32.14	36.311	48.187	44.862	49.355	75.703	87.015	75.193	78.769
	Min	17.793	15.835	18.09	14.647	19.245	19.785	20.031	21.808	14.696	15.462
F3	Mean	0.225	0.12	0.194	0.352	0.541	0.786	1.075	1.954	1.786	2.687
	Median	0.173	0.118	0.17	0.333	0.444	0.689	0.897	1.547	1.518	2.377
	St. Dev.	0.135	0.0396	0.127	0.23	0.33	0.599	0.759	1.344	1.06	1.96
	Max	0.677	0.215	0.676	0.98	1.476	3.289	3.867	5.091	5.182	10.011
	Min	0.0644	0.0606	0.0602	0.0651	0.182	0.148	0.185	0.279	0.502	0.662
F4	Mean	35.095	51.837	31.802	25.663	23.362	24.081	26.437	39.936	38.425	47.304
	Median	36.71	35.856	27.923	24.866	19.231	22.194	24.655	30.775	25.308	34.935
	St. Dev.	7.703	61.316	10.826	8.161	11.673	11.693	10.545	32.309	45.671	37.711
	Max	50.626	278.63	63.32	46.765	52.57	62.249	49.941	159.215	253.085	165.822
	Min	21.005	17.81	19.43	12.335	5.256	7.975	10.563	10.031	11.412	15.176
F5	Mean	9.612	7.264	6.457	5.115	4.808	5.971	6.765	9.348	11.471	13.927
	Median	8.996	7.13	5.608	4.655	4.164	5.612	6.379	8.44	10.66	13.997
	St. Dev.	1.909	1.57	3.174	2	1.917	2.43	2.229	3.688	4.496	3.7
	Max	14.58	11.285	15.01	9.785	10.047	12.955	12.44	19.533	23.016	24.324
	Min	7.583	3.659	2.959	2.065	2.554	2.484	3.464	4.835	3.943	7.464
F6	Mean	4.32	2.956	3.114	2.947	3.082	4.542	4.522	7.065	6.341	11.167
	Median	4.786	1.737	2.88	1.79	2.287	3.802	3.996	5.046	5.684	8.393
	St. Dev.	2.269	2.077	2.008	2.32	2.233	3.002	2.927	6.749	3.584	10.578
	Max	8.468	7.445	8.032	8.694	10.484	11.426	11.92	34.783	16.447	48.445
	Min	1.351	0.859	0.681	0.339	0.34	0.898	0.385	1.595	1.467	1.716
F7	Mean	11.837	4.942	4.344	3.956	5.297	6.34	7.2	10.97	15.146	15.235
	Median	8.278	3.741	3.64	3.401	4.757	5.725	6.292	8.962	14.19	13.594
	St. Dev.	8.306	3.212	2.51	2.047	2.283	3.376	3.364	4.856	6.45	7.061
	Max	28.698	15.734	15.202	12.654	12.421	15.696	16.467	22.779	29.657	33.469
	Min	3.42	2.008	2.177	1.801	2.473	1.897	3.59	4.719	5.745	6.128
F8	Mean	3.042	3.739	4.84	8.62	12.253	16.9	25.211	29.685	44.321	72.062
	Median	3.003	3.753	4.642	8.358	11.642	13.932	22.718	24.861	39.235	62.882
	St. Dev.	0.412	0.613	1.211	3.161	4.643	8.337	11.127	16.096	25.792	43.719
	Max	4.045	5.225	8.246	19.286	22.605	35.574	51.769	84.762	104.87	189.495
	Min	2.085	2.68	2.962	3.577	5.206	6.571	7.748	9.966	6.762	13.875
F9	Mean	0.338	0.396	0.503	0.767	1.393	1.659	2.932	3.621	6.816	10.029
	Median	0.293	0.259	0.396	0.738	1.1	1.242	2.5	3.132	3.217	7.774
	St. Dev.	0.197	0.348	0.297	0.423	1.254	1.345	1.604	2.578	7.089	8.632
	Max	1.07	1.445	1.681	1.747	7.139	6.867	7.538	13.081	29.342	33.576
	Min	0.183	0.163	0.264	0.236	0.47	0.425	0.96	0.511	1.794	2.089

bol α_i where $i = 1, 2, \ldots 10$ is used for denoting techniques with $\alpha = 1, 2, \ldots 10$, respectively.

A one-sided t-test(left tail t-test) is conducted comparing methods α_i to α_j where $i, j = 1, 2, \ldots 10$. A 10×10 matrix is obtained with entries 1 and 0. If entry in (i, j) slot of matrix is 1, then method α_i is better than method α_j, if the entry is 0, then method α_i is statistically similar to method α_j. Such matrices are obtained for each of nine functions, and these nine matrices are superimposed. The summarized result is given in Table 5. Each entry in the slot (i, j) indicates the number of functions for which the performance of method α_i is better than performance of method α_j. The last column and row in Table 5, denoted by P_B and P_W, respectively, gives the sum of the entries in that row and column. The entry in the column P_B indicates the number of cases in which the performance of method α_i is better than all other methods α_j where $i, j = 1, 2, \ldots 10, j \neq i$. The last value in the jth column indicates the number of cases in which the performance of method α_i is worse than all other methods α_j where $i = 1, 2, \ldots 10, i \neq j$.

In Table 5, the highest value of P_B is 50 for α_4 The value indicates that for 50 cases, the performance of α_4 is better than all other methods. Also from Table 5, the lowest value of P_W is 9 for α_4. The value indicates that the performance is worse for least number of cases as compared to other algorithms. The values of P_B and P_W indicates that the performance of α_4 is by far the best as compared to other algorithms. The optimal value is obtained by subtracting the later from former. The results obtained are recorded in Table 6.

On comparison of values obtained in Table 6, it is observed that the highest value is 41 for α_4 which is equivalent to value of $\alpha = 4$. It can therefore be concluded based on the results that the optimal value of $\alpha = 4$ for at least these nine function approximation problems.

Table 5 Left tailed t-test. $\alpha_1, \alpha_2, \ldots \alpha_{10}$ denote weight initialization method corresponding to $\alpha = 1, 2, \ldots 10$, respectively. P_B denotes the number of cases for which performance of α_i is better than α_j where $i, j = 1, 2, \ldots 10, i \neq j$. P_W denotes the number of cases for which performance of α_i is worse than α_j where $i, j = 1, 2, \ldots 10, i \neq j$

	α_1	α_2	α_3	α_4	α_5	α_6	α_7	α_8	α_9	α_{10}	P_B
α_1	0	1	2	3	3	3	3	4	5	7	31
α_2	6	0	2	3	3	4	5	6	6	8	43
α_3	4	2	0	3	3	5	5	6	8	9	45
α_4	6	4	1	0	4	5	6	7	8	9	50
α_5	5	4	4	0	0	3	6	7	8	9	46
α_6	5	4	1	0	0	0	2	6	7	9	34
α_7	5	3	0	0	0	0	0	5	7	9	29
α_8	2	2	0	0	0	0	0	0	6	6	16
α_9	0	0	0	0	0	0	0	0	0	5	5
α_{10}	0	0	0	0	0	0	0	0	0	0	0
P_W	33	20	10	9	13	20	27	41	55	71	

Table 6 Optimal Performance Matrix. $P_B - P_W$ indicates the number of cases for which the performance of method α_i is optimal

	$P_B - P_W$
α_1	−2
α_2	23
α_3	35
α_4	41
α_5	33
α_6	14
α_7	2
α_8	−25
α_9	−50
α_{10}	−71

6 Proposed Weight Initialization Algorithm

Results in Table 6 show that the optimal value of α is 4. Henceforth, a new weight initialization technique is proposed in which magnitude of weights to layer L is 0.7×4, i.e., 2.8.

$$\|W\| = 2.8 \tag{22}$$

The biases to a node in layer L are initialized in the range $(-\|W\|, \|W\|)$, i.e., $(-2.8, 2.8)$. This range also the useful range of hyperbolic tangent function used as the activation function at the hidden layer. Useful range is the range where the value of derivative of activation function is significant. The performance of the proposed method α_4 (corresponding to $\alpha = 4$) is better than other methods ($\alpha = 1, 2, \ldots 10$, $\alpha \neq 4$) for atleast these nine function approximation tasks.

7 Conclusion

A variable parameter α is identified from a well-established weight initialization method (Nguyen–Widrow weight initialization method), and the value of α is varied from 1 to 10. On comparing these 10 techniques using one-sided t-test, it can be concluded that the optimal value of α of these 10 values is 4. Thus, a new weight initialization method is derived where the magnitude of weights to a layer L is 2.8, and all the biases to a layer L are initialized in the range $(-2.8, 2.8)$. The range $(-2.8, 2.8)$ is active range of hyperbolic tangent function which is activation function at the hidden layer.

Acknowledgements This publication is an outcome of the R&D work undertaken project under the Visvesvaraya PhD Scheme of Ministry of Electronics & Information Technology, Government of India, being implemented by Digital India Corporation.

References

1. Bottou, L.: Reconnaissance de la parole par reseaux multi-couches. In: Proceedings of the International Workshop Neural Networks Application, Neuro-Nimes, vol. 88, pp. 197–217 (1988)
2. Breiman, L.: The ii method for estimating multivariate functions from noisy data. Technometrics **33**(2), 125–143 (1991)
3. Chandra, P., Singh, Y.: Feedforward sigmoidal networks-equicontinuity and fault-tolerance properties. IEEE Trans. Neural Netw. **15**(6), 1350–1366 (2004)
4. Chen, C.L., Nutter, R.S.: Improving the training speed of three-layer feedforward neural nets by optimal estimation of the initial weights. In: 1991 IEEE International Joint Conference on Neural Networks, 1991. pp. 2063–2068. IEEE (1991)
5. Cherkassky, V., Gehring, D., Mulier, F.: Comparison of adaptive methods for function estimation from samples. IEEE Trans. Neural Netw. **7**(4), 969–984 (1996)
6. Cherkassky, V., Mulier, F.M.: Learning from Data: Concepts, Theory, and Methods. Wiley (2007)
7. Cybenko, G.: Approximation by superpositions of a sigmoidal function. Math. Control., Signals Syst. **2**(4), 303–314 (1989)
8. Drago, G.P., Ridella, S.: Statistically controlled activation weight initialization (scawi). IEEE Trans. Neural Netw. **3**(4), 627–631 (1992)
9. Funahashi, K.I.: On the approximate realization of continuous mappings by neural networks. Neural Netw. **2**(3), 183–192 (1989)
10. Hagan, M.T., Menhaj, M.B.: Training feedforward networks with the marquardt algorithm. IEEE Trans. Neural Netw. **5**(6), 989–993 (1994)
11. Haykin, S., Network, N.: A comprehensive foundation. Neural Netw. **2**, 2004 (2004)
12. Hornik, K., Stinchcombe, M., White, H.: Multilayer feedforward networks are universal approximators. Neural Netw. **2**(5), 359–366 (1989)
13. Irie, B., Miyake, S.: Capabilities of three-layered perceptrons. In: IEEE International Conference on Neural Networks, vol.1, p. 218 (1988)
14. Jones, L.K.: Constructive approximations for neural networks by sigmoidal functions. Proc. IEEE **78**(10), 1586–1589 (1990)
15. Kim, Y., Ra, J.: Weight value initialization for improving training speed in the backpropagation network. In: 1991 IEEE International Joint Conference on Neural Networks, 1991. pp. 2396–2401. IEEE (1991)
16. Maechler, M., Martin, D., Schimert, J., Csoppenszky, M., Hwang, J.: Projection pursuit learning networks for regression. In: Proceedings of the 2nd International IEEE Conference on Tools for Artificial Intelligence, 1990, pp. 350–358. IEEE (1990)
17. Mittal, A., Chandra, P., Singh, A.P.: A statistically resilient method of weight initialization for sfann. In: 2015 International Conference on Advances in Computing, Communications and Informatics (ICACCI), pp. 1371–1376. IEEE (2015)
18. Nguyen, D., Widrow, B.: Improving the learning speed of 2-layer neural networks by choosing initial values of the adaptive weights. In: 1990 IJCNN International Joint Conference on Neural Networks, 1990, pp. 21–26. IEEE (1990)
19. Riedmiller, M., Braun, H.: A direct adaptive method for faster backpropagation learning: The rprop algorithm. In: IEEE International Conference On Neural Networks, 1993, pp. 586–591. IEEE (1993)

20. Sodhi, S.S., Chandra, P.: A partially deterministic weight initialization method for sffanns. In: 2014 IEEE International Advance Computing Conference (IACC), pp. 1275–1280. IEEE (2014)
21. Sodhi, S.S., Chandra, P., Tanwar, S.: A new weight initialization method for sigmoidal feedforward artificial neural networks. In: 2014 International Joint Conference on Neural Networks (IJCNN), pp. 291–298. IEEE (2014)
22. Yam, J.Y., Chow, T.W.: A weight initialization method for improving training speed in feedforward neural network. Neurocomputing **30**(1), 219–232 (2000)

Impact Analysis of LFM Jammer Signals on Stepped Frequency PAM4 RADAR Waveforms

K. Keerthana and G. A. Shanmugha Sundaram

Abstract Recognizing targets becomes a difficult process when the jamming signal tries to overwhelm the signals detected at a radar receiver. In radar systems, the most common type of signals used are the continuous and pulsed waveforms. For enhanced target detection and range resolution, continuous changes in the waveforms are required in pulsed waveforms. For continuous waveforms, such changes are not required, and both transmitter and receiver are working as two separate antenna systems as in a monostatic radar, which can provide a high signal-to-noise ratio (SNR). In this work, the linear frequency modulated (LFM) waveform is introduced as a strong jamming signal. The radar signals of the compound pulsed radar modulation type are considered here, with the combination of stepped frequency waveform (SFW) and a four-level pulse amplitude modulation (PAM4). The SFW is capable of high range resolution (HRR), while PAM4 can provide a lower bandwidth. This new approach shows characteristics such as better target detection, good range resolution, and proper range measurements, even with the presence of the LFM jamming signal. Simulation is done using a monostatic radar system to analyze the radar operations and performance. Results are investigated in terms of different radar parameters like range-Doppler response, FFT spectrum, target returns, burn-through range, cross-over range, and signal-to-jammer-noise ratio.

Keywords Stepped frequency waveforms · PAM4 · LFM · RADAR jammer · Monostatic radar · Range-Doppler response · Target return · Burn-through range · Cross-over range

---

K. Keerthana · G. A. Shanmugha Sundaram (✉)
Center for Computational Engineering and Networking, Amrita School of Engineering, Amrita Vishwa Vidyapeetham, Coimbatore, India
e-mail: ga_ssundaram@cb.amrita.edu

G. A. Shanmugha Sundaram
Department of Electronics and Communications Engineering, Amrita School of Engineering, Amrita Vishwa Vidyapeetham, Coimbatore, India

G. A. Shanmugha Sundaram
SIERS Research Laboratory, ASE Coimbatore, Amrita University, Coimbatore, India

S. M. Thampi et al. (eds.), *Intelligent Systems, Technologies and Applications*, Advances in Intelligent Systems and Computing 910,
https://doi.org/10.1007/978-981-13-6095-4_12

1 Introduction

Rapid developments have been noticed in radar systems in recent years that have to do with target detection, discrimination, and characterization [1], while the diverse applications of radar systems made for more intense research. Modern radar systems have in addition become much more perfect in terms of collecting information from the receiver side perspective.

Radar systems are generally classified into two specific categories in terms of the locations of the transmit and receive modules: They are the monostatic and bistatic types. The presence of spatially separated system in a bistatic configuration makes it designing much complex and high cost. Monostatic radars can be designed as less complex, low-cost system with proper target tracking and detection, and with reduced false alarm rate [2]. A monostatic radar is capable of a high signal-to-noise ratio (SNR) [3]. Current innovations in monostatic radar considered it as a big platform for research-oriented works.

Target tracking and detection become undesirable when there is the presence of reflections from land, sea, ground surfaces and features, etc. These reflections are termed as clutter. A received radar signal also includes clutter and jammer components. These interferences are cleverly hidden using space-time adaptive processing (STAP) techniques [4].

Recent developments in a radar system have focused on electronic countermeasure (ECM) to mask the target from the radar receivers. Presence of targets can be masked by radio frequency signals coming from an external source which is referred to as the radar jammer process. One of the best innovative techniques used in masking of targets is by deploying digital radio frequency memory (DRFM) [5, 6]. Radar systems cannot identify the targets precisely because DRFM jammers operate in a time-constrained, radar signal spamming mode.

In this work, a linear frequency modulated (LFM) waveform is considered for the jamming signal. LFM was considered because an external weighting factor is needed to reduce the side lobes so as to provide low SNR and range resolution, while the presence of multiple frequencies can provide a difficult task for the receiver to distinguish the target from the jammer [7, 8].

Various types of modulation techniques are considered in radar systems for proper target tracking and detection, each resulting in a different type of waveform, such as non-linear frequency modulation (NLFM), LFM, Barker, Polyphase codes, etc. These types of modulation schemes are being implemented to achieve pulse compression, that provides simultaneous high range resolution (HRR) and high SNR. Using the X-band operating frequency is also a main factor for obtaining HRR [9]. The principle modulation scheme chosen here is the SFW. Integrating SFW and PAM4 results in a compound modulation scheme, and this is advanced as an innovative technique for desirable target detection characteristics in the presence of an LFM jammer signal. Analysis of LFM jammer impact on this compound modulation scheme is performed in terms of different norms like the range-Doppler response, cross-over range, burn-through range, and the signal-to-jammer-plus-noise ratio.

The work presented here is formulated as follows—Sect. 2 reports about the conceptual and mathematical background of SFW, PAM4, monostatic radar, and LFM radar jamming. Section 3 reports about Methodology and Simulation, while Sect. 4 reports on the radar receiver side results obtained with SFW and PAM4 in the presence of an LFM radar jammer signal. The overall conclusions from this work are reported in Sect. 5.

2 Conceptual and Mathematical Background

2.1 *Stepped Frequency Waveform (SFW)*

Higher range resolution (HRR) can be accomplished by minimizing the range bins. All together to get high range determination, pulse width must be shorter; henceforth, the transmitted power would need to be generally lower per duty cycle, which, however, lowers the SNR value on the beneficiary side. Such issues can be positively fathomed with pulse compression technique; certain methods utilized are PAM4 and LFM and frequency modulation schemes [9].

An SFW that involves a sequence of N narrowband pulses is depicted in the schematic Fig. 1. By a fixed sum f, the frequency is extended from step to step [10, 11] to obtain an SFW, a well-known pulse compression technique that is close to other radar waveforms, since it helps achieve a higher range resolution, without giving up the signal power requirement for target identification [11].

Higher SNR values can be obtained even with a low operating bandwidth of the receiver, constructed out of narrow instantaneous bandwidths, which an SFW helps

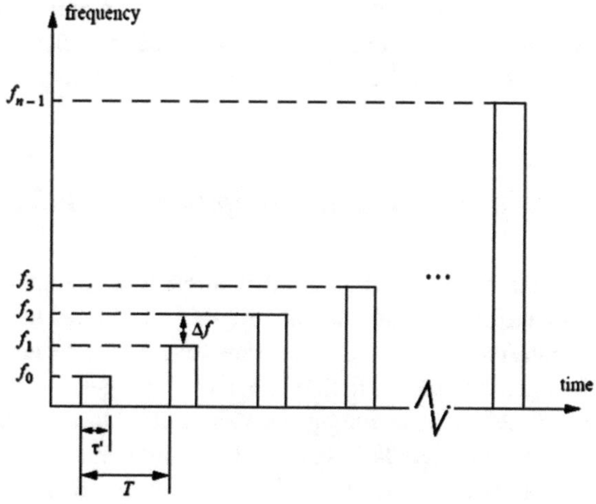

Fig. 1 Schematic of an SFW with constant Δf

realize. The SFW signals are transmitted at a low sample rate and give an adaptive frequency control during transmission of signals [11].

In general, the center frequency for the step is given by Paulose [11]:

$$f_i = f_o + i\Delta f \tag{1}$$

where f_o is the starting carrier frequency, Δf is the step size and i varies from $i =$ 0 to $n-1$, and n is a positive integer. The transmitted waveform for the ith step is formulated as [11]:

$$S_i(t) = \left(\frac{C_i \cos 2\pi f_i t + \theta_i ; iT \leq t \leq iT + \tau'}{O; \text{elsewhere}} \right) \tag{2}$$

where θ_i is the relative phases and c_i are constants.

Range resolution is given by:

$$\Delta R = \frac{C}{2B} \tag{3}$$

And ΔR can be rewritten as:

$$\Delta R = \frac{C}{2N\Delta f} \tag{4}$$

while the bandwidth gets defined by:

$$B = N\Delta f \tag{5}$$

The spectrum of an SFW reveals the frequencies defined changes to signal intensity over every time interval. As the frequency increases, the progression estimate expands straightly to a specific point, and after that, it begins once more [12] and a constant Δf is added every time to the center frequency.

2.2 Four-Level Pulse Amplitude Modulation (PAM4)

In order to transmit digital information, four distinct pulse amplitude levels are used to quantize an analog signal, and this constitutes the PAM4 modulation technique. Various modulation techniques are used to shift the length, the width, the frequency of the signal and the position of the individual signals arranged in specific arrangements, and PAM4 is one among the recent amplitude modulation techniques [13]. One advantage of PAM4 is that, without any bandwidth redundancy, the required data rate can be increased [9]. PAM4 captures the least significant bit (L) signal and adds

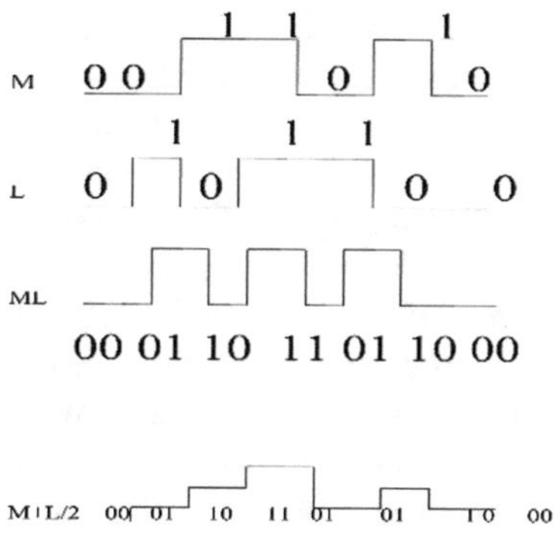

Fig. 2 PAM4 symbol representation waveform diagrams

to the Most Significant Bit (M) signal. Therefore, the four signal levels get converted to two with each signal level corresponding to a two-bit symbol.

A PAM4 symbol trace shown in Fig. 2 has at the bottom a 00 which is followed by 01, 10, 11, which actually is indicative of the four levels in signal amplitude. The two bits are transmitting in parallel, and thereby, the data rate increases. PAM4 is more of a bandwidth efficient modulation scheme compared to various conventional modulation schemes such as QAM or OFDM [13].

Based on the number of amplitude levels, PAM4 is titled as M-ary, where M stands for amplitude, which is usually represented as 2 N where $N = 1, 2, 3$, etc. Higher spectral efficiency is one advantage of PAM4. PAM4 can provide higher SNR and better bandwidth with the understanding that it reduces the expected bandwidth capacity to half for a given data rate, comparing with the PAM2 NRZ [13]. It is well known that PAM4 can reduce intersymbol interference (ISI) and help obtain a higher data rate [14]. The performance analysis is in terms of the bit error rate (BER) and comparison was studied on other modulation scheme like quadrature phase shift keying (QPSK) and binary phase shift keying (BPSK) [14]. With the same data rate more received power is needed for PAM4 when compared with other modulation, since the reduced bandwidth sets a penalty which is less when the data rate is kept constant [15]. The power required for PAM4 to obtain the required SNR is given by:

$$P_{M-\text{PAM}} = 10\log_{10}\left[(M-1)/\left(\log_2 M\right) \wedge 0.5\right] \tag{6}$$

The received power can be reduced using Eq. (6), using the same bit rate needed for other modulation scheme in order touch the same BER.

2.3 *Monostatic Radar*

A monostatic radar is a type of radar in which both transmitter and receiver are co-located. In order to transmit and receive the echo signals, a monostatic radar uses the same antenna [2]. Improved range detection and better range measurements can be obtained when using a monostatic radar antenna system.

The monostatic radar is designed mainly to detect the targets which are in the far range and includes non-fluctuating targets with a particular radar cross-section (RCS) and range resolution. Probabilities of detection and false alarm rates should be carefully considered so as to ensure that the echo signals are not compromised with false information. Range detection can be done in two ways, viz. coherent and non-coherent detection. Generally, non-coherent detection is preferred because it does not need complex computational processing associated with phase angle preservation across the transmit/receive modules in the radar system [2].

For designing a monostatic radar, the main factor considered here is an antenna array. Targets get identified by the radar when it enters the main antenna beam. Usually, when a radar searches for targets it generates an error in the direction of the target. While considering the power factor for the radar plan, a radar target takes a segment of energy; however, reradiates it in the radar course [5]. The quality of the power is resolved by the factor called RCS. The peak power at the radar receiver input (P_{r}) can be defined by Sai Shiva et al. [2]:

$$P_{\mathrm{r}} = P_{\mathrm{t}} G_{\mathrm{t}} G_{\mathrm{r}} \left[\frac{\sigma_c}{(4\pi)^3} * \frac{1}{f^2} * \frac{1}{R^4} \right] \tag{7}$$

where P_{t} is the transmitter power, G_{t} is the transmitter antenna gain, G_{r} is the receiver antenna gain, is the RCS in m^2, R is the distance from the transmitter and f is the pulse repetition frequency (PRF). The wavelength of the target signal being $\lambda = \frac{c}{f}$, the radar range equation calculates the range at which the target cannot be detected accurately or its critical parameters estimated. In the received echoes from a target, the minimum detectable signal power (Smin) is given by Sai Shiva et al. [2]:

$$R\mathrm{max} = \sqrt[4]{\frac{P_{\mathrm{t}} G_{\mathrm{t}} G_{\mathrm{r}} \lambda^2 \sigma}{(4\pi)^2 S_{\mathrm{min}}}} \tag{8}$$

The most important parameter while designing a monostatic radar is the PRF, which is the number of pulses of a periodic signal in a specific time unit that is measured in pulses per second [16]. PRF is chosen based on the maximum unambiguous range, given the need to have a minimum sample rate that is twice the signal bandwidth as per the Nyquist sampling criterion.

Using the probability density function (PDF), the probability of detection can be estimated as using input signal power to noise power ratio [2]:

$$P(x, y) = \frac{1}{y} * \mathrm{e}^{x/y} \quad (x \geq 0) \tag{9}$$

where x is the input signal to noise power ratio and y is the average of x overall fluctuation of the target observed. If the target fluctuations are random, then x is assumed to be a random variable. The PRF is calculated based on the maximum range. The series of steps used here are [2]:

$$\text{Pulse bandwidth(BW)} = \text{Propagation speed}/(2 * \text{range resolution}) \tag{10}$$

$$\text{Pulse width} = 1/(\text{BW}) \tag{11}$$

$$\text{PRF} = 3 * 10^8/(2 * \text{maximum range}) \tag{12}$$

The required SNR is calculated using Albersheim equation [2], applied to non-fluctuating targets and practical results are compared using the expression:

$$\text{SNR}_{\text{dB}} = -5\log_{10}(n) + \left[6.2 + (4.54/\sqrt{n+0.44})\log(A + 0.12AB + 1.7B)\right] \tag{13}$$

for n pulses, where $A = \log_e\left(\frac{0.62}{\text{PFA}}\right)$, and $B = \log_e\left(\frac{\text{PD}}{1-\text{PD}}\right)$.

Here, the probability of false alarm (PFA) is in the order of 10^{-6}, and the probability of detection (PD) ranges from 0.8 to 0.9, and n is the number of samples needed for the detection test.

2.4 *LFM as a Jammer Signal*

Jamming of radar receivers can be modeled in the form of transmit signal distortions or interference components. In this work, LFM is considered as the jamming signal, and it is passed along besides the echo signal. LFM has been the most preferred jamming signal here, compared to random noise and other interferences, since it is found to resolve the targets in range and Doppler domains in the presence of channel noise and radar side lobes. The LFM signal has been considered [7, 8] in the form of a noise in the radar waveform.

The LFM signal is depicted in Fig. 3. The frequency scales linearly across the pulse width, either as an up-chirp (upward) or a down-chirp (downward). Here, the signal is generated as up-chirp signal with specific values for the instantaneous frequency and the center frequency.

LFM chirp waveform complex exponential version is given by Parwana and Sanjay [7]:

$$s(t) = \exp(j\theta(t)) \tag{14}$$

where $\theta(t)$ is the instantaneous phase angle of the signal:

$$\theta(t) = 2\pi(f_o + kt^2) \tag{15}$$

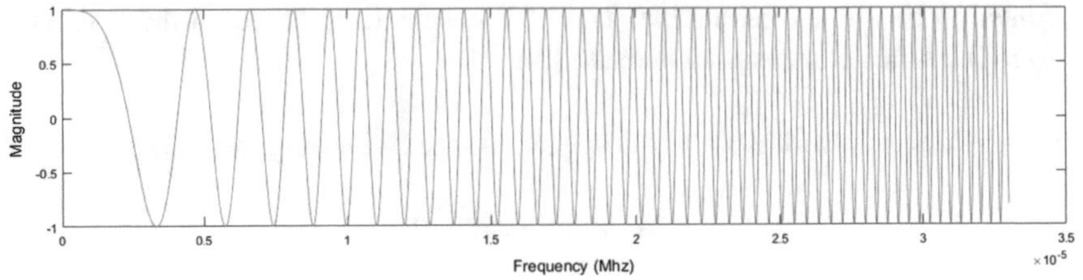

Fig. 3 Up-chirp LFM signal representation

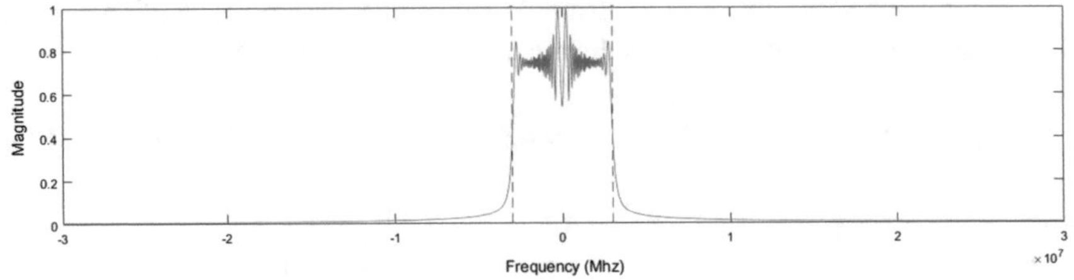

Fig. 4 Spectrum of an LFM signal

where f_o is the radar center frequency.

The instantaneous frequency as a linear function of time is given by Parwana and Sanjay [7]:

$$f_{\mathrm{i}} = \frac{1}{2\pi}\left(\frac{\mathrm{d}\phi}{\mathrm{d}t}\right) \tag{16}$$

$$\text{i.e., } f_{\mathrm{i}} = f_o + kt \tag{17}$$

where κ is the slope and t is the instantaneous time.

The spectrum of LFM is shown in Fig. 4. The frequency components are defined from the characteristics in the spectrum. The spectrum spreads as the pulse bandwidth reduces, while the spectral information reduces when the oversampling rate decreases. Comparing with the normal barrage jammer, an LFM jammer signal can provide stronger interferences [8]. Instead of using a parameter such as the SNR, the SNR-to-interference ratio is used to determine the detection ability of targets by the radar receiver in the midst of a jammer operation. Factors like radar peak power, radar antenna gain, radar operating frequency, target cross-section, radar operating bandwidth, radar losses, jammer peak power, jammer bandwidth, jammer signal's antenna gain, and jammer signal's losses are primarily used for obtaining the jamming signal characteristics. Variation in such factors makes corresponding changes in the performance of the radar as far as target detection is concerned.

The effective radiated power (ERP) is the most important element used for determining the jammer signal's characteristics and this is proportional to the jammer transmitted power P_{j} [17]:

$$\text{ERP} = \frac{P_j G_j}{L_j} \tag{18}$$

where G_j is the jamming signal's antenna gain, L_j is the total jamming signal's losses.

Self-protecting jammers are considered as the main beam jammers, which belong a class of ECM systems and they are carried on the vehicle in which are carrying [16]. Considering a radar with an antenna gain G, wavelength aperture A_r, bandwidth B_r, receiver losses L, and peak power P_t, the radar range equation defines the scattered power available at the receiver for a target at a distance of R km [17]:

$$S = P_t G^2 \lambda^2 \sigma \tau / (4\pi)^3 R^4 L \tag{19}$$

where τ is the pulse width. The power received by the radar from self-protecting jammers at the same range is:

$$J = \frac{P_j G_j}{4\pi R^2} \frac{Ar}{B_j L_j} \tag{20}$$

where P_j, G_j, B_j, L_j are the jammer peak power, antenna gain, operating bandwidth, and losses. Therefore, S/J is given by Mahafza [17]:

$$(S/J) = P_t \tau G \sigma B_j / (\text{ERP})(4\pi) R^2 L \tag{21}$$

The jammer transmission with reference to the radar receiver is a one-way process, the power reaches the radar beside the radar transmitted waveform which is a two-way process, and the target echo reaches the radar, usually with the jamming signal power that is much larger than the target's signal power.

As the target progressively gets nearer to the radar, there will be a particular range in which the (S/J) factor becomes equivalent to unity. This is known as the cross-over range [17]. When the S/J ratio is relatively larger than unity, it defines the detection range, given the receiver bandwidth B_r. To find the cross-over range, the ratio of S/J is fixed to be unity resulting in:

$$(R_{co})_{SSJ} = \sqrt{P_t G \sigma B_J / 4\pi B_r L (\text{ERP})} \tag{22}$$

The radar burn-through range is defined as the range at which the radar can detect and perform proper range measurements with a particular $\frac{S}{J+N}$ [17]:

$$S/(J+N) = \left[\frac{P_t G \sigma A_r \tau}{(4\pi)^2 R^4 L}\right] \Big/ \left[\frac{(\text{ERP}) A_r}{4\pi R^2 B_j} + kT_0\right] \tag{23}$$

To calculate the probability of detection instead of SNR, $\frac{S}{J+N}$ is used, while the SNR is replaced with $\frac{S}{J+N}$ when the detection is based on a coherent or non-coherent integration.

3 Methodology and Simulation

A complete radar system has been implemented as a simulation activity using the Matlab-Simulink-Phased Array toolbox is represented in Fig. 5 [2]. The model consists of a transceiver, array, channel, a target, a jamming signal, and a clutter functional block. In the current work, the microwave X-band operating frequency (10 GHz) is considered. X-band frequencies are used extensively in civilian and military airspace monitoring activities. This region of the spectrum is mainly used in modern radar systems and they employ both continuous and pulsed waveform parameters.

In the transceiver section, the different operational units comprise as SFM block and a PAM4 block, transmitter block, narrowband transmit array, narrowband receiver array, and the related signal processing blocks. The SFW block creates the SFW signal corresponding to a known pulse width, pulse bandwidth, PRF, and the number of frequency steps. The PAM4 block produces the signal based on the M-ary amplitude modulation of the radar waveform. Both these signals are multiplexed and the resultant signal arrives at the transmitter block. The transmitter block amplifies the signals and sends it to the receiver block over a propagation channel [2]. The receiver preamplifies the signals across a narrowband receiver array and passes it to the signal processing block. The signal processing block performs a matched filtering operation, followed by the pulse integration process to enhance the genuine signal component. Matched filtering is usually performed to increase the SNR on the receiver side [18], while the integration process is of the incoherent type that reduces complexity and enhances target detectability [2].

In the array section, the signal from the transmitter is passed to the narrowband transmit array, which generates narrowband planes waves in the far field of the array. This is followed by signal propagation across a channel. The narrowband receiver array block collects the signal from the channel, with the corresponding angle information of the narrowband transmit array provided to the receiver array block.

In both the channel blocks, different multipath propagation and channel fading losses are identified. The target block explains the target's position, velocity, and direction of travel. A stationary target has been considered here. The jamming signal block used here is LFM signal, as it provides a strong jamming component. The detection performance is analyzed in the presence of this LFM jamming signal. Finally, the clutter is also included in order to increase the interferences to the genuine target echo signal.

In the present simulation, there are three models used, which are a target model, jammer model, and clutter model. In order to design a target, the following parameters have been considered: range, Doppler shift, the direction of arrival, time of arrival, and RCS. RCS is defined as the detection capability of a radar for a specific target of interest. For each type of radar, the RCS value considered will be different [19].

In the present study, for the design of the jammer block, the main factors considered are the jammer peak transmitting power, a range of the jammer. The LFM signal is taken as the jamming signal to confuse the radar when it comes to defining

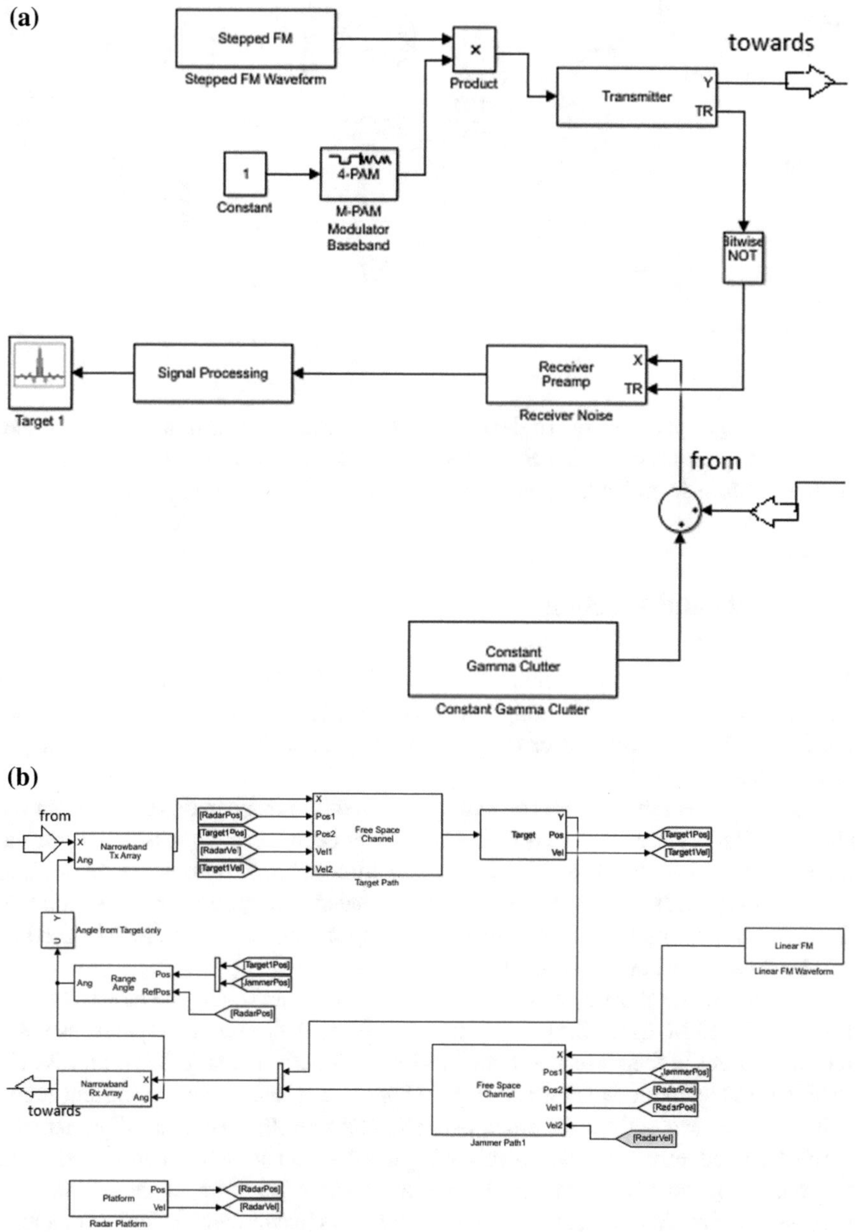

Fig. 5 **a**, **b**, **c** Outline of Simulink model for monostatic radar using SFW, LFM, and PAM4 waveforms

(c)

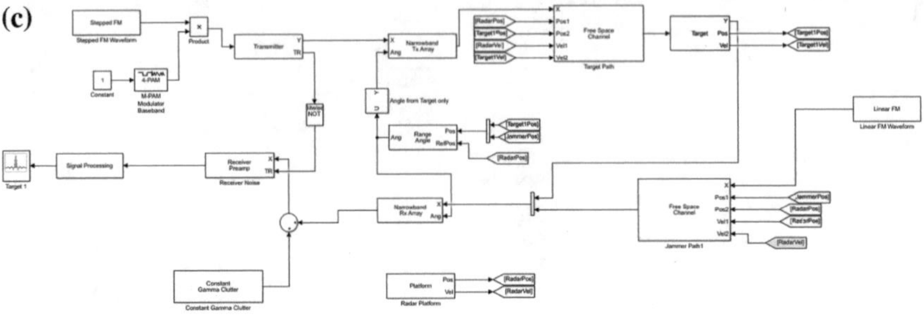

Fig. 5 (continued)

the exact target parameters. To design a clutter, parameters such as grazing angle, horizon range, surface clutter RCS, and the effective Earth radius are considered, with reflections from land, sea getting classified as surface clutter [20].

4 Results and Discussion

The performance due to the combined effect of SFW and PAM4, along with an LFM jamming signal is analyzed through important metrics like range-Doppler response, burn-through range, cross-over range, signal-to-jammer-noise ratio, and the target return plots.

The simulated target returns during operation of the monostatic radar system are shown in Fig. 6. Here, the x-axis represents the range in km and y-axis represents the amplitude value of the target. The precise location of target returns can be inferred through this plot. It is found that the target is located at a location of above 200 km. Here, the target is plotted along with the jamming signal, and its precise location has still been estimated unambiguously.

The analysis for radar system performance is performed using the range-Doppler response for LFM signal [21]. Using these observations, a range-Doppler response plot of a target in the presence of an LFM jamming signal using SFW, and PAM4 modulation technique is shown in Fig. 7. Presence of the target at different range values can be evidently determined using such a plot. In case of multiple targets, their combined influence on the received signals can also be clearly interpreted using the range-Doppler response map, along with the direction of travel and its intent. It process of estimating the properties of several targets that move at different velocities and at different ranges becomes a less complex exercise. The range-Doppler map corresponds to zero Doppler value for a stationary target, if the transmitter platform is also stationary. For targets moving with a particular velocity relative to the transmitter platform, the range-Doppler corresponds to non-zero Doppler values. The response would show both the target as well as the jammer at two different positions. Here, since the target and jammer are considered as stationary, the map is analyzed to a

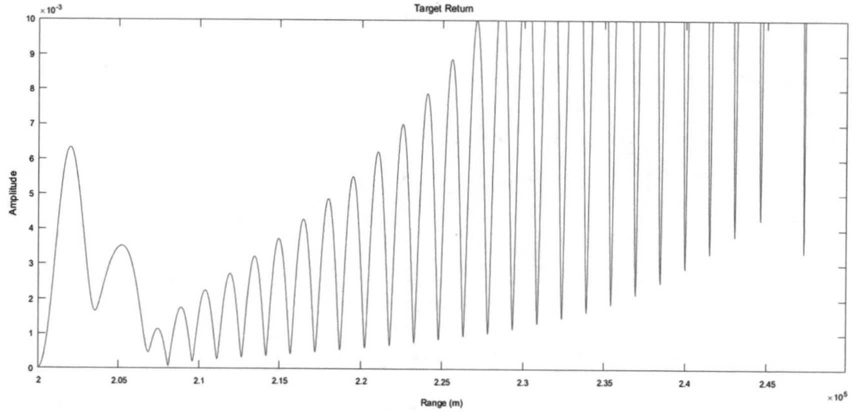

Fig. 6 Target returns indicating its exact position

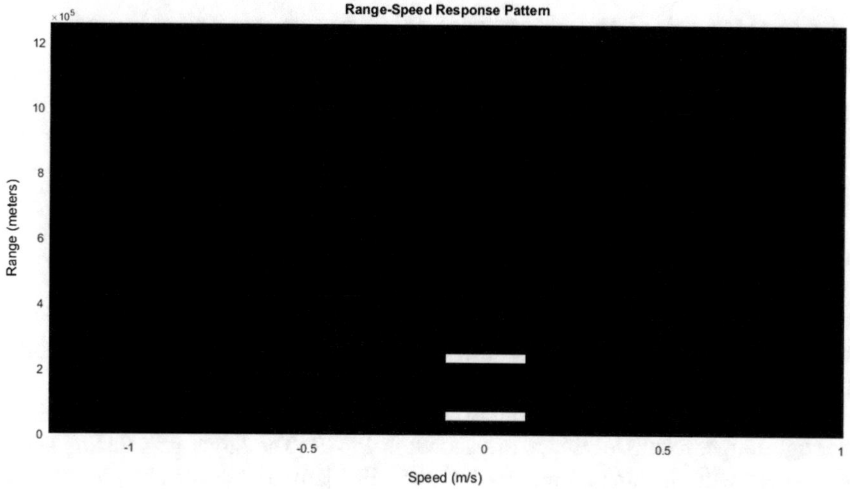

Fig. 7 Range-Doppler response showing a target and a LFM jammer

zero Doppler value. It is a requirement that the receiver operates at peak sensitivity even when the target is observed in the map in the presence of LFM jamming and clutter, to portray a realistic radar detection scenario.

The detection range as a function of signal-to-jammer-noise ratio is plotted in Fig. 8. It is characterized as the range at which the radar can perceive and perform proper estimations for a given $S/(J + N)$ value. As the detection range increases, the signal-to-jammer ratio proportionately diminishes. This procedure can now be applied to distinguish the target at the exact position even with the presence of the LFM jamming signal. Radar peak power, target cross-section, and jammer peak power are few parameters which affect the detection range and signal-to-jammer

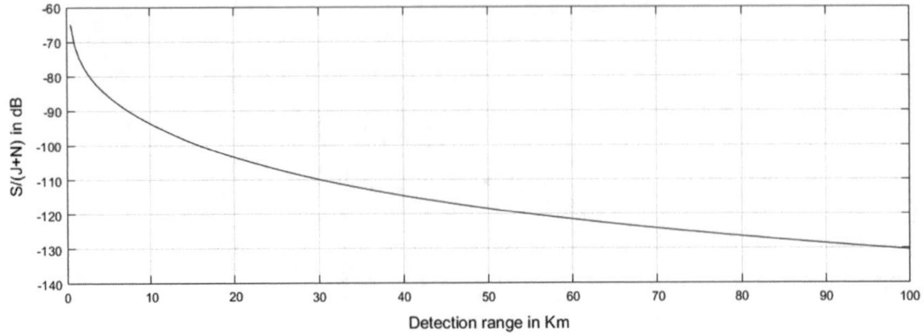

Fig. 8 Changes in detection range as a function of signal-to-jammer-noise ratio

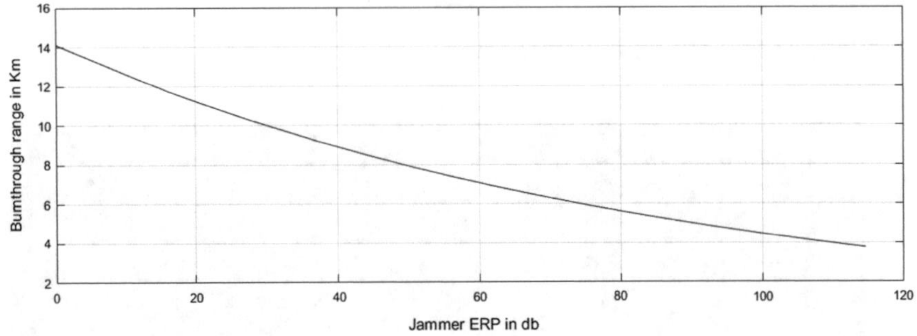

Fig. 9 Variation of ERP as a function of burn-through range

ratio. The plot changes based on the variations in the values of the above stated parameters.

The next performance parameter to be studied is the ERP as a function of the burn-through range, results of which are shown in Fig. 9. The target gets distinguished absolutely with the tandem presence of the LFM jammer signal. The plot demonstrates that as the LFM jammer's ERP increases, the corresponding burn-through range is in a downward trend. At a specific point, the energy from the jammer turns out to become significantly low that it can no more conceal the target echoes. From that specific range onwards, the recognition of a target becomes possible, even while vital radar signal parameters such as the bandwidth, operating frequency, or changes to this frequency are entirely unknown to the jammer. Radar pulse width, jammer antenna gain, radar losses can largely affect the jammer's ERP, and burn-through range. Based on such parameters, the ERP either increases or decreases relative to the burn-through range.

The impact of changes in cross-over range with radar peak power and LFM jammer peak power are shown in Fig. 10. Sub-plot (a) demonstrates that the cross-over range decreases when jammer power increases, while sub-plot (b) shows that the cross-over range remains constant for a particular value of the radar peak power. Thus, the radar peak power is enhanced when the cross-over range increases gradually past a

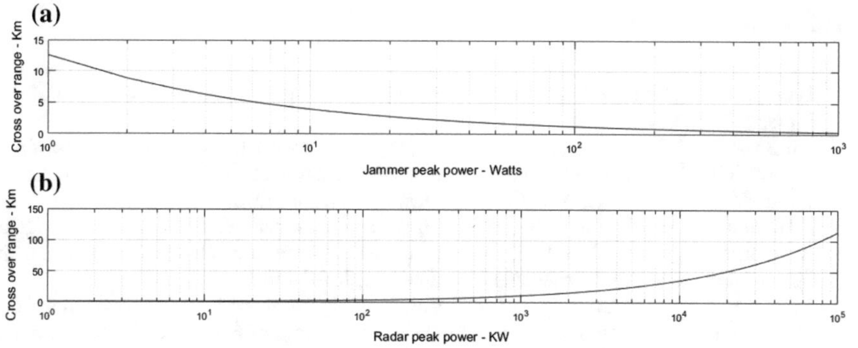

Fig. 10 Changes in cross-over range with LFM jammer peak power (**a**: above) and radar peak power (**b**: below)

specific range of peak power of the radar transmitter signal. Based on the changes of the jammer's bandwidth, radar operating frequency, target's RCS the plot either increases or decreases.

5 Conclusions

In the present work, SFW is taken as the principle modulation scheme for radar systems. Combining SFW with a recent modulation scheme, viz. PAM4, results in a compound modulation technique. The effectiveness of this new technique has been analyzed with a very effective LFM jamming signal. Different criterion such as range-Doppler response, burn-through range, signal-to-jammer-noise ratio and the cross-over range, and the target returns are evaluated to examine this new modulation technique.

Results show that this new modulation technique can be used for better target tracking and detection even with an intense jamming signal as an interference source. This compound modulation technique has been deployed to get accurate results in target detection even at greater ranges and diminished RCS values. The new compound modulation technique also has the potential to be implemented in technologically more complex detection systems such as bistatic radar, passive radars, and MIMO radars, and for application as diverse as imaging, identification, weather monitoring, and projectile tracking.

References

1. Harold, M.: An Introduction to Radar. 2. In: Remote Sensing with Polarimetric Radar. IEEE Press, p. 284 (2007)
2. Sai Shiva, A.V.N.R., Khaled, E., Eman, A.: Improved monostatic pulse radar design using ultrawide band for range estimation. In: Industrial Electronics, Technology & Automation (CT-IETA), Annual Connecticut Conference, pp. 1–7. IEEE (2016)
3. Prakash, D., Shreya, P.: Improving the performance of monostatic Pulse RADAR by changing waveform design using MATLAB In: International Conference on Recent Developments in Science, Engineering and Technology, pp. 345–352. IEEE (2017)
4. Li, X.-M., Ding, l., Chao-yang, Q., Chunsheng, L.: Adaptive generalized DPCA algorithm for clutter suppression in airborne radar system. In: Proceedings of International Conference on Radar. pp. 1093–1098. IEEE (2011)
5. Changyong, J., Gao M., Wang Z., Fu X.: Design of high-speed DRFM system. In: Computer science and information Engineering, pp. 582–586. IEEE (2009)
6. Amuru, S.D., Michael Buehrer, R.: Optimal jamming strategies in Digital communications Impact of modulation. In: Global Communications Conference (GLOBECOM), pp. 1619–1624. IEEE (2014)
7. Parwana, S., Sanjay, K.: Analysis of LFM and NLFM radar wave forms and their performance analysis. Int. Res. J. Eng. Technol. **2**(2), 334–339 (2015)
8. Paik, H., Sastry, N.N., SantiPrabha, I.: Effectiveness of repeat jamming using linear FM interference signal in monopulse receivers. Procedia Comput. Sci. **57**, 296–304 (2015)
9. Ankarao, V., Srivatsa, S., and Sundaram, G.A.S.: Evaluation of pulse compression techniques for X-band radar systems. In: Wireless Communications, Signal Processing and Networking (WiSPNET), International Conference, pp. 1287–1292. IEEE (2017)
10. Gonzalez-Blanco, P., Eduardo de D., Enrique M., Borja E., Ignacio M.: Stepped-frequency waveform radar demonstrator and its jamming. In: International Conference on Waveform Diversity and Design, pp. 192–196. IEEE (2009)
11. Paulose, A.: High radar range resolution with the step frequency waveform. Master Thesis, Naval Post Graduate School Monterey, CA (1994)
12. Mazzaro, G., Koenig, F.: Introduction to stepped-frequency radar. In: 90-minute seminar, US Army Research Laboratory (2013)
13. Breyer, F., Jeffrey, L., Sebastian, R,, Norbert, H.: Comparison of OOK-and PAM4 modulation for 10 Gbit/s transmission over up to 300 m polymer optical fiber. In: Optical Fiber Communication Conference, p. OWB5. Optical Society of America and IEEE (2008)
14. Srivatsa, S., Sundaram, G.A.S.: PAM4-based RADAR counter-measures in hostile environments. In: The International Symposium on Intelligent Systems Technologies and Applications, pp. 390–400. IEEE (2017)
15. Szczerba, K., Westbergh, P., Karout, J., Gustavsson, J.S., Haglund, Å., Karlsson, M., Larsson, A.: 4-PAM for high-speed short-range optical communications. IEEE/OSA J. Opt. Commun. Networking **4**(11), 885–894 (2012)
16. Melvin, W.L., Wicks, M.C., Brown, R.D.: Assessment of multichannel airborne radar measurements for analysis and design of space-time processing architectures and algorithms. In: Proceedings of the IEEE National Radar Conference, pp. 130–135. IEEE (1996)
17. Mahafza, B.R.: Radar Systems Analysis and Design Using MATLAB, 2nd edn. Chapman and Hall/CRC (2005)
18. Salahdine, F., Hassan E.G., Naima K., Wassim F.F.: Matched filter detection with dynamic threshold for cognitive radio networks. In: Proceedings of International Conference on Wireless Networks and Mobile Communications (WINCOM), pp. 1–6. IEEE (2015)
19. Wessling, A.: Radar target modelling based on RCS measurements, Project Report 54. Department of Systems Engineering, Linköping University (2002)

20. Jezak, J.A., Guy, G., Charles, B., Steven, P.L., Donald, M.C.: Configurable clutter models for radar simulation. In: Proceedings of International Conference on Waveform Diversity & Design, pp. 125–1. IEEE (2012)
21. Rossler, C.W., Michael, J.M., Emre, E., Randolph, L.M.: Optimal detectors for multi-target environments. In: Radar Conference (RADAR), pp. 0956–0961. IEEE (2012)

IoT-Enabled Air Monitoring System

Chavi Srivastava, Shyamli Singh and Amit Prakash Singh

Abstract The real-time air monitoring system is today's need for development of good decisive system. This paper proposes to develop IoT-based air monitoring system with the help of Ardunio system. The collection of data has been done through various air monitoring stations. Due to the large involvement of data in this work, this paper proposed to look this data as a big data and data scientist may explore study of air pollution due to smog, emission of methane, and effect of wind flow. This paper introduced the concept of Big Data and how it will be mapped with a data of affecting air pollution. To demonstrate the concept of data analytics, a case study of air monitoring system is presented in this paper.

Keywords IoT · Air pollution · Arduino embedded system

1 Introduction

ICT-based air monitoring systems have been implemented and presented by various researchers and published in [1–3]. These researchers have developed the system using embedded processor and captured the data using sensor. Wireless sensor networks (WSN) have been to deploy sensors and analyze the data at processor end. Dhole et al. have demonstrated the hardware for measuring noise level using Arduino processor. A detailed architecture for hardware and specification of sensors used for measuring noise level is presented in [1]. In this paper, a solution to monitor air and noise pollution levels in a given area using wireless communication and embedded system technology is proposed. The implementation of proposed system is tested for environmental parameters like CO_2, CO, NO_x, CH_4, and noise levels. The collected

C. Srivastava · A. P. Singh (✉)
University School of Information, Communication & Technology, Guru Gobind Singh Indraprastha University, Delhi, India
e-mail: amit@ipu.ac.in

S. Singh
Indian Institute of Public Administration, IP Estate, New Delhi, India

S. M. Thampi et al. (eds.), *Intelligent Systems, Technologies and Applications*, Advances in Intelligent Systems and Computing 910,
https://doi.org/10.1007/978-981-13-6095-4_13

data have been compared with standards specified by the government agencies. In this paper, IoT-based cloud storage software has been used to analyze the data and broadcast for user. This helps to timely broadcast the information of climate change in real-time environment.

2 Air Pollution: Issues and Challenges

Multiple environmental changes lead to Air pollution, these issues are widely studied by scholars of environment and reported in [4, 5]. Various statistical and machine learning techniques have been used to demonstrate the impact of air pollution in human life. A machine learning modeling has been applied to demonstrate weather forecasting in [4]. Air pollution is the presence of contaminations or pollutant substances in the air that interfere with human health or welfare, or produce harmful environmental effects. With the rapid development of communication technology, network technology and remote sensing technology, and air pollution monitoring system can be designed to check the concentration of these pollutants in the air and take appropriate action.

Noise pollution is metro cities is a growing concern as it is slowly affecting our body and mind. Noise pollution is caused due to various factors like industrialization, poor urban-planning household chores, construction activities, social events, and transportation. While this form of pollution may seem harmless, it, in fact, has far reaching consequences. Loud noise can cause ear discomfort, such as ear ringing, ear paints, and hearing loss.

3 Air Monitoring System

This paper demonstrated the approach of real-time monitoring system for air pollution due to traffic, pollutant industry, and construction work. This paper has proposed to design and implement an efficient real-time monitoring system through which the required parameters like CO_2, CO, PM2.5, O_3, noise, etc., are monitored remotely using Internet [5] and the data gathered from the sensors are displayed on a TFT-display screen, stored on an external storage, and projected the estimated trend on ThingSpeak. ThingSpeak is an open-IoTcloud-based platform where the data collected from various sensors are displayed on channels via a Wi-Fi module.

ThingSpeak provides captured data remotely and it also provides various analytics technique for data analysis. The results shown in this paper demonstrate the effect of air pollution.

These data can be used to compare with the threshold values so as to formulate further course of action. The solution provides an intelligent remote monitoring for a particular area of interest. Figure 1 shows the hardware implemented for monitoring of air pollution. Here various sensors have been used as input to Arduino processor.

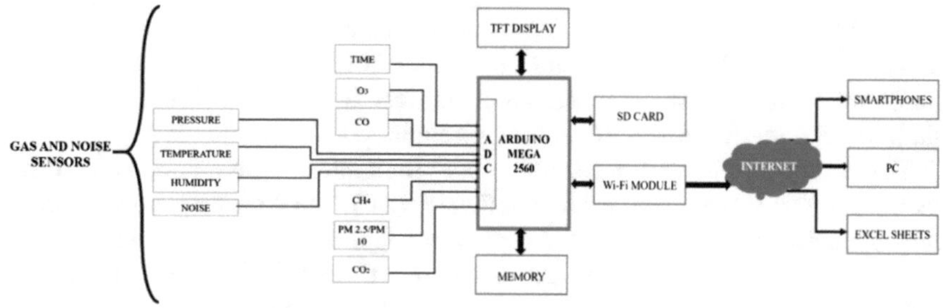

Fig. 1 Architecture of hardware for air monitoring station

This processor also provides interface of Wi-Fi module, which has been used with Internet to transmit and analyze sensors' data. The data is uploaded to ThingSpeak channel from where the required user can take requisite action.

4 Prediction of Air Pollution Using Machine Learning

Baysian theorem is widely used to predict pollution level [6] with the help of knowledge repository. Knowledge repository is developed with the help of data collected through various sensors of the system.

Baysian theory is widely used to train the system. Sensor tends to miss few data, where Baysian theory is widely used to normalize the data and these data are used to learn the system to predict accurate results.

Data collected using sensors are preprocessed and trained using artificial training system like neural network. Network with least error achieved during training is used to predict based on minimum variable. Ke et al. have presented prediction of air pollution in their paper [7] using machine learning. In this paper, it has been shown the procedure of training of the data for air pollution monitoring system.

Air quality forecasting is a final outcome of the air monitoring system. Wani et al. have presented their work for air quality forecasting in [8] using machine learning technique. In his paper, Wani et al. have shown the procedure to train the network based on prior data and these data are used to forecast the air quality.

5 Hardware and Software for IoT

The various sensors including noise detection sensors and gas sensors have been enclosed in a box and this box is placed wherever necessary. The sensors fetch the data from the environment and send them to the microcontroller.

Arduino Mega 2560 is the microcontroller board used, that is, based on the ATmega2560. The Mega 2560 board has a number of facilities for interacting with a computer, another board, or other microcontrollers. It supports TTL communication via UARTS ports, TWI, and SPI communication.

The microcontroller ATmega2560 is pre-programmed with a bootloader. It has been programmed using the Arduino Software (IDE). The Arduino integrated development environment (IDE), is a cross-platform application that has been written in the Java programming language. An Arduino 2.4″ TFT LCD Touch shield has a multi-colored TFT display that has been used to display the data. It also includes an SD card socket. The data is stored on the SD card that is mounted on the shield.

The gas sensors operating belong to the MQ series gas sensors to detect the gases. Such sensors are sensitive for a range of gases and are used indoors at room temperature.

The output is an analog signal and can be read with an analog input of the Arduino. A sound sensor module is used to measure the noise in the environment. Table 1 lists the various sensors used to detect the gases, dust/pollutants, and sound. Figure 2 demonstrates the interconnection of sensors and UART device with Ardunio processor.

Table 1 Details of sensors used to design hardware system

Sr. No.	Data	Sensor
1	Time	DS3231
2	Temperature and humidity	DHT11
3	Ozone gas	MQ131
4	Pressure	BMP280
5	CO gas	MQ7
6	NO_2 gas	MQ135
7	PM2.5/PM10	DSM501A dust sensor
8	CO_2 gas	MG811
9	Methane gas	MQ4
10	Noise	Sound detection sensor module by REES52

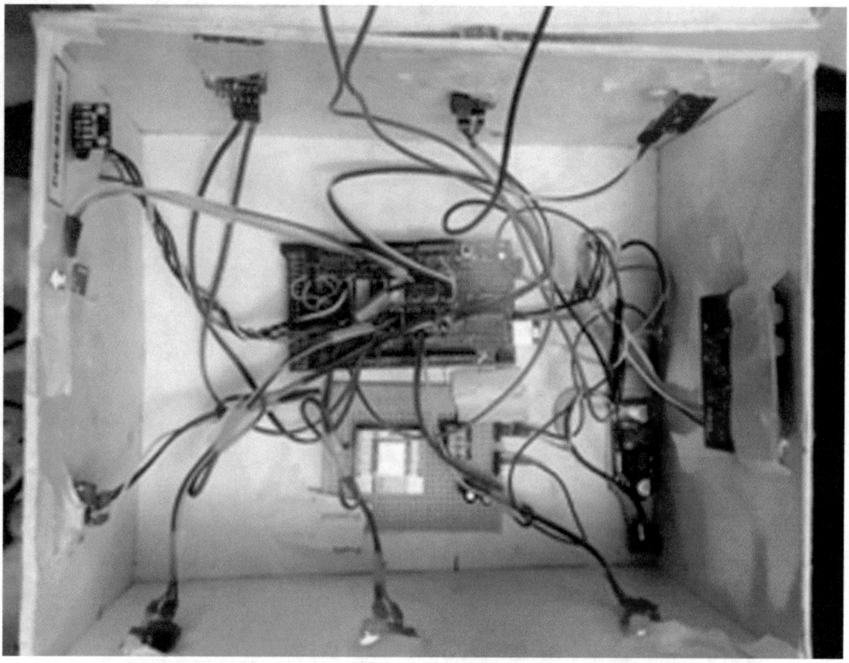

Fig. 2 Air and noise pollution monitoring using embedded System

6 Results and Discussion

Figures 3, 4, 5 and 6 show the results evaluated from various sensors connected with the embedded processor. Arudino system with the help of sensors attached with it, helps to predict real-time output. ThingSpeak portal used to capture the data online and used to transmit on web on real-time basis. Figure 3 shows the O_3 present in the atmosphere, similarly Fig. 4 shown the Particulate Matter, PM2.5 present in the environment. Figure 5 shows the CO present at the experiment site, and similarly noise level has been measured and shown in the Fig. 6. These results are scenario of laboratory; similar experiments can be extended to the site of traffic, construction, and pollutant industry.

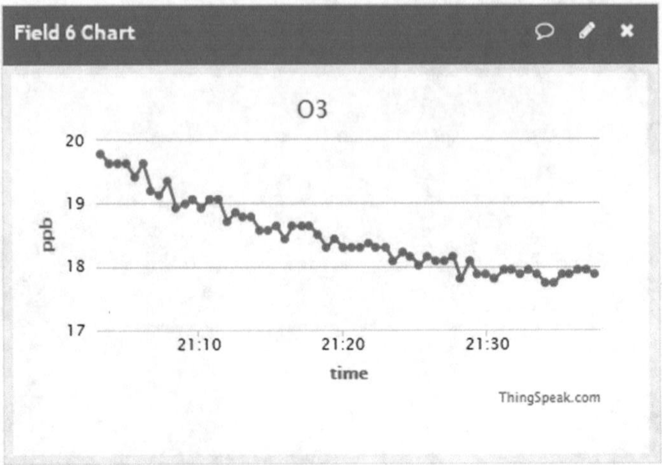

Fig. 3 Concentration of O_3 with respect to time

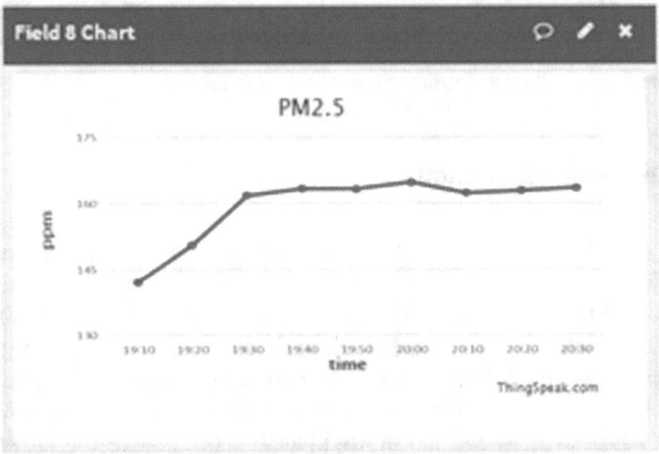

Fig. 4 Concentration of PM2.5 w.r.t. time

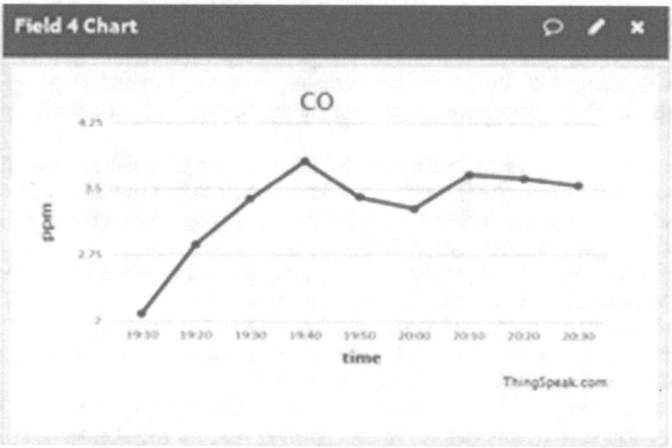

Fig. 5 Concentration of CO w.r.t. time

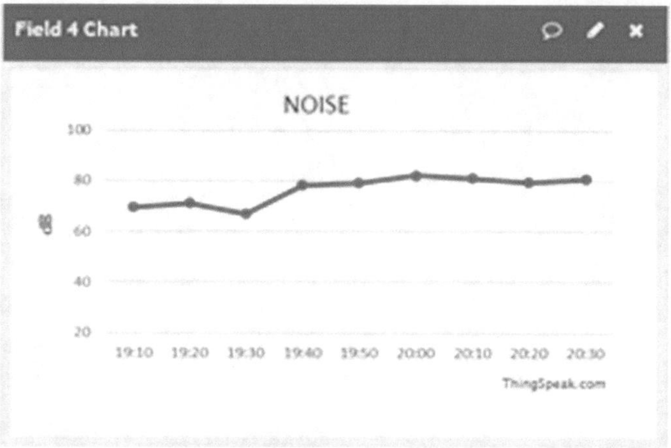

Fig. 6 Noise level

Acknowledgements Authors are thankful to Guru Gobind Singh Indraprastha University, Delhi for providing grants to purchase hardware for this work under FRGS 2017.

References

1. Dhole, R.N., Undre, V.S., Pawale, S.: Arduino based noise detection and image capturing using MATLAB. In: National Conferences in Computing, Networking and Security, pp. 143–146 (2013)
2. Roseline, R.A., Devapriya, M., Sumanthi, P.: Pollution monitoring using sensors and wireless sensor networks: a survey. Int. J. Appl. Innov. Eng. Manag. **2**(7), 119–124 (2013)
3. Shwetal, R.: Calculating pollution in metropolitan cities using wireless sensor network. Int. J. Adv. Res. Comput. Sci. Manag. Stud. **2**(12) (2012)
4. Holmstorm, M., Liu, D.L., Vo, C.: Machine learning applied to weather forecasting. A project Report, Machine Learning Course, Stanford University (2016)
5. Liu, X., Nielse, P.S.: Air quality monitoring system and benchmarking. In: Bellatreche, L., Chakravarthy, S. (eds.) Proceeding of 19th International Conference on Big Data Analytics and Knowledge Discovery. Springer's LNCS, 10440 (2017)
6. Sahu, S.K.: Hierarchical Bayesian models for space-time air pollution data. In: Time Series Analysis, Methods and Applications, vol. 30, pp. 477–495. Elsevier, Amsterdam (2011)
7. Ke, H., Ashfaqur, R., Hari, B., Sivaraman, V.: HazeEst: machine learning based metropolitan air pollution estimation from fixed and mobile sensors. IEEE Sens. J. **17**(11), 3517–3525 (2017)
8. Wani, T., Gilles, N., Christophe, P., Marie-Laure, N., Cyril, V.: Hybridization of air quality forecasting models using machine learning and clustering: an original approach to detect pollutant peaks. Aerosol Air Qual. Res. **16**, 405–416 (2016)
9. https://thingspeak.com/

Phase-Modulated Stepped Frequency Waveform Design for Low Probability of Detection Radar Signals

R. Vignesh, G. A. Shanmugha Sundaram and R. Gandhiraj

Abstract Recent developments in radar electronic warfare in the modern battlefield have led to advancement in various signal processing technologies that have increased the potential to detect and jam the radar signals. Signal detection and jamming results in various problems such as false target detection, false alarm rate, and so on. The introduction of modern electronic support measure (ESM) and radar warning receiver (RWR) provides various threats to the radar operations. To hide and to survive the enemy attack, the radars should have low probability of interception (LPI) or low probability of detection (LPD). In order to improve the robustness of radar signals, in a manner that evades detection, interception, and cognition, complex waveforms which are less susceptible to jammers and intercept receivers, a new technique is presented in the work reported here. Radars are used in target tracking systems to estimate the range and velocity of both stationary and moving objects. Phase-modulated signals help in determining the velocity of the target using Doppler information. Here, a new approach to the pulse compression techniques by means of phase modulating the stepped frequency waveform (SFW) with constant and varying step sizes is advanced. In order to make the signal less detectable, phase modulation techniques such as Barker codes and polyphase codes have been employed in tandem. A random choice of varying frequency step size compresses the ambiguity function estimates of target range, improves covert detection, and reduces the interferences caused by various sources. The comparison between SFW with either a constant or a varying frequency step size is performed. The generated waveforms have wide bandwidth and less peak power which are considered to be integral attributes of LPI signals.

R. Vignesh · G. A. S. Sundaram (✉) · R. Gandhiraj
Amrita School of Engineering, Center for Computational Engineering and Networking, Amrita Vishwa Vidyapeetham, Coimbatore, India
e-mail: ga_ssundaram@cb.amrita.edu

G. A. S. Sundaram
Department of Electronics and Communications Engineering, Amrita School of Engineering, Amrita Vishwa Vidyapeetham, Coimbatore, India

G. A. S. Sundaram
SIERS Research Laboratory, ASE Coimbatore, Amrita University, Coimbatore, India

S. M. Thampi et al. (eds.), *Intelligent Systems, Technologies and Applications*, Advances in Intelligent Systems and Computing 910,
https://doi.org/10.1007/978-981-13-6095-4_14

Also, the waveform properties based on different radar parameters are studied and analyzed using the ambiguity functions plots.

Keywords SFW · Phase modulation · Pulse compression · LPI · Barker codes · Polyphase codes · Ambiguity function · Radar jammer

1 Introduction

Advancements in various electronic countermeasures (ECM) such as electronic support measure (ESM), radar warning receiver (RWR), and anti-radiation missile (ARM) have resulted in serious threats to conventional radar system operations. In order to perform target detection and tracking and simultaneously remain latent to radar signal intercept receivers, the radar systems must be able to 'see without being seen' [15]. This need is addressed by a particular class of radars, viz. low probability of interception (LPI) radars, which possess low peak power, high duty cycle, wide bandwidth, and power management making them difficult to be detected [7]. LPI measures the ability to derive information from the detected signal while LPD measures the ability to detect the presence of signal [3]. The probability of detection and probability of interception is very much dependent on the input signal-to-noise ratio (SNR) of the interceptor receiver [19, 20]. Reducing the peak effective radiated power (ERP) by means of pulse compression techniques is a commonplace LPI radar method [12].

A pulse compression technique is employed in order to generate a train of radar signal pulses having a range resolution that corresponds to that of shorter pulse and a transmitted power that is of a longer pulse [6]. High range resolution radars have the ability to distinguish two closely spaced targets and capture crucial target-related details. The received pulses known as echo signals play a major role in target extraction.

Wideband modulation techniques such as frequency modulation, phase modulation [11], and spread spectrum techniques are employed in increasing LPI properties of the waveform. Frequency coding techniques include linear frequency modulation (LFM), nonlinear frequency modulation (NLFM) or chirp method [13]. Spread spectrum techniques involve frequency hopping spread spectrum and direct sequence spread spectrum (DSSS) [12, 14].

In the present work, the effectiveness of SFW and phase-modulated SFW is rated in terms of critical range parameters [11] in the presence of the jammer signals. Their performances are then compared with the help of ambiguity function plots, also called as uncertainty function, and the two-dimensional autocorrelation functions [4]. Ambiguity function effectiveness is generally measured using the peak sidelobe ratio and integrated sidelobe ratio [4]. The delay and Doppler characteristics of the waveform are also analyzed using the ambiguity function [11].

The contents reported here are organized as follows—Sect. 2 describes the theoretical background with the mathematical basis for SFW, polyphase codes, and jammer signals. Section 3 elucidates the tasks involved in realizing LPI, pulse-compressed radar signals. Section 4 includes the salient results obtained, and the inferences are made from these results. Finally, the important conclusions are made part of Sect. 5.

2 Theoretical Background

In the work reported here, the stepped frequency waveform (SFW) with either a constant or a varying frequency step size is considered as the principle component of the radar signal. This is phase-modulated with phase coding schemes such as Barker codes and polyphase codes, in order to improve the complexity of the waveform and to increase its instantaneous frequency spread, thus making it less detectable and likely to be intercepted, and hence less susceptible to jammer signals.

2.1 Uniform SFW

SFW is used in wideband radar applications [2]. SFW is a type of frequency modulation technique used for better range resolution between targets. SFW achieves high effective bandwidth with narrow instantaneous bandwidth [8]. A sequence of N pulses with a linearly varying carrier frequency and fixed pulse repetition frequency (PRF) is transmitted by an SFW radar. Each set of N pulses is called a sequence or burst. The frequency of each pulse in the sequence is increased uniformly from one pulse to another by a fixed step size (Δf). Modulation occurs across the pulses instead of within the pulses [8]. A carrier frequency (f_0) is considered as the frequency of the first pulse, and subsequent pulses have a difference of step size (Δf) from the preceding pulse, as shown in Fig. 1a, b.

The received signal known as echo signal is sampled, and spectral weighting is applied at the receiver side to reduce the sidelobe levels. The quadrature components for each burst are stored, and an inverse discrete Fourier transform (IDFT) is applied to find out the range profile of the corresponding burst. The carrier frequency waveform for the n-th step, n being positive integers including zero, is given by Paulose [16], Soares [17]:

$$f_n = f_0 + n\Delta f \tag{1}$$

Each pulse in the SFW can have its own modulation scheme in the scheme that is meant to incorporate LPI. The normal modulation scheme present in each pulse is the LFM, while the SFW is derived from LFM. If the transmitted waveform for the n-th pulse is [16]:

$$S_1(t) = A_1 \cos[2\pi(f_0 + n\Delta f)t] \tag{2}$$

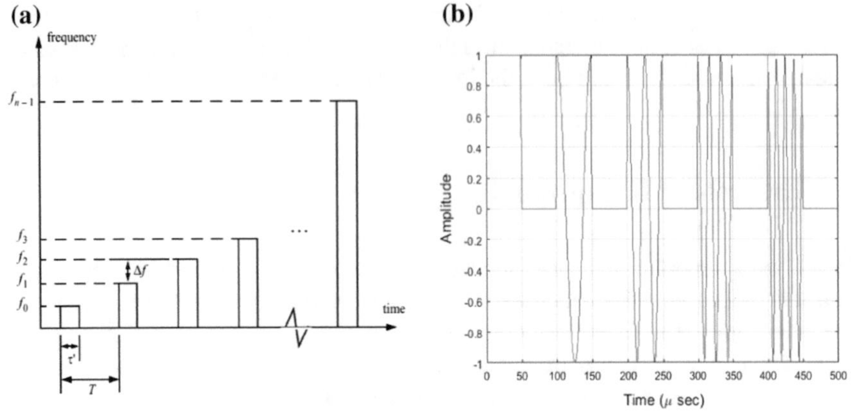

Fig. 1 SFW with **a** constant Δf and **b** with five steps each having a $\Delta f = 20$ kHz

then the received signal from the target after a time delay of 2*R*/*c* is given as [16]:

$$S_2(t) = A_2 \cos[2\pi(f_0 + n\Delta f)(t - 2R/c)] \tag{3}$$

Here, *R* denotes the range of the target in meters, *c* denotes the speed of light in m/s, *N* is the number of pulses, and Δf denotes frequency step size, considering a duration of every pulse as τ and the pulse repetition interval as T. To avoid 'ghost image' phenomenon, the frequency step size Δf should be less than $1/\tau$ [10]. The range resolution (ΔR) is determined from effective bandwidth $B = N.\Delta f$ as [16]:

$$\Delta R = c/(2 * N\Delta f) \tag{4}$$

This implies the range resolution does not depend on instantaneous bandwidth, and its value can be arbitrarily increased by increasing the effective bandwidth. Coherent processing interval considers the number of pulses (*N*) and the pulse repetition interval (PRI) [17] as their product.

2.2 Non-Uniform SFW

Non-uniform SFW is achieved by varying frequency step size (Δf).The carrier frequencies are randomly chosen from a given bandwidth [2]. Non-uniform SFW improves the range-Doppler resolution, suppresses the range ambiguity, and decouples the range and Doppler [5, 6] as shown in Fig. 2a, b. This is observed to be an efficient method that is suitable for electronic counter countermeasures (ECCM), as it reduces interferences between adjacent radar systems [2]. Here, Barker code combinations have been employed in order to choose the frequencies rather than

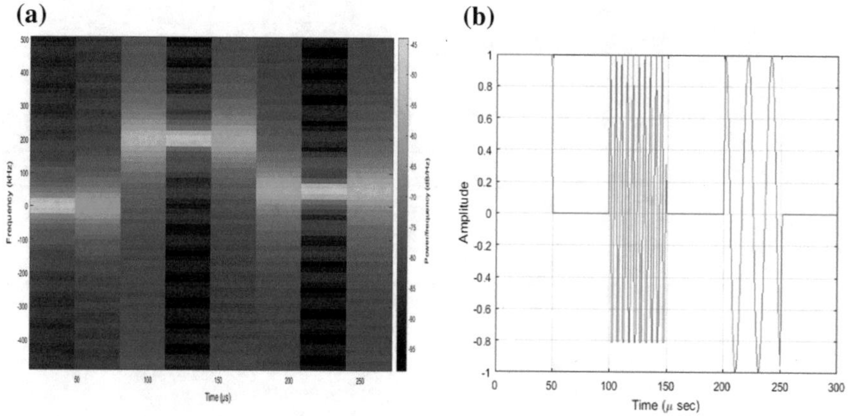

Fig. 2 SFW with **a** non-uniform Δf and **b** with three steps of varying Δf

letting them vary randomly since it helps the transmitter to retain an accurate information about the frequency range transmitted. After choosing the frequencies based on the Barker combinations, the SFW is in turn modulated with a polyphase code (P1) to improve the complexity and thereby incorporate the LPI property in the radar waveform.

The carrier frequency of a random SFW is given by Axelsson [2], Huang et al. [5]:

$$f_n = f_0 + d_n \Delta f \tag{5}$$

where d_n is a random integer between 0, the floor integer is $[B/\Delta f]$, and f_0 is the central or carrier frequency.

2.2.1 Barker Code

Barker code belongs to phase-modulated, pulse compression technique where a phase shift of 0° and 180° occurs [1]. The main advantage of the Barker combinations is that only certain combinations that have equal sidelobe are considered. They exhibit very low-sidelobe performance [8, 11]. The Barker combination over 2–13 bits is listed in Table 1 [4, 9], where '+' stands for a 0° phase shift and '–' for a 180° phase shift.

A five-bit Barker combination has been chosen for selecting the frequency step size, and a total of $2^5 = 32$ combinations are thus possible. When considering combinations having equal amplitude, only five combinations, viz. 11101, 10101, 11011, 11100, 11111, are alone possible. The frequency step size is set for each combination. Based on the input combination entered at the time of radar transmission, the frequency is selected. This is done in order to improve the information regarding the

Table 1 Barker codes

Code length	Code	Range sidelobe level (dB)	Process gain (dB)
2	+ − OR + +	−6.0	3.0
3	+ + −	−9.5	4.8
4	+ + − + OR + + + −	−12.0	6.0
5	+ + + − +	−14.0	7.0
7	+ + + − − + −	−16.9	8.5
11	+ + + − − − + − − + −	−20.8	10.4
13	+ + + + + − − + + − + − +	−22.3	11.1

frequency transmitted non-uniformly on the radar transmitter side as it is needed for decryption of received signal and eventual target information extraction and range and velocity determination by the radar receiver.

2.2.2 Polyphase Code (P1)

Polyphase codes are known for their advanced features such as low range time lobes, low cross-correlation between codes and are compatible with digital implementation [1, 4]. Polyphase code lengths of any size are possible, and an example is the Frank code or the P-code [1, 11]. Here, the P1 code has been considered, which has N^2 elements. With i being the number of sample and j being the frequency number, the phase of ith sample of frequency j is given by Farnane et al. [4]:

$$\varphi_{i,j} = \frac{-\pi}{N} * [N - (2j - 1)][(j - 1)N + (i - 1)] \tag{6}$$

Here, i and j are positive integers excluding zero.

The length of P1 code should be even [4]. The pulse width (τ) that corresponds to a longer pulse is divided into N smaller sub-pulses called the chip width T_c. Each sub-pulse can be in binary format. The phase-coded waveform (P1) is thus modulated with SFW having a varying step size. This increases the complexity of the waveform and, hence, such radar waveforms possess the properties of LPI.

2.3 Radar Ambiguity Function

The radar waveform's ambiguity function describes the interference due to a point target return positioned at non-identical range and velocity from a reference target of interest [8]. It informs of the range-Doppler position of ambiguous responses and hence the range and Doppler resolution [14]. It is defined as the absolute value of the matched filter output envelope. The input signal to the filter is a Doppler-shifted version of the return signal, to which the filter response is matched. The ambiguity function of a waveform $s(t)$ is given as [11, 13]:

$$|X(\tau, f_{\mathrm{d}})| = \left| \int_{-\infty}^{\infty} s(t).s(t-\tau).\mathrm{e}^{j.2\pi.f_{\mathrm{d}}t}\mathrm{d}t \right| \tag{7}$$

where $s(t)$ is the transmitted signal, τ is the delay, and f_{d} is Doppler frequency shift.

2.4 Radar Jamming

Jammers play a major role in the modern electronic battlefield. A waveform is said to possess LPI only when it is less detectable by a jammer system. Jammer equipments are classified as barrage and deceptive jammer devices (repeaters) [11]. Deceptive jamming is a technique where the radar receiver is misguided by the jammer by employing various electronic and computational techniques [11]. A barrage jammer transmits a high strength signal across a wideband frequency to completely mask the target reflection. Barrage jammer design is less complex to construct and operated over a wide portion of electromagnetic spectrum, while deceptive jammer requires only less transmission power [11, 12]. When a strong jammer signal is present, the detection capability is determined by receiver signal-to-noise-plus-interference ratio rather than the conventional signal-to-noise ratio. In general, a jammer is characterized by its operating bandwidth B_{j} and its effective radiated power (ERP), which is proportional to jammer transmitter power P_{j}, related by Mahafza and Elsherbeni [11]

$$\mathrm{ERP} = (P_{\mathrm{j}}/G_{\mathrm{j}}/L_{\mathrm{j}}) \tag{8}$$

Here, G_{j} is the antenna gain for the jammer module, and L_{j} is total loss factor of the jammer. The jammer signal's effect on radar receiver performance is evaluated by the signal-to-jammer ratio (SJR).

3 Methodology and Simulations

The non-uniform SFW is modulated with the P1 polyphase code. The functional blocks in a system simulation of a monostatic SFW radar along with a jammer are shown in Fig. 3. The modulated SFW waveform is transmitted with the help of the radar transmitter having high transmitter gain. The free space propagation path is considered here so that the interferences caused is less. The target can be considered either as stationary or to be moving. The transmitted signals, after hitting the target, reach the receiver end as echo signals which are used to extract vital information about the target. Two important models, such as target model and jammer model, are to be addressed before the radar simulation process is initiated. Model parameters of the target include its radar cross-section (RCS), range, velocity of target, and direction of travel [13]. RCS plays a major role in target detection. The accuracy of target detection varies with its RCS value.

To model a jammer, its signal peak power, range, and its velocity information have been considered. A standoff jammer or a self-screening jammer [12, 18] is typical models applied as the jammer signal. The detection of target information decreases with increase in jammer peak power and is also based on the effect of jammer on the waveform considered.

The received echo signals, shown in Fig. 4, are processed at the receiver side after performing matched filtering and pulse integration, in order to extract target and range information. Various techniques are employed in processing the receiver signals such as the time-frequency analysis methods, Choi-Williams distribution (CWD [21]), or the quadrature mirror [4] filter bank methods. Here, only the transmitter module in the block is considered. Ambiguity function is used for analyzing the matched filter output response of the transmitted signals without applying rigorous signal processing techniques.

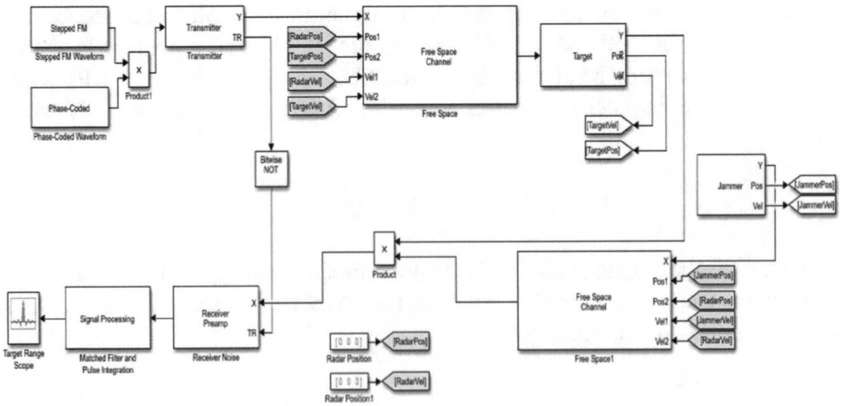

Fig. 3 Functional diagram of simulation of the SFW radar and a jammer system

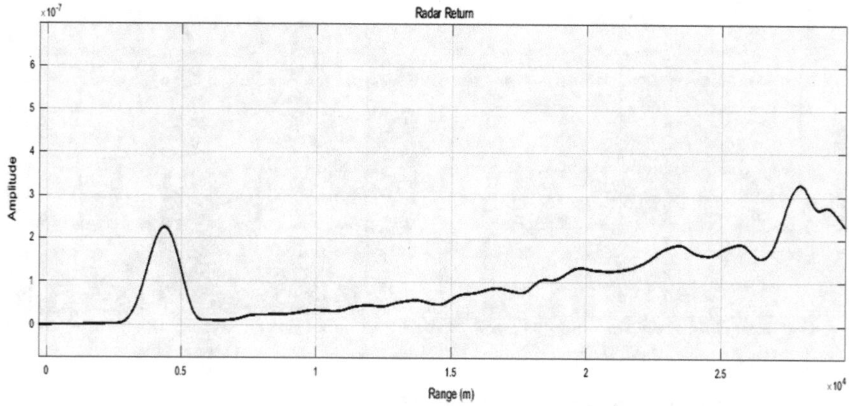

Fig. 4 Detection of target echo signals using non-uniform SFW

4 Results and Discussion

The performance of phase-modulated SFW is estimated using parameters such as crossover range and burn-through range, delay-Doppler and ambiguity functions, with the following specifications: maximum range of 30 km, range resolution of 150 m, and a sampling frequency of 2 MHz The number of frequency steps in the SFW considered here is 4, with a constant step size 200 kHz. The Barker combination for selecting frequency step size is of five-bit dimension.

A section of the ambiguity surface along the delay axis at $f_{\mathrm{d}} = 0$ is depicted in Fig. 5a; it shows a single peak called the mainlobe along with sidelobes having smaller magnitudes. This indicates absence of ambiguity for the waveform when the Doppler velocity profile of the target is known. This plot also gives a null width of 2.5, which agrees with the computation using the formula $2/N\Delta f$, while being the time-domain output of the matched filter. A section of the ambiguity surface along the frequency axis at $\tau = 0$ is shown in Fig. 5b. The null-to-null width is 0.025 obtained using 2/(*NT*), and this denotes the Doppler resolution.

Figure 6a, b shows the delay and Doppler cut of an SFW having a variable step size. The step size is selected based on the Barker combinations that is given as input. Here, varying step sizes of 104, 410, 206, and 308 kHz have been considered. This improves the range resolution $(1/N\Delta f)$ and Doppler resolution (1/*N* * PRI); the values are 0.045 and 0.00425 [8], respectively.

Polyphase codes are finite length, discrete time complex sequences with constant magnitude and variable phase [8, 11]. The compression ratio is given by N^2. The delay and Doppler cut of the P1 phase-coded waveform is shown in Fig. 7a, b. The P1 code is generated with the following specifications: number of chips = 16 fs of 2 MHz, chipwidth of 1 μs and PRF of 5 kHz. The polyphase codes are known for their reduced sidelobe levels. From Fig. 7a, b, the first null along the time domain occurs at 4 ms, and the first null along Doppler cut occurs at 0.0625 MHz implying

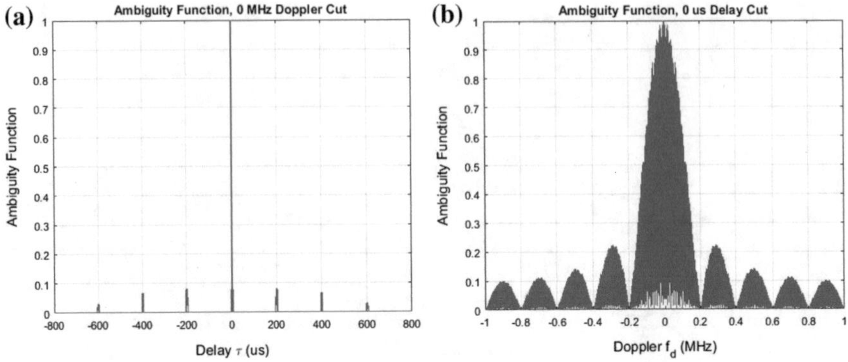

Fig. 5 **a** Doppler cut and **b** delay cut of uniform SFW

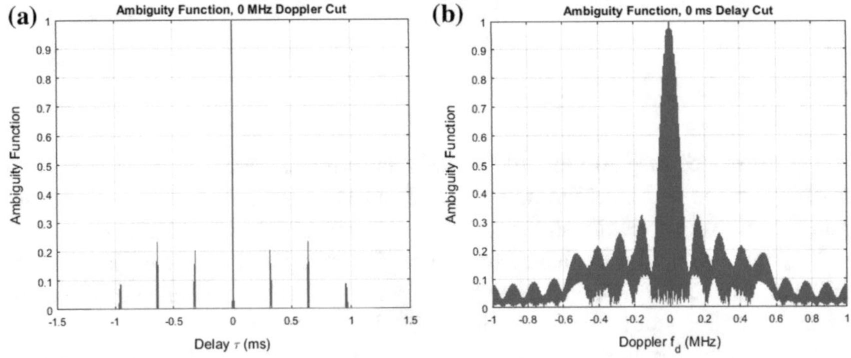

Fig. 6 **a** Doppler cut and **b** delay cut of non-uniform SFW

that the targets should be separated by at least 4 ms in their separation profiles, and 0.0625 MHz in their velocity profiles for an accurate classification.

The non-uniform SFW and P1 polyphase codes are modulated to improve the pulse compression and to spread the energy equally in the waveform thus reducing the peak power and increasing the bandwidth making the waveform an LPI. The specifications mentioned above are considered for generation of the modulated waveform. A section of the ambiguity surface along the delay axis is shown in Fig. 8a, where the first null occurs at 0.038 ms; while Fig. 8b depicts a section of the ambiguity surface along the frequency axis at $\tau = 0$. The null-to-null width is 0.0085 MHz along frequency domain. The range resolution and Doppler resolution values obtained here imply that the performance of the waveform has improved compared to that for a uniform SFW. The sidelobe is of a high value as observed from Fig. 8a. This indicates a reduction in the SNR ratio at on the receiver side, thus highlighting that the detection of the signal is low. The unambiguous range and delay between the targets are also reduced in the modulated waveform compared that with a uniform SFW.

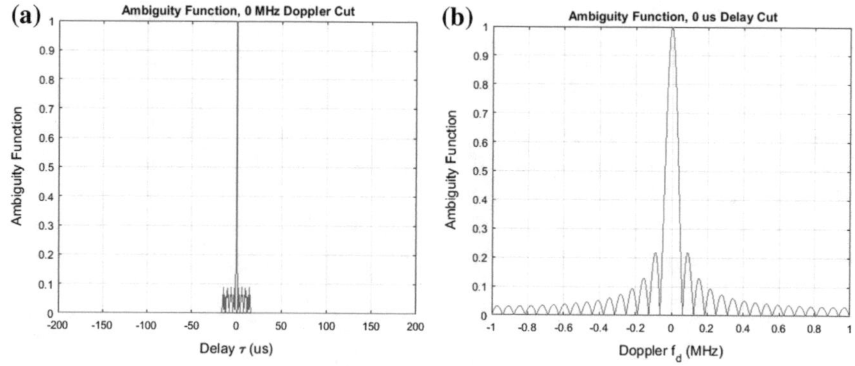

Fig. 7 P1 code **a** Doppler cut, and **b** Delay cut

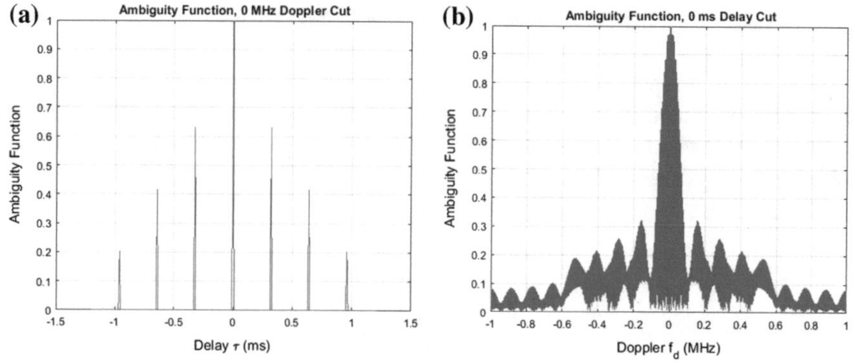

Fig. 8 Phase-modulated SFW: **a** Doppler cut and **b** delay cut

The ambiguity function plots of uniform SFW and phase-modulated SFW are shown in Fig. 9a, b. In general, the ambiguity function is plotted as a relation between three parameters, viz. time delay, Doppler, and signal power density. The detection of a radar target is good only in a particular region where the power density is more, i.e., at the point where the time delay and Doppler values are null. As the delay and Doppler value increases, the detection capability reduces.

The performance of the waveform in the presence of the jammer can be best analyzed by parameters such as the burn-through range and the crossover range. Crossover range is one where the signal power from reflections off the target equates with that of the jammer. The radar crossover range R_{co} for self-screen jammer is calculated by Mahafza and Elsherbeni [11], Srivatsa and Sundaram[18]:

$$R_{co} = \sqrt{\left(P_t G \sigma B_j / 4\pi B_r L(\mathrm{ERP})\right)} \tag{9}$$

Here, P_t is the radar peak power at transmitter, σ is the radar RCS, G is the transmitting gain of the antenna, B_j is the effective bandwidth of the jammer, B_r the

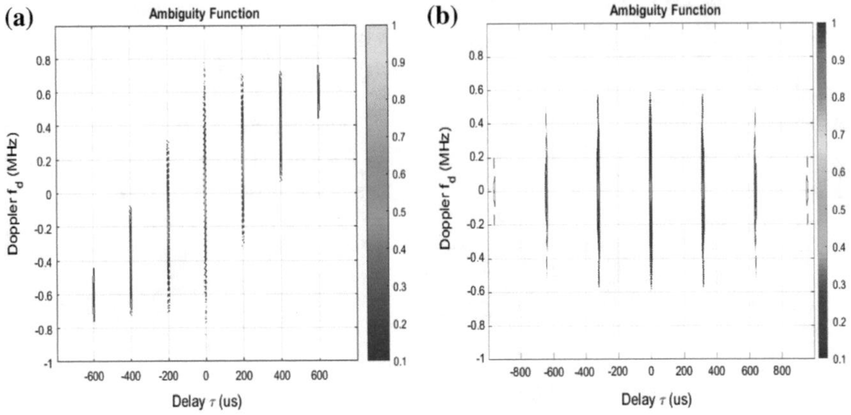

Fig. 9 **a** Uniform and **b** phase-modulated: SFW ambiguity function plots

operating bandwidth of the radar, L is the atmospheric loss, and ERP is the effective radiated power which determines strength of the jamming signal. The received signal-to-jammer-plus-noise ratio is given by Mahafza and Elsherbeni [11], Srivatsa and Sundaram[18]:

$$\frac{S}{J+N} = \frac{\left[\frac{P_t G \sigma A_r \tau}{(4\pi)^2 R^4 L}\right]}{\left[\frac{(\mathrm{ERP}) A_r}{4\pi R^2 B_j} + K T_0\right]} \tag{10}$$

The S-to-$(J + N)$ ratio can be applied instead of SNR in the radar equation and the probability of target detection. It is also used in the process of coherent and non-coherent pulse integration in the radar receiver-side signal processing [11]. The range at which a radar can detect and perform accurate measurements for a given signal-to-jammer-plus-noise ratio is defined as the burn-through range. In terms of the effectiveness of the jammer, it is the least distance of separation between the target and radar in excess of which the target is entirely masked by the jammer [18].

A comparison of the jammer's effective radiated power with the burn-through range for phase-modulated SFW is depicted in Fig. 10. The study has been performed so as to estimate the radar system performance associated with target detectability in the presence of a high-intensity jammer signal that has a spatio-temporal and spectral overlap with the radar parameters. Variations of crossover range with respect to radar peak power and jammer peak power are depicted in Fig. 11a, b.

From the indications in these plots, it is observed that phase-modulated non-uniform SFW performs far better than a simple LFM signals in terms of their ability to remain low in profile, while not compromising on target detectability. The waveform also possesses significant LPI properties, as inferred from the high bandwidth and low peak power. Also, the time-bandwidth product of the modulated waveform is found to be greater. The signal is also less susceptible to jamming and interference.

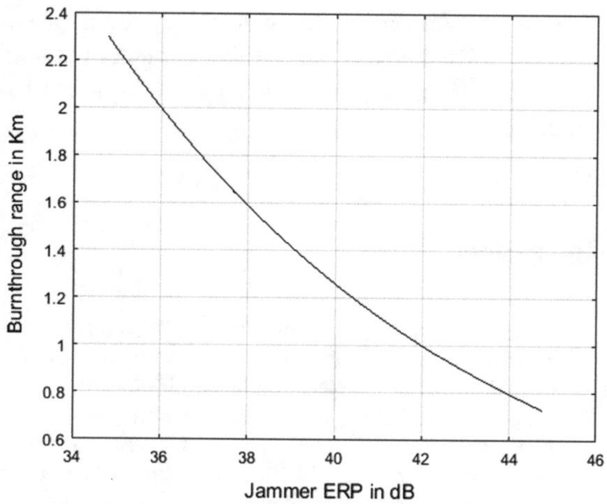

Fig. 10 Jammer ERP versus burn-through range

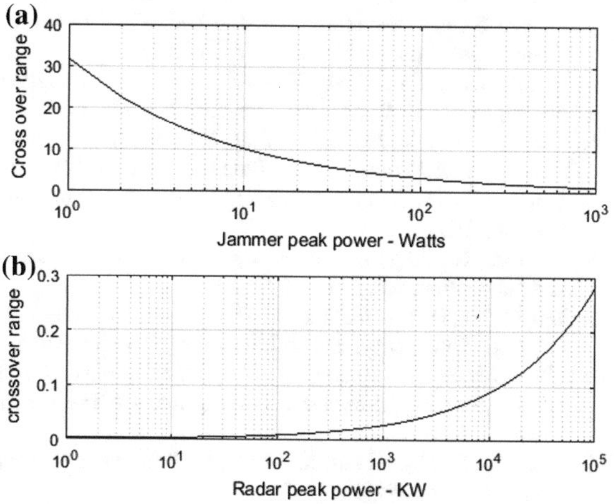

Fig. 11 Crossover range as a function of **a** jammer peak power and **b** radar peak power

Thus the P1-modulated SFW proves to be less detectable and less susceptible to jamming, while also offering an enhanced range resolution.

5 Conclusion

A new waveform design scheme that modulates an SFW with polyphase code P1 is introduced here. The characteristics of the phase-modulated SFW are analyzed with the help of ambiguity functions diagrams that include separate investigations of their Doppler and delay domain components, followed by a comparison with a uniform SFW for performance assessment. The SFW complexity is increased by using varying step size, and this is then modulated with an LPI code such as the

P1-type, thus decreasing the detectability and improving the stealthiness of the radar waveform. The performance of this complex modulated waveform in the presence of a jammer signal is also analyzed using parameters such as the burn-through range and crossover range.

References

1. Ankarao, V., Srivatsa, S., Sundaram, G.A.S.: Evaluation of pulse compression techniques for X-band radar systems. In: Proceedings of International Conference on Wireless Communications, Signal Processing and Networking. IEEE/SSN (2017)
2. Axelsson, S.R.J.: Analysis of random step frequency radar and comparison with experiments. IEEE Trans. Geosci. Remote Sens. **45**, 890–904 (2007)
3. Dybdal, R., Soohoo, K.: LPI/LPD detection sensitivity limitations. In: Proceedings of 2014 IEEE Military Communications Conference. IEEE (2014)
4. Farnane, K., Minaoui, K., Rouijel, A., Aboutajdine, D.: Analysis of the ambiguity function for phase-coded waveforms. In: Proceedings of 12th International Conference of Computer Systems and Applications. IEEE/ACS (2015)
5. Huang, T., Liu, Y., Li, G., Wang, X.: Randomized stepped frequency ISAR imaging. In: Proceedings of IEEE Radar Conference. IEEE (2012)
6. Huang, T., Liu, Y., Meng, H., Wang, X.: Cognitive random stepped frequency radar with sparse recovery. IEEE Trans. Aerosp. Electron. Syst. **50**, 858–870 (2014)
7. Lange, J.B.: The Relative Nature of Low Probability of Detection Radar. Defence R&D Canada-Ottawa (2012)
8. Levanon, N.: Stepped-frequency pulse-train radar signal. In: IEE Proceedings—Radar, Sonar and Navigation, vol. 149, p. 297 (2002)
9. Liu, G., Gu, H., Su, W., Sun, H.: The analysis and design of modern low probability of intercept radar. In: Proceedings of CIE International Conference on Radar. IEEE (2001)
10. Liu, Y., Meng, H., Zhang, H., Wang, X.: Eliminating ghost images in high-range resolution profiles for stepped-frequency train of linear frequency modulation pulses. IET Radar Sonar Navig. **3**, 512 (2009)
11. Mahafza, B., Elsherbeni, A.: MATLAB Simulations for Radar Systems Design. Chapman and Hall/CRC, Boca Raton (2003)
12. McRitchie, W.K., McDonald, S.E.: Detection and Jamming of LPI Radars. MC Countermeasures Inc., Ottawa (1999)
13. Skolnik, M.I.: Introduction to Radar Systems, 2nd edn., pp. 607–609. McGraw-Hill, New York (2001)
14. Ning, B., Li, Z., Guan, L., Zhou, F.: Probabilistic frequency-hopping sequence with low probability of detection based on spectrum sensing. IET Commun. **11**, 2147–2153 (2017)
15. Pace, P.E.: Detecting and Classifying Low Probability of Intercept Radar, 2nd edn. Artech House, London (2009)
16. Paulose, A.T.: High radar range resolution with the step frequency waveform. Ph.D. thesis, Naval Post Graduate School, Monterey, CA (1994)
17. Soares, P.A.: Step frequency waveform design and analysis using the ambiguity function. Master thesis, Naval Post Graduate School, Monterey, CA (1996)
18. Srivatsa, S., Sundaram, G.A.S.: PAM4-Based RADAR Counter-measures in hostile environments. In: Proceedings of International Conference: Advances in Intelligent Systems and Computing Intelligent Systems Technologies and Applications, pp. 390–400. Springer (2017)
19. Urkowitz, H.: Energy detection of unknown deterministic signals. Proc. IEEE **55**, 523–531 (1967)

20. Weeks, G., Townsend, J., Freebersyer, J.A.: Method and metric for quantitatively defining low probability of detection. In: IEEE Military Communications Conference Proceedings MILCOM 98. IEEE (1998)
21. Upperman, T.L.O.: ELINT signal processing using Choi-Williams distribution on reconfigurable computers for detection and classification of LPI emitters. Master thesis, Naval Post Graduate School, Monterey, CA (2008)

Effect of Waveform Coding on Stepped Frequency Modulated Pulsed Radar Transmit Signals

G. Priyanga and G. A. Shanmugha Sundaram

Abstract The choice of a waveform in a radar system plays a predominant role in characterizing the radars' ability to differentiate two closely spaced targets in both range and Doppler domains. RADAR waveforms are broadly categorized into two types, viz. pulsed or continuous waveform. A waveform can be analyzed in both time and frequency domains. The frequency domain analysis informs about the range and Doppler resolution of the radar target. The ambiguity of the transmitted waveform in both range and Doppler domain is used to understand the critical parameters used in the design and performance of any radar system. Precise radar measurements like range and Doppler calculation need higher resolution in both domains. But, in a pulsed radar system, there is an important factor of trade-off between range resolution and sensitivity, that necessitates pulses of smaller width, while the pulse energy in the transmitted waveform requires a longer pulse duration. An approach to overcome this trade-off is pulse compression which modulates the pulses to be transmitted. In the work reported here, a new approach to pulse compression technique is provided by modulating the transmitted waveform of a stepped frequency waveform (SFW) with orthogonal frequency-division multiplexing (OFDM). The reported work evaluates the performance of frequency spacings that are either uniform or non-uniform in the SFW waveform in terms of their maximum detection capability, observed from the graphs of delay and Doppler cuts in the ambiguity function plot. Various phase coding techniques are also used to define the non-uniform spaced SFW. The robustness of

G. Priyanga · G. A. Shanmugha Sundaram (✉)
Amrita School of Engineering, Center for Computational Engineering and Networking, Amrita Vishwa Vidyapeetham, Coimbatore, India
e-mail: ga_ssundaram@cb.amrita.edu

G. A. Shanmugha Sundaram
Department of Electronics and Communications Engineering, Amrita School of Engineering, Amrita Vishwa Vidyapeetham, Coimbatore, India

G. A. Shanmugha Sundaram
Sponsored by the National Instruments Inc. (USA), as part of the NI Academic Research Grant 2017, and conducted at the SIERS Research Laboratory, ASE Coimbatore, Amrita University, Coimbatore, India

S. M. Thampi et al. (eds.), *Intelligent Systems, Technologies and Applications*, Advances in Intelligent Systems and Computing 910,
https://doi.org/10.1007/978-981-13-6095-4_15

radar waveforms in terms of their static and dynamic resolution parameters, as well as their ambiguities, is also evaluated.

Keywords SFW · Pulse compression · OFDM · Range resolution · Doppler resolution · Ambiguity

1 Introduction

Recent advances in radar technology, along with complex operational environments and development in electronic countermeasure (ECM) technologies, have created a demand in radar engineering toward design of newer types of signals on the transmit side of the radar system, in order to remain evasive, while also making precise radar target estimate [1]. Although there exist a wide variety of radar types in terms of the operating parameters, the basic working principle has remained the same. A known signal is transmitted, and the reflected echo signal from the target is analyzed [2]. Selection of waveforms has become a critical parameter in developing new radar systems. Generally, radar systems are characterized in terms of ambiguity in both range and Doppler, with its corresponding resolution and accuracy [3]. The minimum separable distance between the two targets is known as range resolution. A system with low range resolution cannot differentiate between two closely spaced targets. High range resolution necessitates the use of a shorter pulse width. On the other hand, the maximum unambiguous range that can be detected by the radar system depends on the pulse repetition frequency (PRF) [3]. Next to PRF, the maximum range of a pulsed-type radar depends on the average transmit power [4]. Increasing the transmit power leads to improved signal-to-noise (SNR) [2]. One alternative way for power amplification is by extending the pulse duration [5]. Pulse compression techniques overcome this issue by extending the pulse duration while keeping the pulse width short [6]. Here, the signal to be transmitted with the desired pulse duration is modulated either in frequency or phase (so as to obtain a higher time-bandwidth product) [4]. Pulse compression techniques help in maximizing the operating range of radars by increasing the transmit duration, while also enabling a very acceptable range resolution. Moreover, it also offers better immunity against noise and jamming, by increasing the SNR. The main enablers in pulse compression perform either waveform modulation or coding or both to the transmitted waveform. Modulation schemes, such as orthogonal frequency-division multiplexing (OFDM), frequency modulation—linear (LFM), or stepped (SFW), can be found in the literature [7] and are practically implemented in many radar systems [8]. In the work reported here, uniformly and non-uniformly spaced SFW are modulated with OFDM [9] in order to derive the transmitted waveforms. Hamming codes are used to vary the step size of the latter type that results in an innovative type of radar signal with superior characteristics attributed to target detection and resistance to conventional ECMs.

An SFW modulated with OFDM is chosen to be the transmitted waveform, and its performance with uniform and varying frequency step sizes is compared by means of the following criterion: ambiguity function diagram, range plot, and Doppler plot. The effectiveness in the use of various pulse compression techniques is established as a comparative analysis of such radar waveforms. Section 2 briefly explains pulse compression techniques based on frequency modulation, while the following section describes the SFW and Hamming codes techniques. Section 4 discusses the results from comparison of the modulated waveform described in Sects. 2 and 3, and the major conclusions are confined to Sect. 5.

2 Pulse Compression Based on Frequency Modulation

Modulating a signal increases the bandwidth of the transmitted waveform and produces a narrow peak for detection by the matched filter [6]. In a regular frequency modulation (FM), autocorrelation of the signal gives high-energy sidelobes. So, an FM signal in which the signal frequency varies linearly with time may be used to obtain highly suppressed sidelobes, so that a good pulse compression may be achieved [3]. In regard to the frequency–time relationship, FM can be either linear or nonlinear and are briefly discussed here. The autocorrelation output of a rectangular pulse is shown in Fig. 1. After matched filtering of the received pulses, the output occupies a complete time period. Two or more targets lying close enough along a

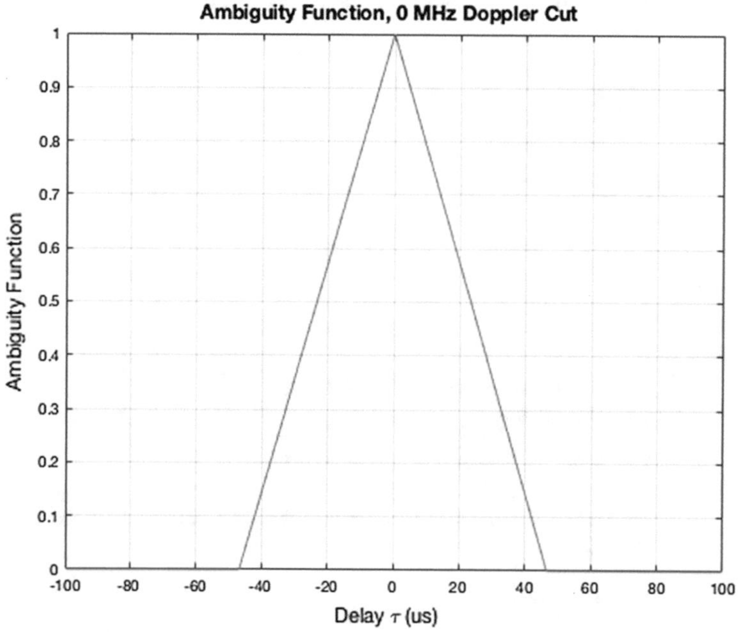

Fig. 1 Autocorrelation of rectangular waveform

given line of sight would appear within a coherent processing interval (CPI), and they shall not be distinguished by the radar receiver [3]. But, if the output of the matched filter has a narrow peak, then multiple targets can be detected that are also very close enough, depending on the time bandwidth of the narrow peak [4].

3 Stepped Frequency Waveform (SFW)

The advantage of using FM is that they can be readily generated with various techniques. A rectangular pulse modulated with a sine wave may be represented in general by Mahafza [10]:

$$x(t) = \text{rect}(t/T_r)A\ \cos(2\pi f_i(t)t + \varphi) \tag{1}$$

where A represents the amplitude, *fo* represents the frequency, and φ represents the initial phase, and rect is the rectangular function defined as [3]:

$$\text{rect}\left(\frac{t}{T_r}\right) = \begin{cases} 1, |t| \le T_r/2 \\ 0, |t| > T_r/2 \end{cases} \tag{2}$$

An SFW is generated by transmitting N narrowband pulses in series. A burst represents each group of N narrowband pulses. The frequency step between pulses

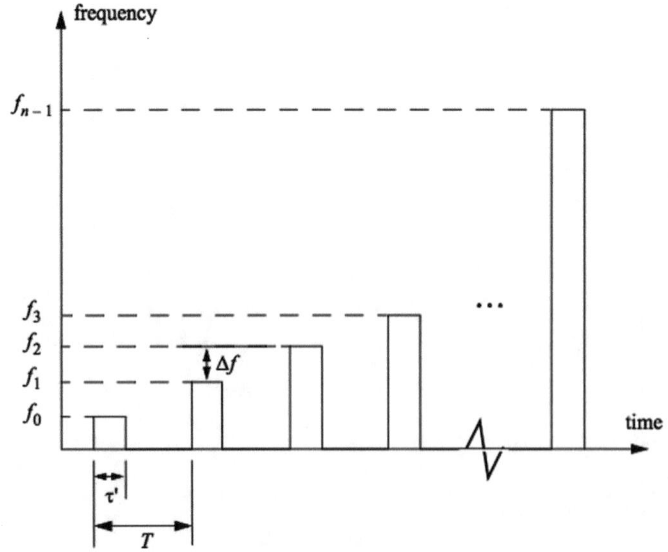

Fig. 2 SFW burst from a radar transmitter [10]

scales in terms of a fixed frequency step Δf, as shown in Fig. 2. T denotes the pulse repetition interval (PRI), and τ denotes the width of the pulse. Each pulse can be modulated in turn using a specific digital modulation technique. In the work reported here, the transmitted waveform is modulated using OFDM [7].

The range resolution of SFW [10] is given by:

$$\Delta R = c/2N \cdot \Delta f \tag{3}$$

where c is the velocity of light, B represents the bandwidth, and the number of pulses is given by N. The maximum unambiguous range of SFW is calculated using:

$$Ru = c/2\Delta f \tag{4}$$

The maximum transmitted power is estimated using radar equation [5], and minimum SNR is calculated using the Albersheim equation [5].

4 Uniformly Spaced SFW

In uniformly spaced SFW, the frequency step (Δf) of the waveform is kept constant, as shown in Fig. 3. It can be mathematically expressed as [11]:

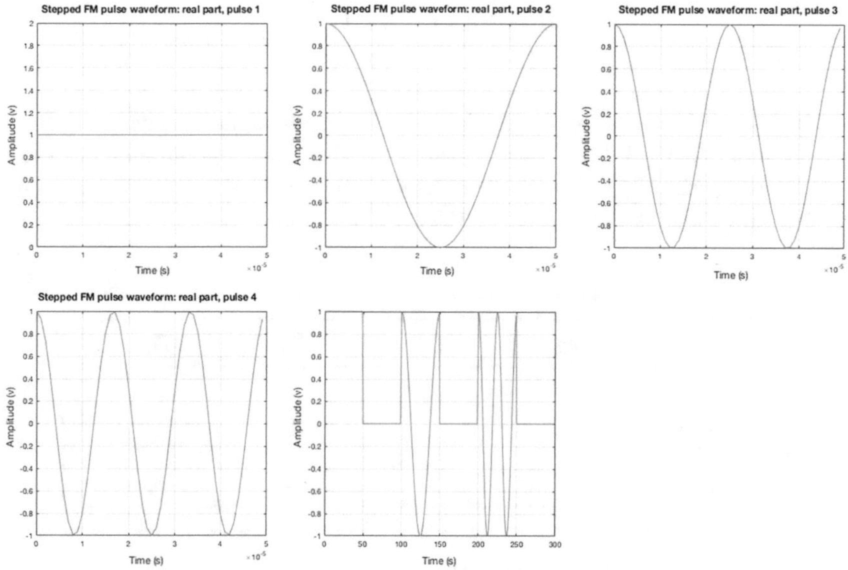

Fig. 3 Representation of uniformly spaced SFW

$$fi = fo + i\Delta f \tag{5}$$

Representation of each burst at the 'i'th step can be given by Skolnik [11]:

$$s_i(t) = \begin{pmatrix} C_i \cos 2\pi f_i t + \theta_i;\ iT \leq t \leq iT + \tau' \\ 0 \qquad\qquad \text{elsewhere} \end{pmatrix} \tag{6}$$

The subfigures plotted above show the real part of a pulse at each step. It is inferred that there is an equal Δf increase between successive pulses.

5 Non-uniformly Spaced SFW Using Hamming Codes

In the case of a non-uniformly spaced SFW (Fig. 5), the frequency step (Δf) is no more a constant. Hamming codes have been used here in order to obtain a variable step size. Hamming codes are a set of error-correcting codes that provides a parity check bits with each of the output data stream [12, 13]. The most widely used Hamming code is the (7, 4) code, where four data bits (D1, D2, D3, and D4) are encoded into seven bits by adding three parity check bits (P1, P2, and P3) at the end of the bit stream. The parity check matrix H that has been used for this work is depicted in Fig. 4:

$$H = \begin{bmatrix} 1\,0\,0\,0\,1\,1\,1 \\ 0\,1\,0\,1\,0\,1\,1 \\ 0\,0\,1\,1\,1\,0\,1 \end{bmatrix}_{3,7}$$

The transmitted waveform is chosen to be the bit stream. At the end of each bit stream, a parity check bit is added. The parity check bits are pre-initialized to a particular step size (Δf). Now, the step size (Δf) between the pulses depends on the parity check which is added at the end of each bit stream.

	D_1	D_2	D_3	D_4
P_1	×	×		×
P_2	×		×	×
P_3		×	×	×

Fig. 4 Data bits with its associated parity check bit

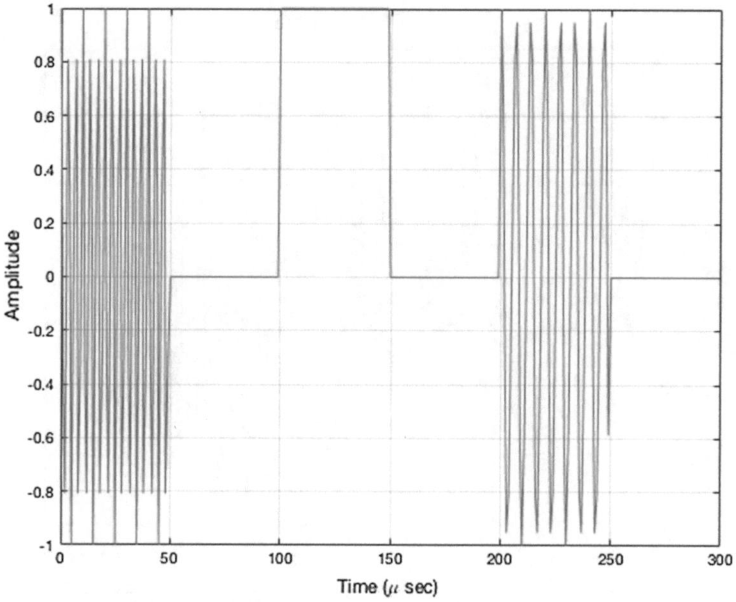

Fig. 5 Non-uniformly spaced SFW

6 OFDM Modulation

OFDM is a special form of the frequency-division multiplexing (FDM) multicarrier modulation technique [10] in which the information to be transmitted is split into multiple smaller chunks and transmitted independently. The total bandwidth is divided equally among all the carrier waveforms, thus constituting subcarriers. In order to avoid intercarrier interference (ICI), the traditional FDM uses guard bands between carriers where no information can be transmitted, although it would result in wastage of the available spectrum [8]. To overcome this drawback, OFDM uses subcarriers which are orthogonal to each other, which is depicted in Fig. 6a. This makes OFDM bandwidth-efficient modulation scheme [1]. To generate OFDM, the data to be transmitted is first channel coded and modulated. Among the traditional modulation schemes such as QAM, PAM, QPSK [4], QPSK is chosen as the modulation scheme. OFDM adds a cyclic prefix before each symbol to avoid intersymbol interference (ISI). This preserves the orthogonality of the transmitted waveform [8]. The serial data stream of symbols is then split into N parallel streams, where N corresponds to the number of subcarriers used in the system. An N-point inverse fast Fourier transform (IFFT) is then used to create the OFDM symbols, of the type shown in Fig. 6b; the use of the FFT/IFFT makes the implementation very computationally efficient.

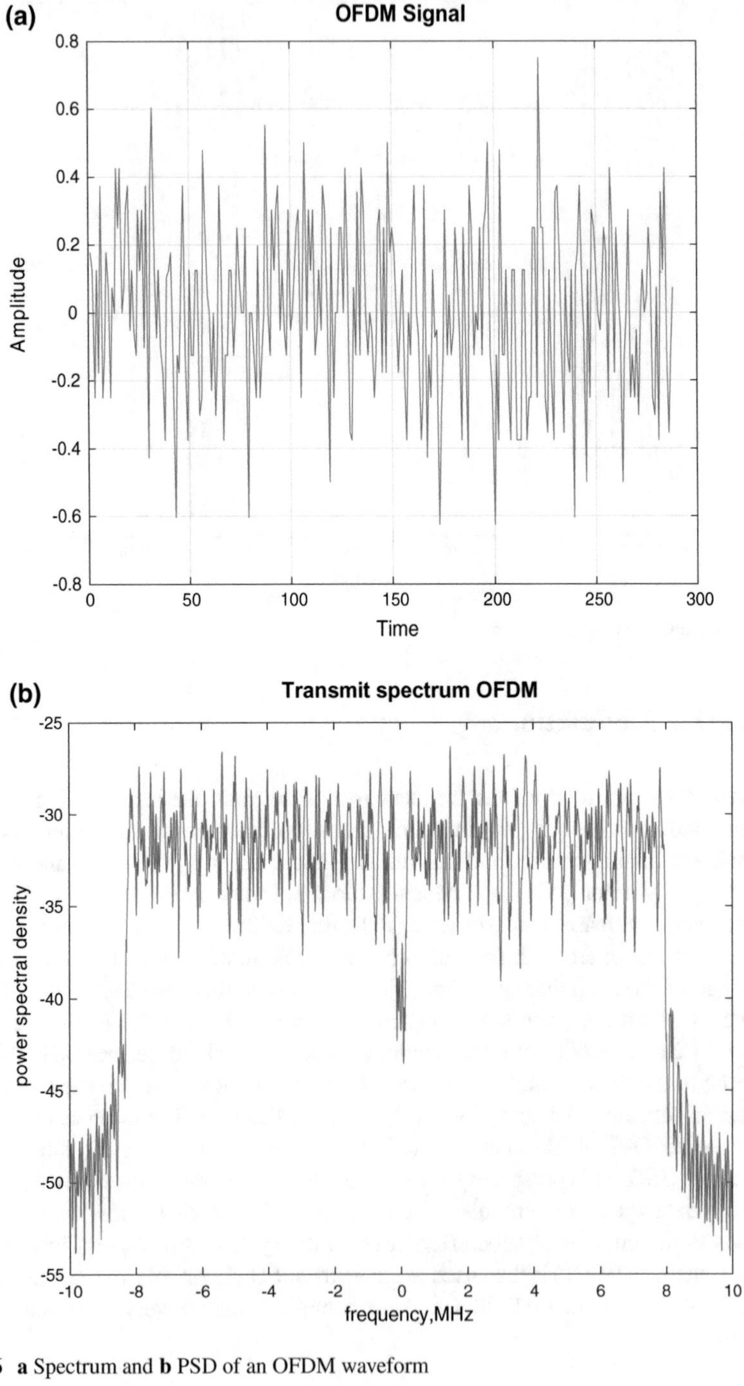

Fig. 6 **a** Spectrum and **b** PSD of an OFDM waveform

The main purpose to use OFDM in the work reported here is to prevent the ICI caused by Doppler shift. The generated OFDM symbol is modulated with SFW signal using QPSK modulation.

7 RADAR System Implementation

A Simulink block has been modeled to generate the two types of SFW. The generated waveform is then transmitted through free space. The waveforms are made to incident on a radar target at a known location, and the reflected echoes are collected at the receiver.

The received echoes are processed to test the detection capability of the transmitted waveform. The block diagram for the radar's transmit side has been modeled using Simulink as shown in Fig. 7. The range plot of the target is depicted in Fig. 8, from which it is inferred that the SFW can differentiate two closely located target and, thus, preserve the range and Doppler resolution parameters.

The following Simulink block is designed with the following specifications: maximum range of 10 km, a PRF of 30 MHz and a pulse width of 33 μs, an operating frequency that is set to 10 GHz, a range resolution of 50 m, and a target RCS as 1.5 m^2. The targets are located at the 2000 and 3500 m ranges away from the radar transmitter.

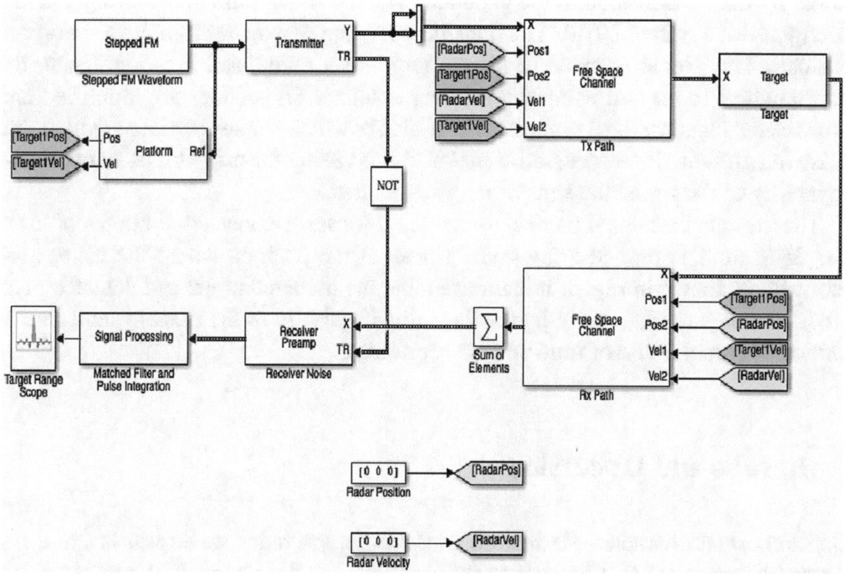

Fig. 7 Simulink model of a radar system for generation, transmission, and processing of SFW

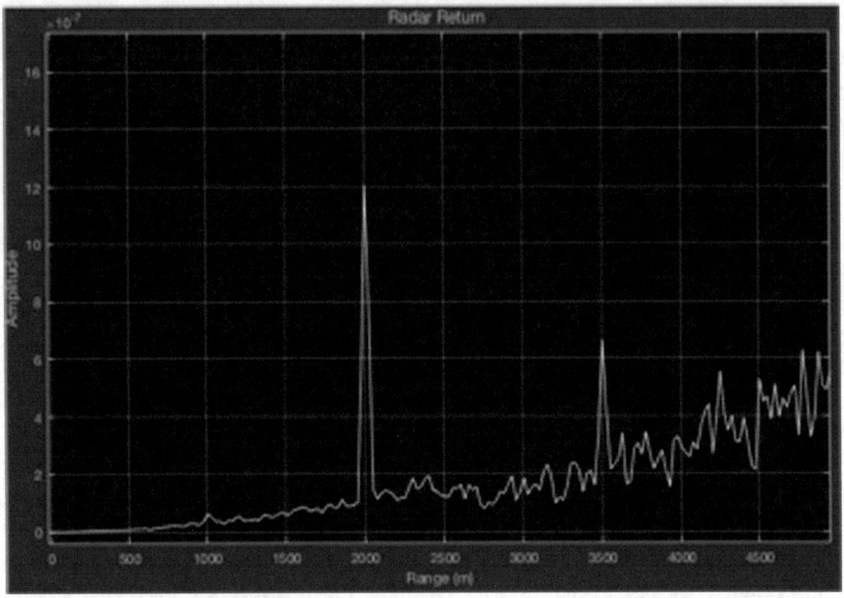

Fig. 8 Detection capability and robustness of SFW

The transmitter block generates the SFW signal of a specific type. This is followed by the transmission of the generated waveform with the transmitting antenna configured at a gain of 20 dB. The transmitted signal propagates through a free space channel. This signal is made to hit the target at a predefined location. Then, the return echo is collected using the receiver antenna. The echoes are amplified and processed using the signal processing sub-blocks which consist of matched filter and pulse integration. The processed signal is plotted using the array block to check the capability of the signal in terms of range estimates.

The various functional blocks are modeled for targets located at ranges of 2005 and 3500 m. The plotted array gives a peak at the position where the echoes are received, as shown in Fig. 8. It is inferred that the modeled target and detected peak are in agreement, thus verifying the detection capability of the radar system for the particular combination of transmitted waveforms.

8 Results and Discussion

One important characteristic feature of the complex radar waveform is the time-bandwidth product [3]. The basic radar waveform has time-bandwidth product in the order of unity because they is no level of coherent pulse integration. The matched filter response plot in Fig. 9a shows that most of the energy is concentrated within a time interval that corresponds to duration of the pulse, and delay interval corresponds to the bandwidth of the pulse. Since the time-bandwidth product is in the order of unity, the ambiguity plot is concentrated in a unit size, thus resulting in corresponding delay duration and Doppler resolution [14]. Existence of side lobe is based on the shape of the pulse. In this section, the effect of Doppler shift and delay on the ambiguity plot will be discussed. The main lobe of the ambiguity plot will have lower peak and energy gets more concentrated in the sidelobes. The resistance of a pulse compression technique to the Doppler effect is known as Doppler tolerance. A radar system with a 10 kHz is considered in this study. Assuming a duty cycle of 20 dB, that gives pulse width (τ) = 1 μs, the maximum unambiguous range R_{max} = $cT_p/2$, which results in 15 km. The range resolution obtainable without any pulse compression is $R = c\tau/2 = 150$ m. For a radar operating frequency of 10 GHz, a target with a velocity of 660 m/s produces a maximum Doppler shift of 22 kHz.

The delay and Doppler plots of conventional modulated waveform such as rectangular and LFM types are plotted in Figs. 9 and 10, and these are then compared with the SFW to show the robustness of the transmit-side waveforms. Rectangular waveforms, also known as single frequency waveform, are the simplest of all conventional radar waveforms.

From Fig. 9 it is inferred that the zero delay has a broad profile. The first null appears at 100 kHz corresponding to an equivalent Doppler shift. If two targets are at the same range, then a minimal Doppler shift of 100 kHz is required that corresponds to a speed of 30 m/s so that they are separately identified. Since the obtained speed is a large value it can no longer separate two targets in Doppler domain.

The first null in the zero-Doppler cut occurs at 10 m/s which corresponds to a distance of 1.5 km, thus preserving the ranger resolution. In case of the delay plot, the first null occurs at 10 μs preserving the range resolution. As in case of Doppler first null occurs at is 20 kHz which is one-fifth of that of the previous waveform.

The response of a uniformly varying frequency step between pulses is shown in Fig. 11. From the plot, it is inferred that there is a uniform (Δf) increase between two pulses, and the corresponding delay and Doppler cuts for a uniformly spaced SFW are determined in terms of the magnitudes and positions of the primary power patterns and sidelobes.

From the plots of ambiguity, delay, and Doppler shown in Fig. 12 it can be inferred that changes have occurred over the main lobe. The width of the main lobe relative to the sidelobes is seen to reduce with the Doppler shift. It also affects the amount of pulse compression achieved, since introduction of Doppler shift or delay increases the time width of the main lobe. In case of uniform SFW, the first null occurs at 10 μs in the delay plot, preserving the range resolution. The sidelobes disappear because one pulse is different from the other. As with the case of Doppler shift, the first null occurs at 2 kHz.

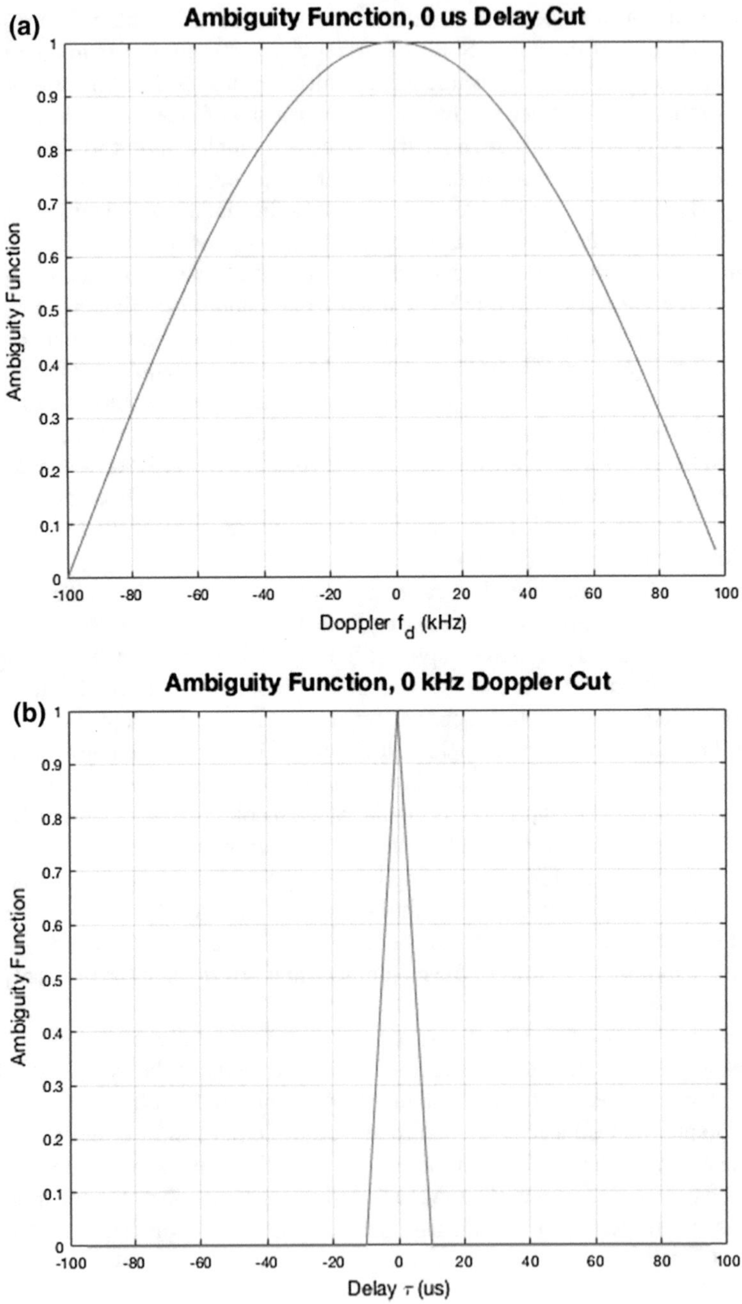

Fig. 9 **a** Delay plot and **b** Doppler plot of a rectangular waveform

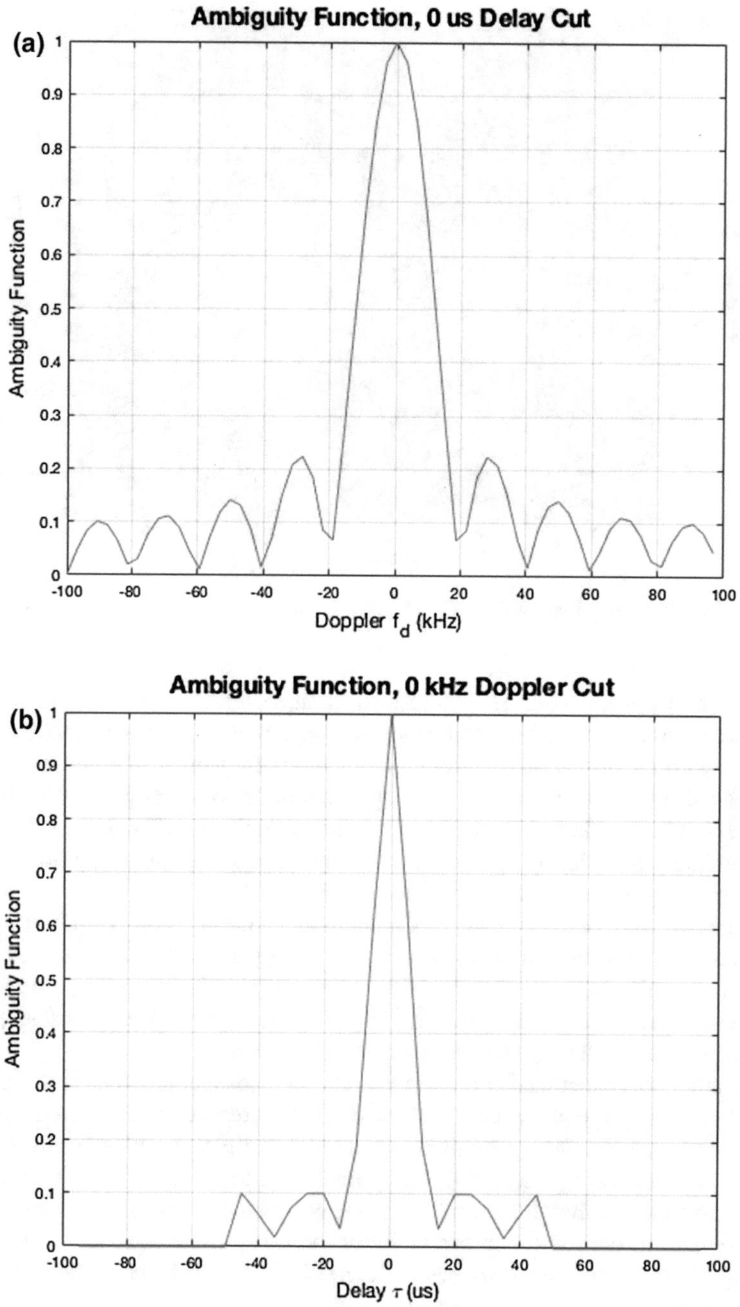

Fig. 10 **a** Delay plot and **b** Doppler plot of an LFM waveform

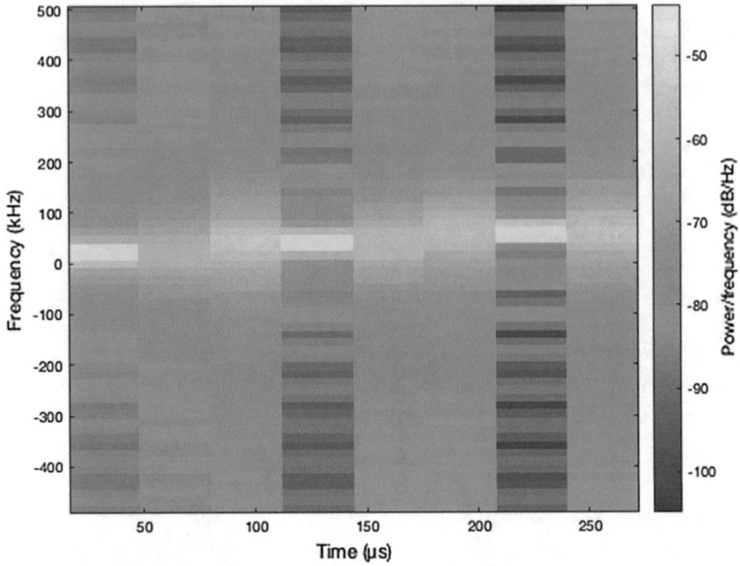

Fig. 11 Response plot of uniformly spaced SFW

A novel approach to pulse compression technique, which is presented in the work discussed here, is to vary the frequency step in a non-uniform manner. Based on the parity check matrix of the Hamming code, the frequency step between pulses is assigned. Figure 13 represents the response plot of non-uniform SFW. This plot is used to infer the detection capability of the non-uniform SFW when it comes to differentiating two closely spaced targets, where sidelobe positions, power pattern positions, and widths become a significant factor.

The main lobe of the ambiguity plot sections along the delay and Doppler domains shown in Fig. 14 reveal a higher peak, and the energy gets more concentrated in the mainlobe. In the Doppler plot, the side lobes are significant, because one pulse is different from other pulse, again highlighting the significance of various power pattern parameters in target detection and classification.

The ambiguity function diagrams in Fig. 15 consist of subplot (a) that is referred to as the bed of nail-type plot which is caused due to the presence of extensive sidelobes. It is to be noted that the unambiguity in the plot of SFW is greatly increased, yet its quality metrics are preserved for the given PRF. In the subplot (b), the sidelobes are suppressed and the plot is no longer one of a bed of nails. There is an evident enhancement to the resolution capability of the waveform on account of the least amount of distortions and sidelobes in the non-uniform SFW type. The ambiguity, delay, and Doppler plots inferences are summarized in Table 1.

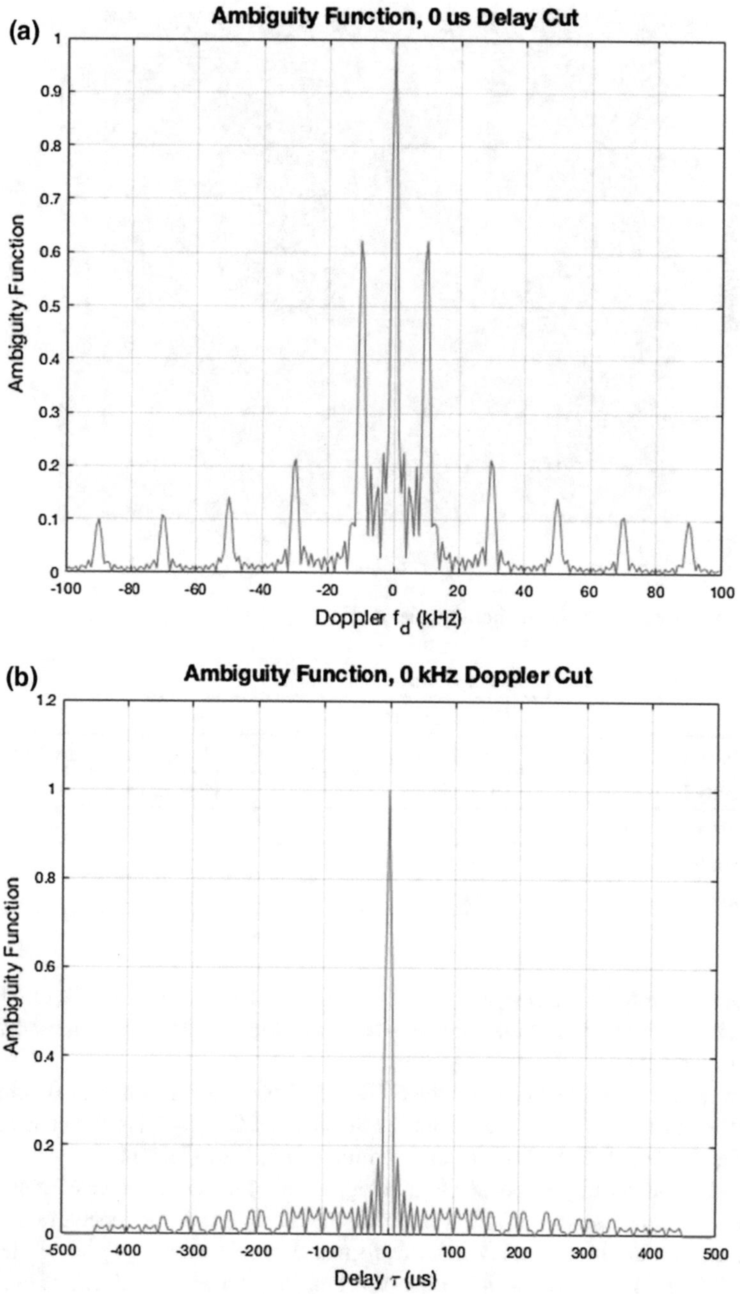

Fig. 12 **a** Delay plot and **b** Doppler plot for a uniformly spaced SFW

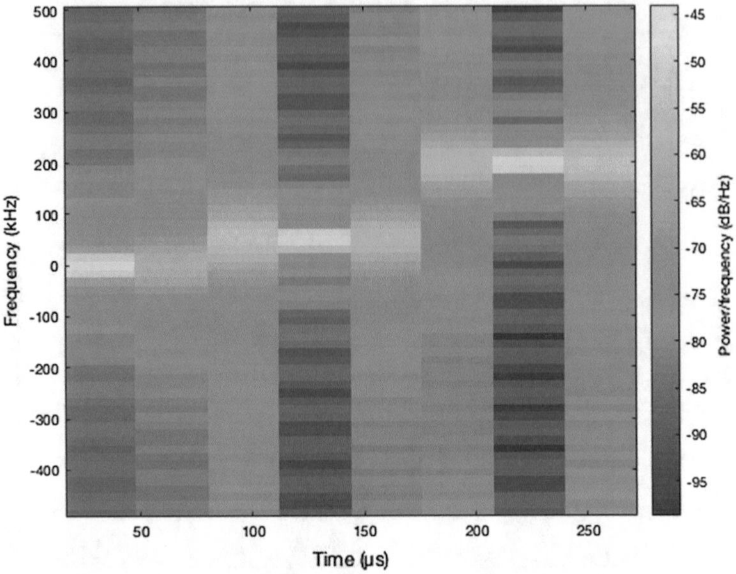

Fig. 13 Response plot of non-uniformly spaced SFW

Table 1 Characteristics of various radar waveforms

	Basic waveform	SFM
Time-bandwidth product	Unity	Large compared with unity
Ambiguity function	Thumbtack	Bed of nails
Resolution cell size	Unity	$1/T * B$
Ambiguities	No	Spike
Sidelobes	Low	Low

The spectra of an OFDM signal and the SFW signal modulated with OFDM are shown in Fig. 16 as part of the comparative exercise used to frequency spread among these waveforms.

The spectrum of the generated the SFW signal is shown in subplot (a), while the spectrum of the SFW signal modulated with the an OFDM signal is shown in subplot (b). These depict the locations of their primary frequency elements.

From the ambiguity response plots in Fig. 17 of either a normal SFW or an SFW-OFDM combined signal, it is inferred that the ambiguities are concentrated about unity, as with the traditional radar waveforms. From the response plot of the SFW with an OFDM component, it is observed that the response is of unit size as like conventional waveform but has a time-bandwidth product greater than unity, thus preserving the range resolution.

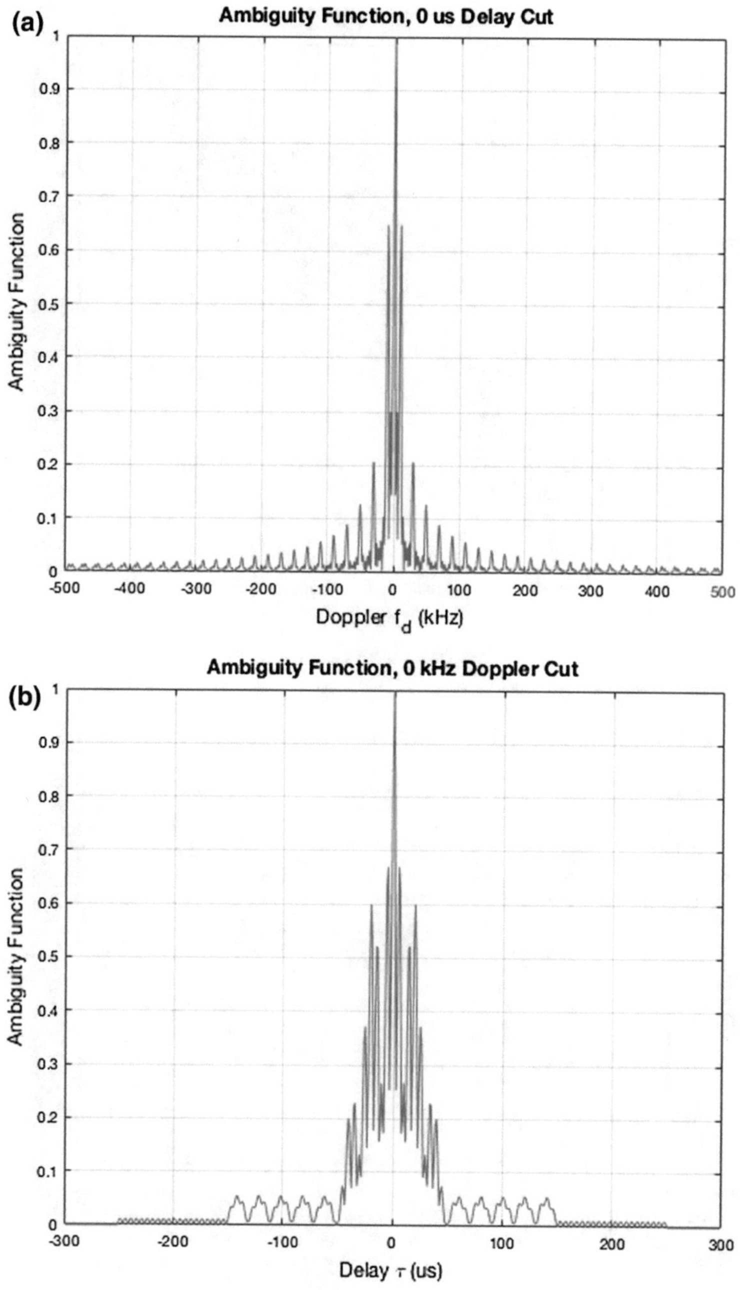

Fig. 14 **a** Delay plot and **b** Doppler plot for a non-uniformly spaced SFW

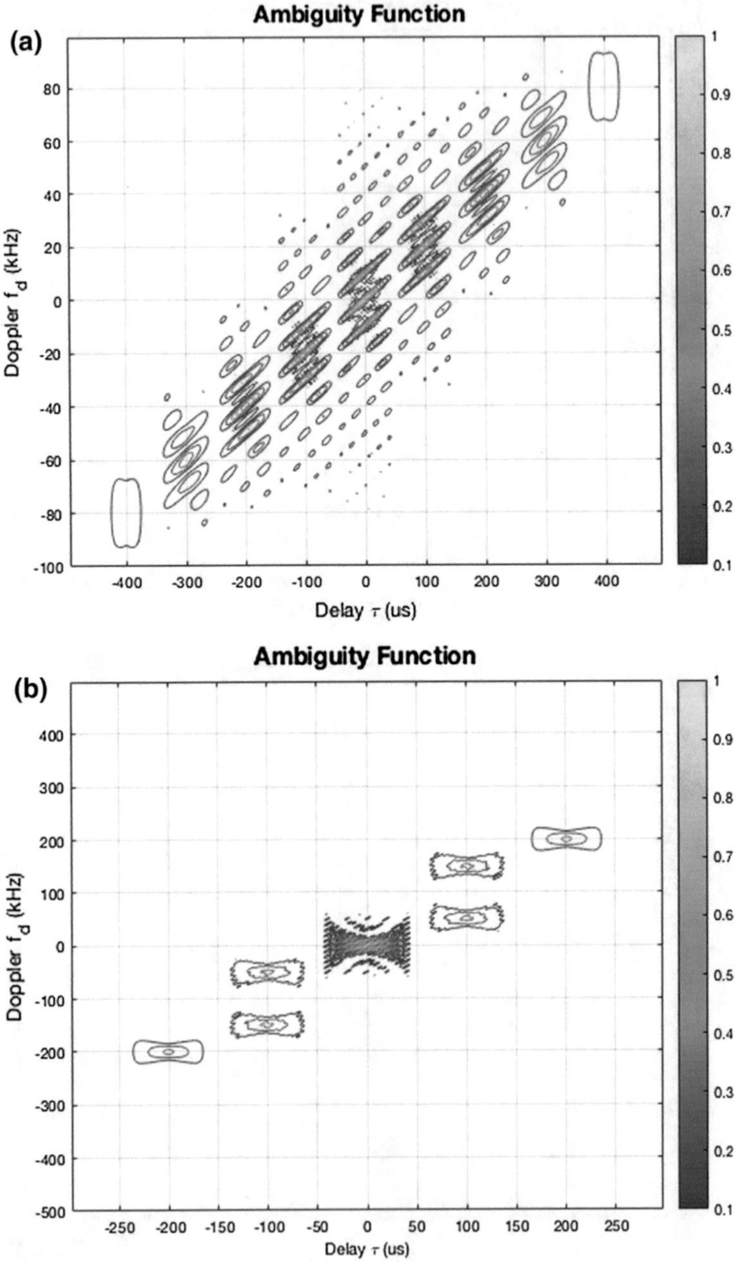

Fig. 15 Ambiguity plot of **a** an uniformly spaced SFW and **b** a non-uniformly spaced SFW

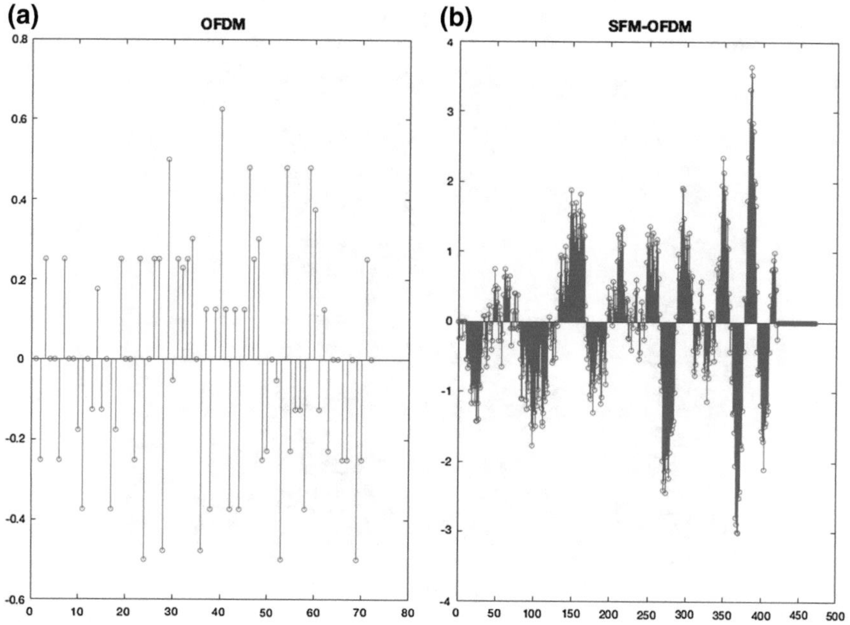

Fig. 16 Spectrum of SFW (**a**: left) and SFW modulated with OFDM (**b**: right)

9 Conclusion

The need for pulse compression in transmitted waveform to obtain better resolution has been discussed here. Comparison of uniformly spaced SFW-OFDM with that of non-uniform spaced SFW-OFDM complex radar waveform-based pulse compression is presented in terms of delay and Doppler shift. A new pulse compression technique is presented which is based on the OFDM and Hamming codes. From the analysis of these techniques in terms of delay, Doppler, and ambiguity function diagram, it is inferred that SFW with non-uniform step size preserves both range and Doppler resolution, whereas the SFW with uniform step size fails to preserve either of the resolution parameters, proving an evident demonstration of the former waveform type as superior in performance in terms of target detection and resistance to ECM signals when employed as a transmit signal in programmable radar systems.

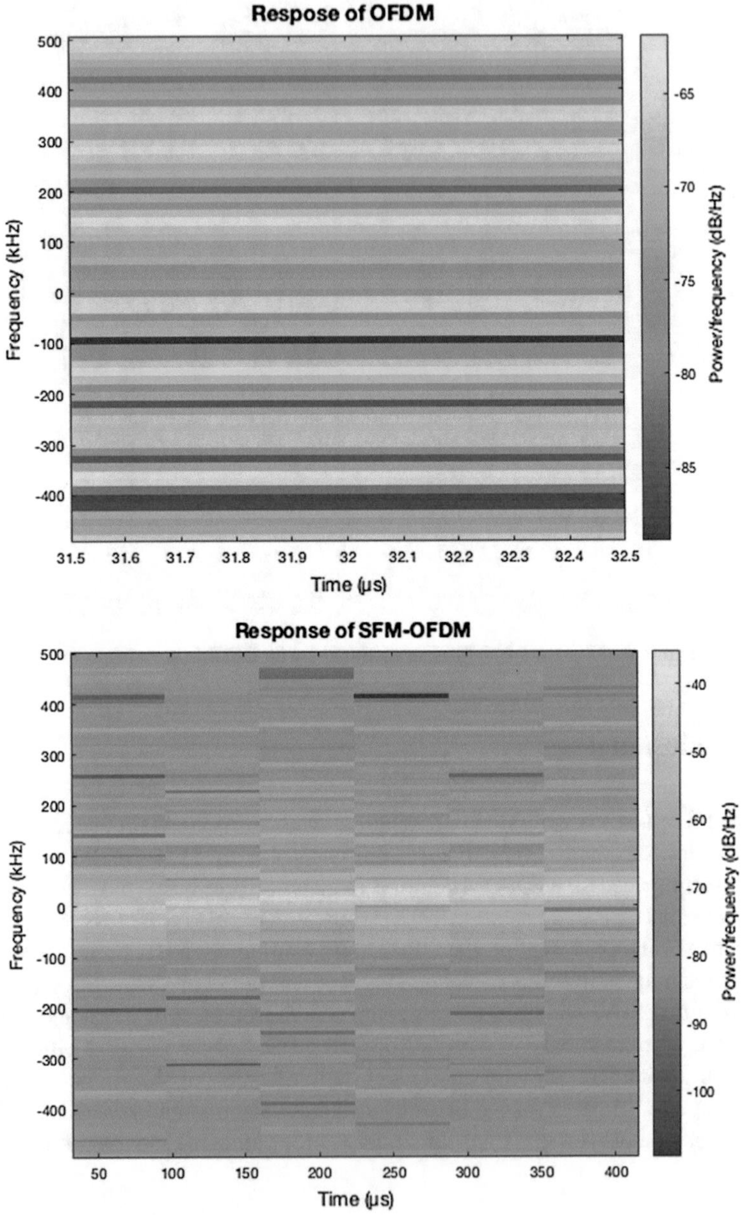

Fig. 17 **a** Ambiguity response plot of OFDM and **b** SFW with an OFDM

References

1. Rihaczek, A.W.: Principles of High-Resolution Radar. McGraw-Hill, New York (1969)
2. Levanon, N., Mozeson, E.: Radar Signals, 1st edn. Wiley, New York (2004)
3. Pierre, S., Siclet, C., Lacaille, N.: Analysis and design of OFDM/OQAM systems based on lterbank theory. IEEE Trans. Sig. Process. **50**(5), 1170–1183 (2002)
4. Ankarao, V., Srivatsa, S., Shanmugha Sundaram, G.A.: Evaluation of pulse compression techniques in X-band radar systems. In: Proceedings of International Conference on Wireless Communications Signal Processing and Networking. IEEE (2017)
5. Carpentier, M.H.: Evolution of pulse compression in the radar field. In: Proceedings of 9th European Microwave Conference. IEEE (1979)
6. Pramudita, I.R.: Implementasi Radar Doppler dengan Sinyal orthogonal frequency division multiplexing (OFDM) pada Perangkat berbasis software define radio (SDR). Doctoral dissertation. Institut Teknologi Sepuluh Nopember (2017)
7. Rihaczek, A.W.: Radar waveform selection-a simplified approach. IEEE Trans. Aerosp. Electron. Syst. **6**, 1078–1086 (1971)
8. Frederiksen, F.B., Prasad, R.: An overview of OFDM and related techniques towards development of future wireless multimedia communications. In: Proceedings of Radio and Wireless Conference. IEEE (2002)
9. Li, G., Meng, H., Xia, X.G., Peng, Y.N.: Range and velocity estimation of moving targets using multiple stepped-frequency pulse trains. Sensors **8**(2), 1343–1350 (2008)
10. Mahafza, B.R.: Radar Systems Analysis and Design Using MATLAB, 3rd edn. CRC Press, Boca Raton (2013)
11. Skolnik, M.I.: Introduction to Radar Systems. McGraw Hill Book Company Inc., New York (1962)
12. Huimin, L., Jingya, Z.: Analysis of a combined waveform of linear frequency modulation and phase coded modulation. In: Proceedings of 11th International Symposium on Antennas, Propagation and EM Theory. IEEE (2016)
13. Hamming, R.W.: Error detecting and error correcting codes. Bell Labs Tech. J. **29**(2), 147–160 (1950)
14. Farnane, K., Minaoui, K., Rouijel, A., Aboutajdine, D. Analysis of the ambiguity function for phase-coded waveforms. In: Proceedings of 12th International Conference on Computer Systems and Applications. IEEE (2015)

A Decision-Making Approach Based On L-IVHFSS Setting

AR. Pandipriya, J. Vimala, Xindong Peng and S. Sabeena Begam

Abstract In 2018, we have propounded the notion of lattice-ordered interval-valued hesitant fuzzy soft set and some of its basic properties were discussed. In this present work, we inherit some essential definitions such as $\mathcal{L}$-optimistic interval-valued fuzzy soft set, $\mathcal{L}$-pessimistic interval-valued fuzzy soft set, $\mathcal{L}$-neutral interval-valued fuzzy soft set, $\mathcal{L}$-threshold interval-valued fuzzy set and Contra-$\mathcal{L}$-threshold interval-valued fuzzy set. Also a new technique has been implemented to solve a real-life problem by applying these definitions over lattice-ordered interval-valued hesitant fuzzy soft set.

Keywords $Max_{\mathcal{L}}$-threshold · $Mid_{\mathcal{L}}$-threshold · $Min_{\mathcal{L}}$-threshold

Subject Classification 06D72

1 Introduction

In 1965, Zadeh [1] extended the notion of crisp sets into fuzzy sets for concerned with numerous domains as modelling, engineering, medical science and biological science. Later the hybrid model of soft sets has been enlarged by Molodtsov for solving the problem of lack of describing parameters. Maji granted soft sets in real-life problems by transforming to fuzzy soft sets. The fuzzy sets have been elongated as hesitant fuzzy sets as the response of unsettled thoughts of human by Torra and Narukawa. Subsequently, Xu et al. and Chen et al. Babitha and John [2] merged soft sets with hesitant fuzzy sets and in 2013 properties of this new hybrid model were examined. In 2016, the conception of generalized hesitant fuzzy soft sets was emerged by Chen [3]. Moreover, hesitant fuzzy soft sets have been executed successfully in

AR. Pandipriya · J. Vimala (✉) · S. S. Begam
Department of Mathematics, Alagappa University, Karaikudi, Tamilnadu, India
e-mail: vimaljey@alagappauniversity.ac.in

X. Peng
Shaoguan University, Shaoguan, People's Republic of China

S. M. Thampi et al. (eds.), *Intelligent Systems, Technologies and Applications*, Advances in Intelligent Systems and Computing 910,
https://doi.org/10.1007/978-981-13-6095-4_16

decision making [4]. Xu and Da provoked the notion of OWA operator as a tool for handling intervals.

Further, Yang proposed interval-valued fuzzy soft sets and its various operations with their applications were studied by Feng in 2010. The mechanism of decision making on interval-valued intuitionistic fuzzy soft sets [5, 6] was inspected by some researchers. Chen et al. [7], Peng and Yang extended the notion of interval-valued intuitionistic hesitant fuzzy soft sets.

In 2015, Zhang derived interval-valued hesitant fuzzy soft sets and Peng et al. and Yang et al. [8] explored the strategy for decision-making problems. Also, Jyoti [9] found more operators on it. In 2014, Birkhoff [10] evaluated lattice theory. Further, Vimala et al. discovered the concepts of lattice-ordered fuzzy soft group and multi-sets on lattice. Later, Pandipriya et al. [11] have conferred L-IVHFSS with real-life applications in 2018.

In recent days, researchers are quite interested in dealing the situation of uncertainties through fuzzification of sets instead of crisp sets. Also interval-valued hesitant fuzzy soft set has been introduced because of describing membership functions in intervals by replacing specified values to indicate the tentative decisions of human beings. In this paper, $\mathcal{L}$-optimistic IVFSS, $\mathcal{L}$-pessimistic IVFSS and $\mathcal{L}$-neutral IVFSS are derived from $\mathcal{L}-\mathcal{IVHFSS}$. Further $\mathcal{L}$-level hesitant fuzzy soft set, contra $\mathcal{L}$-level hesitant fuzzy soft set are inherited from a newly introduced $\mathcal{L}$-threshold IVFS and contra-$\mathcal{L}$-threshold IVFS, respectively. Eventually, $\mathcal{L}$-top-level Hfss, $\mathcal{L}$-mid-level Hfss and $\mathcal{L}$-low-level Hfss are defined. Later, the real-life problem has been solved by applying these new techniques of $\mathcal{L}-\mathcal{IVHFSS}$.

This work is constructed in the following way. In Sect. 2, definition of lattice-ordered interval-valued hesitant fuzzy soft set is given. In Sect. 3, some new definitions are established and a new algorithm has been designed to solve a real-life problem via L-IVHFSS.

2 Preliminaries

Definition 1 [11] Let U be a universal set and P be a parameter set. Let $(\mathcal{F}, \mathcal{P})$ be an interval-valued hesitant fuzzy soft set. We say that $(\mathcal{F}, \mathcal{P})$ a lattice-ordered interval-valued hesitant fuzzy soft set ($\mathcal{L}-\mathcal{IVHFSS}$) if $\mathcal{F}(e_i) \subseteq \mathcal{F}(e_j)$ whenever $e_i \leq e_j, \forall e_i, e_j \in \mathcal{P}$ and is shortly written as $\mathcal{L}-\mathcal{IVHFSS}$.

3 A Decision-Making Approach Based on L-IVHFSS Setting

In this part, we define some new definitions required for solving real-life problem on $\mathcal{L}-\mathcal{IVHFSS}$.

Definition 2 Let $(\mathcal{F}, \mathcal{P})$ be an $\mathcal{L} - \mathcal{IVHFSS}$ over U. Then, $\mathcal{L}$-optimistic interval-valued fuzzy soft set $(\mathcal{F}_{\mathcal{L}+}, \mathcal{P})$ is defined as

$$\begin{aligned}(\mathcal{F}_{\mathcal{L}+}, \mathcal{P}) &= \{< x, \mathcal{F}_{\mathcal{L}+}(e)(x) > |x \in U\} \\ &= \{< x, \vee \gamma_{\mathcal{L}}^{\sigma(k)} > |x \in U\},\end{aligned}$$

$\forall e \in \mathcal{P}$ and $\forall k = 1, 2, \ldots, n$, where $\gamma_{\mathcal{L}}^{\sigma(k)} = [\gamma_{\mathcal{L}}^{\sigma(k)(L)}, \gamma_{\mathcal{L}}^{\sigma(k)(U)}] \in \mathcal{F}(e)(x)$.

Definition 3 Let $(\mathcal{F}, \mathcal{P})$ be an $\mathcal{L} - \mathcal{IVHFSS}$ over U. Then, $\mathcal{L}$-neutral interval-valued fuzzy soft set $(\mathcal{F}_{\mathcal{L}\sim}, \mathcal{P})$ is defined as

$$\begin{aligned}(\mathcal{F}_{\mathcal{L}\sim}, \mathcal{P}) &= \{< x, \mathcal{F}_{\mathcal{L}\sim}(e)(x) > |x \in U\} \\ &= \{< x, \sum_{k=1}^{n} \frac{\gamma_{\mathcal{L}}^{\sigma(k)}}{n} > |x \in U\},\end{aligned}$$

$\forall e \in \mathcal{P}$ and $\forall k = 1, 2, \ldots, n$, where $\gamma_{\mathcal{L}}^{\sigma(k)} = [\gamma_{\mathcal{L}}^{\sigma(k)(L)}, \gamma_{\mathcal{L}}^{\sigma(k)(U)}] \in \mathcal{F}(e)(x)$.

Definition 4 Let $(\mathcal{F}, \mathcal{P})$ be an $\mathcal{L} - \mathcal{IVHFSS}$ over U. Then, $\mathcal{L}$-pessimistic interval-valued fuzzy soft set $(\mathcal{F}_{\mathcal{L}-}, \mathcal{P})$ is defined as

$$\begin{aligned}(\mathcal{F}_{\mathcal{L}-}, \mathcal{P}) &= \{< x, \mathcal{F}_{\mathcal{L}-}(e)(x) > |x \in U\} \\ &= \{< x, \wedge \gamma_{\mathcal{L}}^{\sigma(k)} > |x \in U\},\end{aligned}$$

$\forall e \in \mathcal{P}$ and $\forall k = 1, 2, \ldots, n$, where $\gamma_{\mathcal{L}}^{\sigma(k)} = [\gamma_{\mathcal{L}}^{\sigma(k)(L)}, \gamma_{\mathcal{L}}^{\sigma(k)(U)}] \in \mathcal{F}(e)(x)$.

Definition 5 Let $(\mathcal{F}, \mathcal{P})$ be an $\mathcal{L} - \mathcal{IVHFSS}$ over U and $(I, \mathcal{P})$ be an IVFS. Then,
$\alpha^{\sigma(k)} = \begin{cases} 1 & \text{if } \gamma^{\sigma(k)} \geq I(e) \\ 0 & \text{if } \gamma^{\sigma(k)} < I(e) \end{cases}$,
$\forall e \in \mathcal{P}, \forall \gamma^{\sigma(k)} \in \mathcal{F}(e)$ and $\forall k = 1, 2, \ldots, n$.
Then, $(I, \mathcal{P})$ is called an $\mathcal{L}$-threshold interval-valued fuzzy set. Also an $\mathcal{L}$-level hesitant fuzzy soft set $(\mathcal{F}_{\mathcal{L}I}, \mathcal{P})$ with respect to $(I, \mathcal{P})$ is defined as

$$\begin{aligned}(\mathcal{F}_{\mathcal{L}I}, \mathcal{P}) &= \{< x, \mathcal{F}_{\mathcal{L}I}(e)(x) > |x \in U\} \\ &= \{< x, \{\alpha^{\sigma(k)}\} > |x \in U\},\end{aligned}$$

$\forall e \in \mathcal{P}, \forall k = 1, 2, \ldots, n$

Definition 6 Let $(\mathcal{F}, \mathcal{P})$ be an $\mathcal{L} - \mathcal{IVHFSS}$ over U and $(J, \mathcal{P})$ be an IVFS. Then,
$\beta^{\sigma(k)} = \begin{cases} 1 & \text{if } \gamma^{\sigma(k)} \leq J(e) \\ 0 & \text{if } \gamma^{\sigma(k)} > J(e) \end{cases}$,
$\forall e \in \mathcal{P}, \forall \gamma^{\sigma(k)} \in \mathcal{F}(e)$ and $\forall k = 1, 2, \ldots, n$.
Then $(J, \mathcal{P})$ is called a contra-$\mathcal{L}$-threshold interval-valued fuzzy set. Also a contra-$\mathcal{L}$-level hesitant fuzzy soft set $(\mathcal{F}_{\mathcal{L}J}, \mathcal{P})$ with respect to $(J, \mathcal{P})$ is defined as

$$(\mathcal{F}_{\mathcal{L}J}, \mathcal{P}) = \{< x, \mathcal{F}_{\mathcal{L}J}(e)(x) > | x \in U\}$$
$$= \{< x, \{\beta^{\sigma(k)}\} > | x \in U\},$$

$\forall e \in \mathcal{P}, \forall k = 1, 2, \ldots, n.$

Definition 7 Let U be an universal set and $\mathcal{P}$ be a parameter set. Let $(\mathcal{F}, \mathcal{P})$ be $\mathcal{L} - \mathcal{IVHFSS}$. Then,

(i) $\vee_{x \in U}\{\mathcal{F}(e)(x)\} = \mathcal{F}(e)(x_k)$ *if* $\mathcal{F}(e)(x_i) \subseteq \mathcal{F}(e)(x_j) \subseteq \mathcal{F}(e)(x_k)$
(ii) $\wedge_{x \in U}\{\mathcal{F}(e)(x)\} = \mathcal{F}(e)(x_i)$ *if* $\mathcal{F}(e)(x_i) \subseteq \mathcal{F}(e)(x_j) \subseteq \mathcal{F}(e)(x_k)$
(iii) $\vee \wedge_{x \in U}\{\mathcal{F}(e)(x)\} = \mathcal{F}(e)(x_j)$ *if* $\mathcal{F}(e)(x_i) \subseteq \mathcal{F}(e)(x_j) \subseteq \mathcal{F}(e)(x_k)$.

Definition 8 Let $(\mathcal{F}, \mathcal{P})$ be an $\mathcal{L} - \mathcal{IVHFSS}$ over U and $(\mathcal{F}_{\mathcal{L}+}, \mathcal{P})$ be an $\mathcal{L}$-optimistic IVFSS over U. Then, $Max_{\mathcal{L}}$-threshold of $\mathcal{L} - \mathcal{IVHFSS}$ is defined as $\mathcal{F}_{Max-\mathcal{L}}(e) = \vee \mathcal{F}_{\mathcal{L}+}(e)(x), \forall e \in \mathcal{P}$, where $x \in U$. The $\mathcal{L}$-top-level Hfss of $(\mathcal{F}, \mathcal{P})$ with respect to $Max_{\mathcal{L}}$-threshold is defined as $(\mathcal{F}_{Max-\mathcal{L}}, \mathcal{P}) = \{< x, \mathcal{F}_{Max-\mathcal{L}}(e)(x) > | x \in U\}$.

Definition 9 Let $(\mathcal{F}, \mathcal{P})$ be an $\mathcal{L} - \mathcal{IVHFSS}$ over U and $(\mathcal{F}_{\mathcal{L}\sim}, \mathcal{P})$ be an $\mathcal{L}$-neutral IVFSS over U. Then, $Mid_{\mathcal{L}}$-threshold of $\mathcal{L} - \mathcal{IVHFSS}$ is defined as $\mathcal{F}_{Mid-\mathcal{L}}(e) = \vee \wedge \mathcal{F}_{\mathcal{L}\sim}(e)(x), \forall e \in \mathcal{P}$, where $x \in U$. The $\mathcal{L}$-mid-level Hfss of $(\mathcal{F}, \mathcal{P})$ with respect to $Mid_{\mathcal{L}}$-threshold is defined as $(\mathcal{F}_{Mid-\mathcal{L}}, \mathcal{P}) = \{< x, \mathcal{F}_{Mid-\mathcal{L}}(e)(x) > | x \in U\}$.

Definition 10 Let $(\mathcal{F}, \mathcal{P})$ be an $\mathcal{L} - \mathcal{IVHFSS}$ over U and $(\mathcal{F}_{\mathcal{L}-}, \mathcal{P})$ be an $\mathcal{L}$-pessimistic IVFSS over U. Then $Min_{\mathcal{L}}$-threshold of $\mathcal{L} - \mathcal{IVHFSS}$ is defined as $\mathcal{F}_{Min-\mathcal{L}}(e) = \wedge \mathcal{F}_{\mathcal{L}-}(e)(x), \forall e \in \mathcal{P}$, where $x \in U$. The $\mathcal{L}$-low-level Hfss of $(\mathcal{F}, \mathcal{P})$ with respect to $Min_{\mathcal{L}}$-threshold is defined as $(\mathcal{F}_{Min-\mathcal{L}}, \mathcal{P}) = \{< x, \mathcal{F}_{Min-\mathcal{L}}(e)(x) > | x \in U\}$.

3.1 Technique of Solving a Real-Life Problem on Lattice-Ordered Interval-Valued Hesitant Fuzzy Soft Set

Consider x_1, x_2 and x_3 represent three sets of diet and e_1, e_2, e_3, e_4 and e_5 represent protein, carbohydrates, fat, water and vitamins, respectively. Thus, the given $\mathcal{L} - \mathcal{IVHFSS}(\mathcal{F}, \mathcal{P})$ represents the amount of protein, carbohydrates, fat, water and vitamins present in the three sets of diets and this data is collected from two nutrionists. Now, the problem is that the decision maker has to select the best diet set among the three given sets of diets. For this situation, We will choose the most correct decision by constructing an algorithm as follows:

Step:1

Write a given $\mathcal{L} - \mathcal{IVHFSS}(\mathcal{F}, \mathcal{P})$

Given $(\mathcal{F}, \mathcal{P}) = \{\mathcal{F}(e_1) = \{< x_1, \{[0.4, 0.6], [0.4, 0.65]\} >,$
$< x_2, \{[0.2, 0.31], [0.3, 0.5]\} >, < x_3, \{[0.45, 0.5], [0.4, 0.6]\} >\},$
$\mathcal{F}(e_2) = \{< x_1, \{[0.5, 0.7], [0.4, 0.5]\} >, < x_2, \{[0.35, 0.55], [0.2, 0.3]\} >,$
$< x_3, \{[0.4, 0.5], [0.55, 0.65]\} >\}, \mathcal{F}(e_3) = \{< x_1, \{[0.44, 0.6], [0.51, 0.63]\} >,$
$< x_2, \{[0.25, 0.5], [0.33, 0.4]\} >, < x_3, \{[0.47, 0.6], [0.5, 0.6]\} >\},$
$\mathcal{F}(e_4) = \{< x_1, \{[0.4, 0.55], [0.58, 0.7]\} >, < x_2, \{[0.3, 0.5], [0.35, 0.45]\} >,$
$< x_3, \{[0.5, 0.6], [0.5, 0.7]\} >\}, \mathcal{F}(e_5) = \{< x_1, \{[0.51, 0.6], [0.5, 0.67]\} >,$
$< x_2, \{[0.2, 0.4], [0.5, 0.6]\} >, < x_3, \{[0.5, 0.65], [0.55, 0.7]\} >\}\}$

Step:2

(i) *Find $Max_{\mathcal{L}}$-threshold of $(\mathcal{F}, \mathcal{P})$.*
$\mathcal{F}_{Max-\mathcal{L}}(e) = \{< e_1, [0.4, 0.65] >, < e_2, [0.5, 0.7] >, < e_3, [0.51, 0.63] >,$
$< e_4, [0.58, 0.7] >, < e_5, [0.55, 0.7] >\}$

(ii) *Find $\mathcal{L}$-top-level hesitant fuzzy soft set $(\mathcal{F}_{Max-\mathcal{L}}, \mathcal{P})$ of $(\mathcal{F}, \mathcal{P})$.*
$(\mathcal{F}_{Max-\mathcal{L}}, \mathcal{P}) = \{\mathcal{F}(e_1) = \{< x_1, \{0, 1\} >, < x_2, \{0, 0\} >, < x_3, \{0, 0\} >\},$
$\mathcal{F}(e_2) = \{< x_1, \{1, 0\} >, < x_2, \{0, 0\} >, < x_3, \{0, 0\} >\},$
$\mathcal{F}(e_3) = \{< x_1, \{0, 1\} >, < x_2, \{0, 0\} >, < x_3, \{0, 0\} >\},$
$\mathcal{F}(e_4) = \{< x_1, \{0, 1\} >, < x_2, \{0, 0\} >, < x_3, \{0, 0\} >\},$
$\mathcal{F}(e_5) = \{< x_1, \{0, 0\} >, < x_2, \{0, 0\} >, < x_3, \{0, 1\} >\}\}$

(iii) *Assign weight w_{j+} for each parameter e_j, where $j = 1, 2, \ldots, m$ for $(\mathcal{F}_{Max-\mathcal{L}}, \mathcal{P})$.* $w_{1+} = 0.5, w_{2+} = 0.45, w_{3+} = 0.2, w_{4+} = 0.4, w_{5+} = 0.5$

(iv) *Draw a table for $\left(\mathcal{F}_{Max-\mathcal{L}}, \mathcal{P}, w_{j+}\right)$.*

U	$(\mathcal{P}, w_{i+})$					p_i
	$e_1, w_{1+} = 0.5$	$e_2, w_{2+} = 0.45$	$e_3, w_{3+} = 0.2$	$e_4, w_{4+} = 0.4$	$e_5, w_{5+} = 0.5$	
x_1	{0, 1}	{1, 0}	{0, 1}	{0, 1}	{0, 0}	1.55
x_2	{0, 0}	{0, 0}	{0, 0}	{0, 0}	{0, 0}	0
x_3	{0, 0}	{0, 0}	{0, 0}	{0, 0}	{0, 1}	0.5

Step:3

(i) *Find $Mid_{\mathcal{L}}$*-threshold of $(\mathcal{F}, \mathcal{P})$.
$\mathcal{F}_{Mid-\mathcal{L}}(e) = \{< e_1, [0.4, 0.625] >, < e_2, [0.475, 0.575] >, < e_3, [0.485, 0.6] >, < e_4, [0.49, 0.625] >, < e_5, [0.505, 0.635] >\}$

(ii) *Find $\mathcal{L}$-mid-level hesitant fuzzy soft set $(\mathcal{F}_{Max-\mathcal{L}}, \mathcal{P})$ of $(\mathcal{F}, \mathcal{P})$.*
$(\mathcal{F}_{Mid-\mathcal{L}}, \mathcal{P}) = \{\mathcal{F}(e_1) = \{< x_1, \{0, 1\} >, < x_2, \{0, 0\} >, < x_3, \{0, 0\} >\},$
$\mathcal{F}(e_2) = \{< x_1, \{1, 0\} >, < x_2, \{0, 0\} >, < x_3, \{0, 1\} >\},$
$\mathcal{F}(e_3) = \{< x_1, \{0, 1\} >, < x_2, \{0, 0\} >, < x_3, \{0, 1\} >\},$
$\mathcal{F}(e_4) = \{< x_1, \{0, 1\} >, < x_2, \{0, 0\} >, < x_3, \{0, 0\} >\},$
$\mathcal{F}(e_5) = \{< x_1, \{0, 1\} >, < x_2, \{0, 0\} >, < x_3, \{1, 1\} >\}\}$

(iii) *Calculate weight $w_{j\sim} = \frac{w_{j+}}{2}$ for each parameter e_j, where $j = 1, 2, \ldots, m$ for $(\mathcal{F}_{Mid-\mathcal{L}}, \mathcal{P})$.*
$w_{1\sim} = 0.25, w_{2\sim} = 0.225, w_{3\sim} = 0.1, w_{4\sim} = 0.2, w_{5\sim} = 0.25$

(iv) *Draw a table for* $\left(\mathcal{F}_{Mid-\mathcal{L}}, \mathcal{P}, w_{j\sim}\right)$.

U	$(\mathcal{P}, w_{i\sim})$					q_i
	$e_1, w_{1\sim} = 0.25$	$e_2, w_{2\sim} = 0.225$	$e_3, w_{3\sim} = 0.1$	$e_4, w_{4\sim} = 0.2$	$e_5, w_{5\sim} = 0.25$	
x_1	$\{0, 1\}$	$\{1, 0\}$	$\{0, 1\}$	$\{0, 1\}$	$\{0, 1\}$	1.025
x_2	$\{0, 0\}$	$\{0, 0\}$	$\{0, 0\}$	$\{0, 0\}$	$\{0, 0\}$	0
x_3	$\{0, 0\}$	$\{0, 1\}$	$\{0, 1\}$	$\{0, 0\}$	$\{1, 1\}$	0.825

Step:4

(i) *Find* $Min_{\mathcal{L}}$*-threshold of* $(\mathcal{F}, \mathcal{P})$.
$\mathcal{F}_{Min-\mathcal{L}}(e) = \{< e_1, [0.2, 0.31] >, < e_2, [0.2, 0.3] >, < e_3, [0.25, 0.4] >, < e_4, [0.3, 0.45] >, < e_5, [0.2, 0.4] >\}$

(ii) *Find* $\mathcal{L}$*-low-level hesitant fuzzy soft set* $(\mathcal{F}_{Min-\mathcal{L}}, \mathcal{P})$ *of* $(\mathcal{F}, \mathcal{P})$.
$(\mathcal{F}_{Min-\mathcal{L}}, \mathcal{P}) = \{\mathcal{F}(e_1) = \{< x_1, \{0, 0\} >, < x_2, \{1, 0\} >, < x_3, \{0, 0\} >\},$
$\mathcal{F}(e_2) = \{< x_1, \{0, 0\} >, < x_2, \{0, 1\} >, < x_3, \{0, 0\} >\},$
$\mathcal{F}(e_3) = \{< x_1, \{0, 0\} >, < x_2, \{0, 0\} >, < x_3, \{0, 0\} >\},$
$\mathcal{F}(e_4) = \{< x_1, \{0, 0\} >, < x_2, \{0, 1\} >, < x_3, \{0, 0\} >\},$
$\mathcal{F}(e_5) = \{< x_1, \{0, 0\} >, < x_2, \{1, 0\} >, < x_3, \{0, 0\} >\}\}$

(iii) *Calculate* $w_{j-} = 1 - w_{j+}$ *for each parameter* e_j*, where* $j = 1, 2, \ldots, m$ *for* $(\mathcal{F}_{Min-\mathcal{L}}, \mathcal{P})$.
$w_{1-} = 0.5, w_{2-} = 0.55, w_{3-} = 0.8, w_{4-} = 0.6, w_{5-} = 0.5$

(iv) *Draw a table for* $\left(\mathcal{F}_{Min-\mathcal{L}}, \mathcal{P}, w_{j-}\right)$.

U	$(\mathcal{P}, w_{i-})$					r_i
	$e_1, w_{1-} = 0.5$	$e_2, w_{2-} = 0.55$	$e_3, w_{3-} = 0.8$	$e_4, w_{4-} = 0.6$	$e_5, w_{5-} = 0.5$	
x_1	$\{0, 0\}$	$\{0, 0\}$	$\{0, 0\}$	$\{0, 0\}$	$\{0, 0\}$	0
x_2	$\{1, 0\}$	$\{0, 1\}$	$\{0, 0\}$	$\{0, 1\}$	$\{1, 0\}$	2.15
x_3	$\{0, 0\}$	$\{0, 0\}$	$\{0, 0\}$	$\{0, 0\}$	$\{0, 0\}$	0

Step:5

(i) *Calculate* $p_i = \sum_{j=1}^{m} \mathcal{F}_{Max-\mathcal{L}}(e)(x_i)\, w_{j+}$
$p_1 = 1.55, p_2 = 0, p_3 = 0.5$

(ii) *Calculate* $q_i = \sum_{j=1}^{m} \mathcal{F}_{Mid-\mathcal{L}}(e)(x_i)\, w_{j\sim}$
$q_1 = 1.025, q_2 = 0, q_3 = 0.825$

(iii) *Calculate* $r_i = \sum_{j=1}^{m} \mathcal{F}_{Min-\mathcal{L}}(e)(x_i)\, w_{j-}$
$r_1 = 0, r_2 = 2.15, r_3 = 0$

Step:6

Calculate $C_i = \frac{p_i + q_i}{2} - r_i$*, for each* $i = 1, 2, \ldots, n$
$C_1 = 1.2875, C_2 = -2.15, C_3 = 0.6625$
Choose the optimal alternative as x_i *whose* C_i *value is maximum.*
Hence, the optimal alternative is x_1.

Decision:

Thus, the decision maker choose the optimal alternative x_1, as the best diet. In [11], we analysed the decision problem in a neutral view and solved the problem using aggregation method. But here, we examined the problem in different views such as optimistic, pessimistic and neutral. Thus, we inspected the problem in many dimensions to get the precise decision.

4 Conclusion

In this work, the disparate strategy has been implemented for decision-making problem over lattice-ordered interval-valued hesitant fuzzy soft set. We ensure that our forthcoming work is to enlarge the characteristics of lattice-ordered interval-valued hesitant fuzzy soft set in multifarious fields.

Acknowledgements We are grateful to the editors for giving pivotal suggestions to fine-tune this paper.

References

1. Zadeh, L.A.: Fuzzy sets. Inf. Control **8**, 338–353 (1965)
2. Babitha, K.V., John, S.J.: Hesitant fuzzy soft sets. J. New Results Sci. **3**, 98–107 (2013)
3. Bin, C.: Generalized hesitant fuzzy soft sets. Ital. J. Pure Appl. Math. **36**, 35–54 (2016)
4. Wang, Y.M.: Using the method of maximizing deviations to make decision for multiindices. J. Syst. Eng. Electron. **8**, 21–26 (1997)
5. Qin, H., Noor, A.S., Ma, X., Chiroma, H., Herawan, T.: An adjustable reduction approach of interval-valued intuitionistic fuzzy soft sets for decision making. Appl. Math. Inf. Sci. **11**, 999–1009 (2017)
6. Wen, X., Xiao, Z.: Optimistic and pessimistic decision making based on interval-valued intuitionistic fuzzy soft sets. Comput. Model. New Technol. **18**(12), 1284–1290 (2014)
7. Chen, N., Xu, Z., Xia, M.: Interval-valued hesitant preference relations and their applications to group decision making. Knowl. Based Syst. **37**, 528–540 (2013)
8. Yang, Y., Lang, L., Lu, L., Sun, Y.: A new method of multiattribute decision-making based on interval-valued hesitant fuzzy soft sets and its application. Mathematical Problems in Engineering, 8p (2017)
9. Borah, M.J., Hazarika, B.: Some operators on interval-valued hesitant fuzzy soft sets. General Mathematics, 14p. arXiv:1604.00902v1 (2016)
10. Birkhoff, G.: Lattice Theory, vol. XXV. American Mathematical Society Colloqium Publications (1967)
11. Pandipriya, A.R., Vimala, J., Begam, S.S.: Lattice ordered interval-valued hesitant fuzzy soft sets in decision making problem. Int. J. Eng. Tech. **7**, 52–55 (2018)

Battery Assisted, PSO-BFOA based Single Stage PV Inverter fed Five Phase Induction Motor Drive for Green Boat Applications

Y. Suri Babu and K. Chandra Sekhar

Abstract Single-stage solar powered, battery assisted five-phase induction motor drive is proposed to avail benefits such as less cost, higher efficiency, and compact in size. Major goal of this paper is to investigate the performance of five-phase induction motor for solar boat application. Also, such type of solar boats requires battery energy storage systems for continuous operation over the specified period. This paper also investigates the performance of input ripple currents of battery as well as solar PV array in both conventional three-phase and proposed five-phase systems. Simulation-based results proven that input current ripples less in case of five-phase systems, which results in enhanced lifetime of battery as well as DC link capacitor in comparison with three-phase. Also incorporates all the performance parameters such as cost, reliability, power density, device current stresses, and starting currents for both the systems. Results proven that five-phase system offers superior performance compared to existing conventional approach.

Keywords Solar boats · Five-phase induction motor · Solar energy · Power management

1 Introduction

Solar technology is one of the most emerging technologies among all other renewable energy sources. Over the years, because of contamination of environment, increase in per capita energy demand and depletion of fossil fuel causes for the hunting of new alternative sources has become a worldwide concern. Among accessible renewable

Y. Suri Babu (✉)
Acharya Nagarjuna University, Guntur 522 510, India
e-mail: ysuribabu@gmail.com

Y. Suri Babu · K. Chandra Sekhar
Department of Electrical and Electronics Engineering, R.V.R & J.C College of Engineering, Guntur 522019, Andhra Pradesh, India
e-mail: cskoritala@gmail.com

S. M. Thampi et al. (eds.), *Intelligent Systems, Technologies and Applications*, Advances in Intelligent Systems and Computing 910,
https://doi.org/10.1007/978-981-13-6095-4_17

energy sources, solar energy had drawn utmost significance because it is available freely, noise free. Thus, PV systems are playing an important role in energy source to meet electrical energy demand. Stand-alone PV systems have wide use in remote area applications such as water pumping and electric boat (green boat) propulsion. These stand-alone PV system-driven applications necessitate the Inclusion of Energy Storage System (ESS) because solar energy generation is not constant. With ESS, the excess energy is stored in the battery bank and this stored energy is provided to load when there is reduction in source power and/or increase in load demand. This feature allows the PV system to become more reliable. This technology is almost saturated in on-road electrical vehicle applications which are zero emissive and eco-friendly in nature. Currently, most of the researchers are exploring the same concept for water vehicles like small passenger boats [1, 2], In the literature, a basic solar boat was developed in which PV panels are used to drive three-phase asynchronous motor by utilizing the space available on the terrace of boats [3]. Later on, several features such as reduced V_{dc} at partial speeds, applying optimum flux level to improve the performance, reliability and optimize efficiency have been proposed [4]. Hybrid solar boats which powered by photovoltaics and fuel cells [5] are developed in the literature, utilizing diesel generator and lithium battery sources used as backup sources [6]. The major problem with PV fed electric boats is that availability of sun is not predictable and that depends on weather conditions, demands electrical energy storage system such as batteries. Based on power generation by solar PV, either energy need to store/supply the additional energy in the battery subjected to load power greater or lesser than the PV power generation. So, energy management plays a crucial role in solar boat to utilize the energy effectively in all weather conditions, many methods proposed [7–12]. In electric boats three-phase induction motors [4, 13], BLDC motors [14], permanent magnet synchronous motors [15], and DC motors [15] are preferred for propulsion, but induction motors got lot of attention due to its compact in size, economical and rugged in nature. However, three-phase IM is less reliable (not greater fault tolerant) for stand-alone applications. It is necessary to maintain an equal rating of the motor for stand by conditions [13] when fault occurs in any one-phase or drive system. Usage of additional motor for standby operation is not economical and also causes additional baggage. Due to advancements in processing of power electronics data, and multiphase ac electrical motors, multiphase drives become more popular for this application where high-reliability operation is required. Certainly, multiphase ac drives can perform its operation even when phases are lost due to fault. Under such conditions, a minimum of three phases is still available for operation.

From this viewpoint, multiphase machines are finding their niche for several applications such as electric and hybrid electric vehicles (mainly for propulsion), ship propulsion, aerospace or wind power (predominantly for remote off-shore), locomotive traction and high-power industrial applications. In autonomous applications like solar boat, EV/HEV, and electric traction have no restriction on the number of phases to three, opening the opportunity to multi-phase drives [16–18]. This paper presents stand-alone PV powered five-phase induction motor drive for electric boat application system along with the energy storage for backup functionality. To drive the five-phase induction motors, two-stage power conditioning structure is conventionally used. However, suffers from more losses, bulky, high cost, and two control loops required. Hence, single-stage power converter scheme is proposed in this paper which offers more efficiency, compact, less cost compared to two-stage systems. To get higher conversion efficiency of the solar PV system, maximum power point tracking (MPPT) algorithm needs to be used. ESS is incorporated in stand-alone solar PV system to manage system energy to supply uninterrupted power to the connected load. Among all available storage devices such as batteries, fuel cells, and flywheel energy storage, batteries are mostly preferred for PV applications because they have high energy density and performance. Among variants of batteries, lead-acid batteries are employed in this work due to their characteristics are well suited for this application. Moreover, these are more reliable and efficient. The five-phase inverter is shown in Fig. 1, which supplies five-phase induction motor load. To achieve the required speed with fast dynamic response, closed-loop control system is designed. In this paper, two converters are used, namely bidirectional DC-DC converter and five-phase inverter. MPPT, DC link voltage and energy management control techniques are applied to BDC, whereas v/f control strategy is used for inverter control to meet the variable speed requirements. So, to control the proposed system, three independent control loops are designed. MPPT is used to obtain the maximum power available from solar PV module even in varying conditions in nature and load. A control strategy is used to keep DC link voltage constant during change in load or source that by charging and discharging of battery are presented. Battery with this control technique maintains the power equilibrium in the system by varying the DC bus voltage. Solar PV module is interfaced with battery by common DC link as well as bidirectional DC-DCconverter. Here, five-phase VSI with PWM is used to supply constant voltage and frequency to load. To evaluate the proposed system dynamic performance, a step changing load is measured. Proposed stand-alone solar PV system with battery assistance mainly consists of a bidirectional buck-boost converter with its control circuit, ESS, a DC-AC five-phase VSI with its control and five-phase asynchronous motor. These two converters work collectively to supply power to the motor and for propelling the boat under all weather conditions. The structure of paper is as follows, proposed system is described in Sect. 2. In Sect. 3, simulation results and discussions are presented, and finally, in Sect. 4, conclusions are presented.

2 Proposed System

2.1 Solar PV Module Modeling

From [19–24], mathematical characteristic equations are used to model PV array. The equivalent circuit of a PV cell required for modeling and simulation is presented in Fig. 2a. It includes a current source (I_{ph}) across diode, D and shunt resistance, R_{sh} and a series resistance, R_s. The output current of PV module is given by

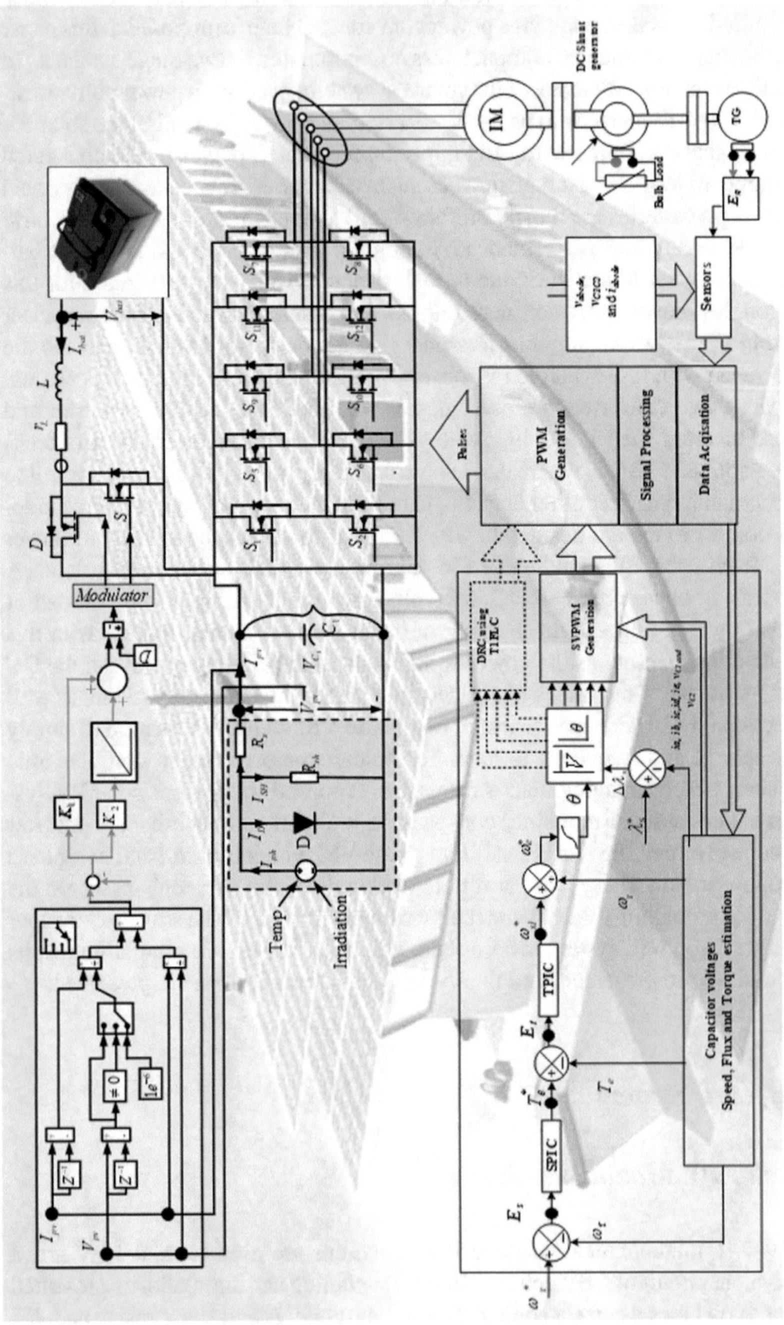

Fig. 1 Battery assisted, solar-powered single-stage power conditioning unit for five-phase induction motor drive

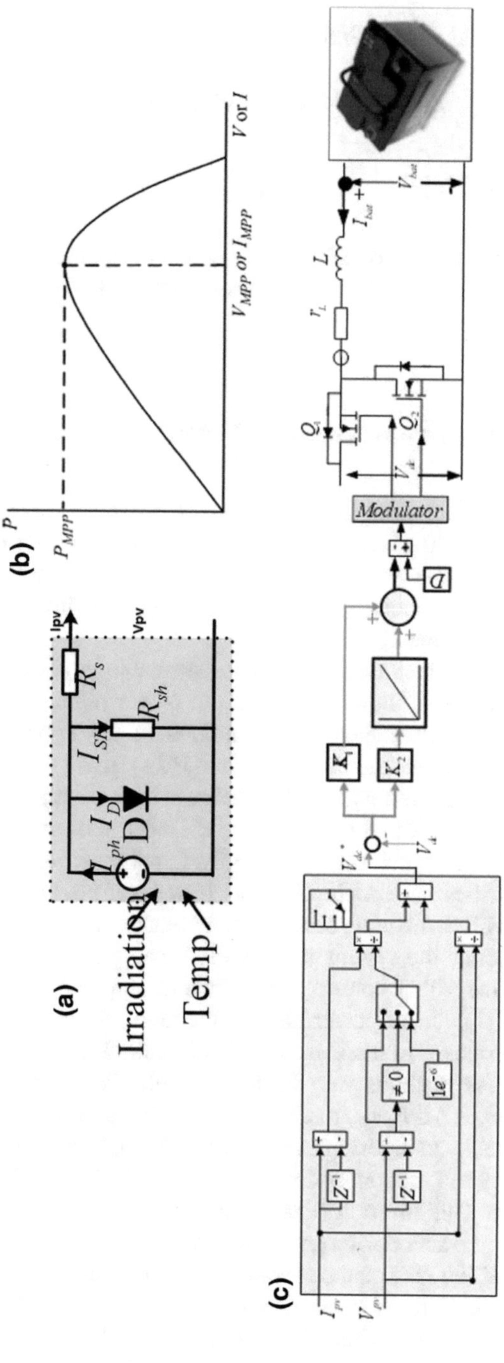

Fig. 2 **a** Equivalent circuit diagram of solar PV cell, **b** PV and PI characteristics of solar PV cell, **c** control circuit of bidirectional DC-DC converter

$$I_{pv} = N_p * I_{ph} - N_p * I_D \tag{1}$$

where I_{ph} the photocurrent and I_D diode current

$$I_{ph} = [I_{SC} + K_I(T_C - T_{Ref})]\lambda \tag{2}$$

$$I_D = I_o\left[\exp\left(\frac{q * (V_{pv} + I_{pv} * R_p)}{N_s AKT}\right) - 1\right] \tag{3}$$

where I_{SC} is short-circuit current of cell, K_I is temperature coefficient of cell, T_{Ref} is reference temperature of cell and λ is the solar insolation in kW/m^2. T_C is the cell's working temperature.

2.2 Intelligent Perturb and Observe MPPT Method

Tracking of maximum power point (MPP) of a solar PV array is generally an important part of a PV system [20]. Figure 2b shows the PV array power curve characteristics. The MPPT technique automatically finds the voltage VMPP or the current IMPP at which a PV array should operate to obtain the maximum power output PMPP under a given parameter.

The optimum point at which maximum power available in power-voltage characteristic of a PV module is known as maximum power point (MPP). MPP not constant and it depends on climate and load conditions. Thus, to operate efficiently MPP tracking (MPPT) methods are necessary for a PV system.

Over the few decades, numerous MPPT techniques have been proposed. Among these P&O, HC and INC are widely used methods. Under uniform solar irradiation basic said MPPT, algorithms perform satisfactorily. In partial shading condition, the PV module characteristics have multiple peaks (few local peaks with one global peak) and so-called MPPT methods cannot differentiate the local peak and global peak, so unable to operate the system at the MPP. When partial shading condition (PSC) arises, change the MPPT controller performance, reducing the output power of the system due to significant power loss. In literature, there are several MPPT methods have been proposed to track accurate MPP under PSC without modifying system hardware. These methods vary in their complexity, accuracy, and speed. Even these methods track MPP accurately, the dynamic response of system is low. By using artificial intelligent algorithm to the MPPT controller, dynamic response of the system increases and accurately determines the MPP under PSC.

The Particle Swarm Optimization (PSO) is a stochastic optimization technique that was inspired from the actions of a group of birds or social insects with restricted individual capabilities. The PSO method shows substantial potential, because of easy implementation and capability to determine the MPP. The bacterial foraging (BF) is grounded upon search and optimal foraging choice making capabilities of the *E. coli* bacteria. The hybrid MPPT algorithm considered in this paper is achieved by combining two artificial intelligent algorithms, namely PSO and BFOA, aims to make use of PSO ability to interchange social information and BF ability in finding a new result by removal and dispersal.

Over the past a few decades, many biologically inspired Evolutionary Algorithms (EAs) have been widely applied to solve different engineering problems. They have been using due to their simplicity and sturdiness. But, Evolutionary Algorithms (EAs) have slow convergence rate. Hybrid EAs need the objective function to be differentiable. Two intelligent algorithms, namely Ant Colony Optimization (ACO) and Particle Swarm Optimizer (PSO), have been developed to attempt applications for which sustainable solutions are difficult or sometimes impossible. Application of PSO to optimization problem which is high-dimensional in large scale is difficult mainly they impose computational load enormously.

The study of bacterial foraging algorithm (BFA) receives large interest in the computational intelligence. BFA was developed as optimization algorithm based on the fundamental phenomena of the surviving of bacteria in a complex environment. In BFOA searching, the feature of self-adaptability of individuals in the group has attracted plenty of interest. Process of BFA could not obtain convincing results in high-dimensional problems [25–30].

To PV system users, so many MPPT techniques available, some of the MPPT methods only apply to particular topologies. It might be clear to decide which one suits better to our application. The major aspects of the MPPT methods are implementation, cost, sensors, multiple local maxima, convergence speed, whether it works with analog or digital or either or both etc. to be taken into consideration. Out of these techniques we choose P&O method, since it satisfies almost all the main aspects of MPPT technique. Figure 3a, b shows control circuit of PSO-BFOA-based MPPT controller and flowchart of bacterial foraging oriented by PSO algorithm, respectively.

2.3 Five-Phase Inverter and Five-Phase Induction Motor

Five-phase induction motor is constructed using ten-phase sets, each set of 36°. Therefore, the spatial displacement among phases is 72°. The rotor is treated as equivalent properties of five-phase stator winding and assumed that the rotor winding has been referred to stator with the use of transformation ratio. A five-phase induction machine modeling equations which are considered in this paper described in [21].

2.4 v/f Control

This sub-section presents a closed-loop constant v/f control scheme to five-phase induction motor drive system. Broadly, control methods of IM can be categorized as scalar and vector control. In the scalar control, only either voltage or current

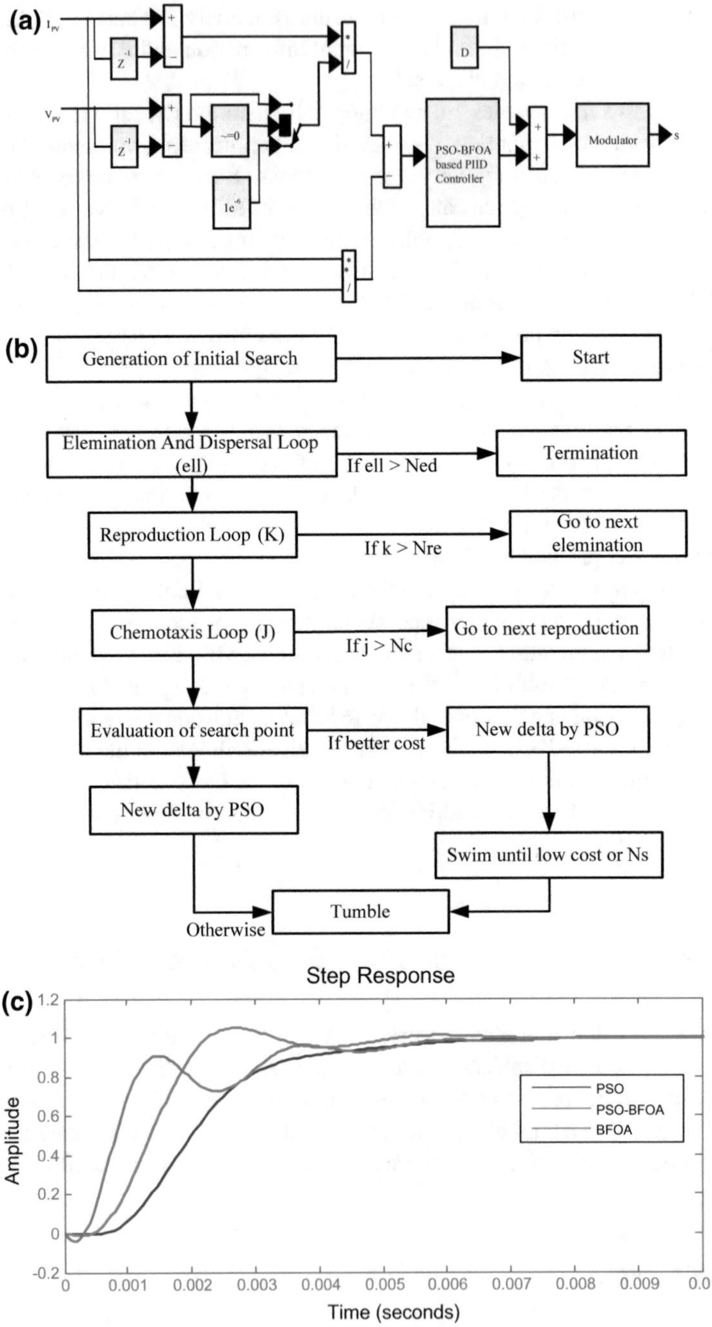

Fig. 3 **a** Control of PSO-BFOA-based MPPT controller, **b** flow chart of BFO by PSO algorithm, **c** how, PSO + BFOA best with simulation circuit

magnitude and frequency are controlled. In vector control, instantaneous positions of voltage and current are controlled also. Among these two schemes, scalar control is best suited where there is no fast response is required, because this method is simple to realize [24].

In many applications where the dynamic performance of the drive is not much significant and rapid change in speed is not considerable. For such cases, the cheap and simple solution is constant *v/f* control scheme. The fundamental principle of this control technique is to maintain constant rated flux at all operating points to retain the torque capability of the drive to its maximum value throughout transition period and avoiding the core saturation losses.

2.5 Control of Bidirectional DC-DC Converters

Control is significant subject to examine the operation of the system, so discussed here. The lead-acid/Lithium ion phosphate battery is widely used to assist while managing the energy. In the proposed system, battery functions in two modes, namely charging and discharging to meet the surplus power. To keep constant DC link voltage and to retain uninterrupted power flow between the DC bus and ESS, bidirectional DC-DC converter which has ability to boost and/or buck is controlled such a way [23]. Control of bidirectional DC-DC converter is shown in Fig. 2c. During charging mode, Switch Q1 is activated to run as buck converter, Q2 is activated to run as boost converter during discharging. So, the Switches Q1 and Q2 turn on in complimentary fashion either in buck mode or in boost mode. The voltage V_{dc} is compared with a reference DC link voltage V_{dcref} to get the error signal, and this error is processed to produce battery reference current I_{batref}. Then, the I_{batref} is compared with the current sensed at battery, comparison gives error signal. The obtained error is processed by the PI controller to get control signal. Now compare this control signal with a high-frequency triangular carrier signal to get the PWM switching pulses. This DC-DC bidirectional converter is switched at a frequency of 20 kHz.

3 Simulations Results and Discussion

To validate the performance of proposed system i.e., battery assisted, solar-powered single-stage power conditioning unit for five-phase induction motor drive have been simulated on Matlab/Simulink environment. Considered simulation parameters have been tabulated in Table 1.

The simulation results of solar PV Voltage; PV current and PV Power; DC link voltage and DC link current; battery input current and voltage and output current and voltages are presented in this Results section. Also were presented simulation results of the motor side i.e., speed, stator currents, nominal current discharge characteristics, and modulation index of the five-phase inverter. PV array of 5 kW rating with

Table 1 Simulation parameters

Solar PV Module Specifications	
Parameter	Value
I_{SC}	9.49 A
V_{OC}	43.20 V
I_{MPP}	8.47 A
V_{MPP}	36 V
P_{MPP}	300 W
N_S	16
Energy Storage Power Converter	
Parameter	Value
Input power	3 kW
Input voltage	200 V
Output voltage	200–600 V
Switching frequency	20 kHz
Power Converters Parameters	
3L-VSI	
Parameter	Value
Input power	2 kW
Input voltage	200-600 V
Output voltage	90 V @ 367 Hz
Switching frequency	20 kHz
Induction Motor	
Parameter	Value
Output power	2 kW
Rated speed	1430 rpm
Rated torque	10 Nm

intelligent MPP tracking technique has been considered and connected in parallel to battery fed bidirectional buck-boost DC-DC converter to form DC bus.

In the simulations, two disturbances in the form of solar irradiation and speed of the induction motor drive are introduced to confirm the dynamic performance of the proposed system. Assuming PV temperature remains constant throughout simulation time. From Fig. 4a it can be observed that at 1.25 s solar irradiation is decreased so consequently IPV and PPV decreased. The PV current and PV voltage are shown in Fig. 4a, b, respectively. Similarly, battery current and voltage are shown in Fig. 4g, h. Figure 4k shows the stator currents during both transient and steady states. Stator currents under steady state are sinusoidal waveform shown in inset. Figure 4n shows the simulation results of five-phase induction motor drive with a torsional load of 10 N-m in solar boat. Reference speed at time 0 s is 1430 rpm and it is changed to 1000 rpm at time 1.5 s.

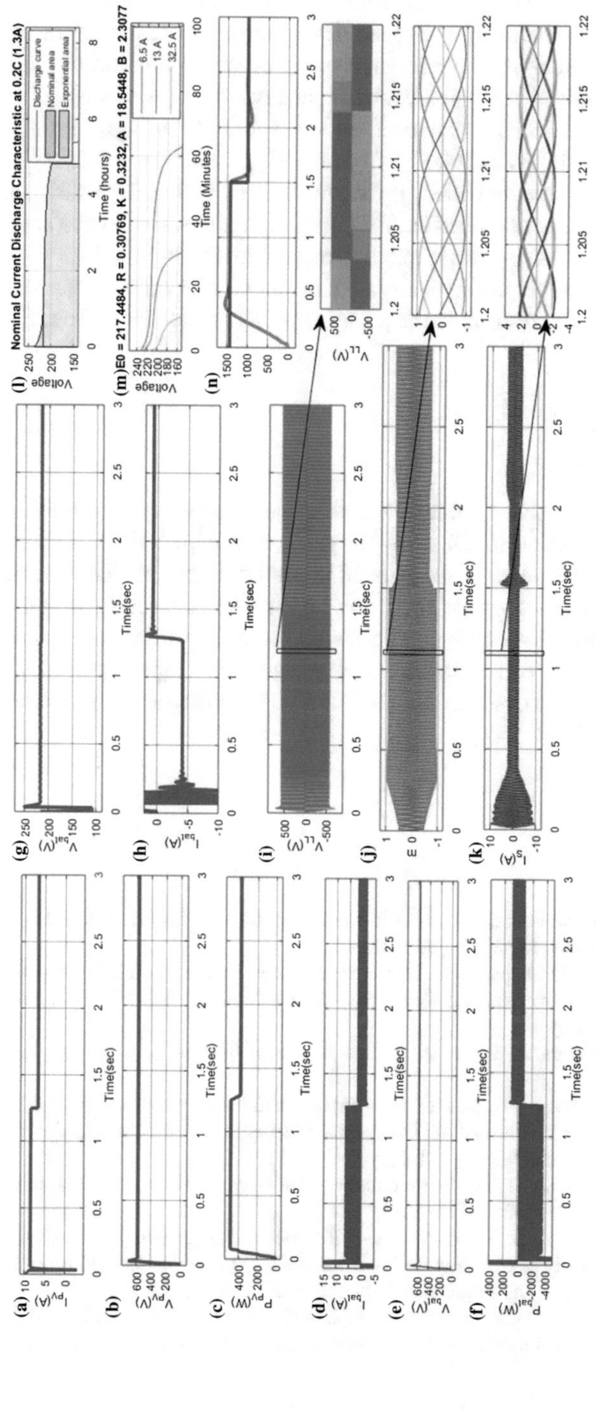

Fig. 4 **a** solar PV current, **b** solar PV voltage, **c** solar PV power, **d** battery fed DC-DC converter output current, **e** battery fed DC-DC converter output voltage, **f** battery power, **g** battery voltage, **h** battery current, **i** line-line voltages, **j** modulation index waveforms of VSI, **k** stator currents, **l** nominal current discharge, **m** battery discharge curves for different loads, **n** motor speed

It can be observed from Fig. 4h that actual speed of the motor exactly follows the reference speed. At 2 s reference speed is not changed but load is increased correspondingly we can observe the speed disturbances momentarily and at steady state actual speed is tracking the reference speed exactly. Variation of modulation index according to speed reference is shown in Fig. 4j at $t = 2$ s as ref speed is decreased; modulation index is adjusted to low value accordingly. From Fig. 4c power available from the PV is 5000 W, from starting instant to $t = 1.25$ s. So excess in power available is stored in battery up to time 1.25 s. At time 1.25 s PV current is reduced since irradiation is reduced but PV voltage remains same, so power available from PV is reduced which is not sufficient to meet load. During $t = 1.25$ s to $t = 3$ s, there is deficiency in power available from PV to meet load requirements. Therefore, deficient power is obtained from battery. The same can be observed from Fig. 4h.

4 Conclusions

In this paper, battery assisted solar-powered single-stage power conditioning unit-based five-phase induction motor drive for green boat applications is successfully validated using simulation results simulated in Matlab/Simulink. These results validated that battery assisted single-stage solar PV system successfully able to feed the power to load in all weather conditions with the help of energy storage unit. Controllers designed for drive are able to trace the required speed perfectly, and on the other hand, controller designed for BDC is able to fulfill the energy management function quintessentially.

References

1. Jurdana, V., Sladic, S.: Green energy for low power ships. In: 57th International Symposium ELMAR, 28–30 September 2015, Zadar, Croatia
2. Schirripa Spagnolo, G., Papalilo, D., Martocchia, A.: Eco friendly electric propulsion boat. In: IEEE 2011 10th International Conference on Environment and Electrical Engineering (EEEIC)—Rome, Italy
3. Schaffrin, C., Benndorf, S., Schneider, S.: The solar boat 'Korona': a complete photovoltaic-powered system
4. Sousa, G.C.D., Simonetti, D.S.L., Norena, E.E.C.: Efficiency optimization of a solar boat induction motor drive. In: Thirty-Fifth IAS Annual Meeting and World Conference on Industrial Applications of Electrical Energy, vol. 3, pp. 1424–1430 (2000)
5. Leiner, R.: Solar radiation and water for emission-free marine mobility. In: IEEE International Energy Conference, pp. 1425–1428 (2014)
6. Bellache, K., Camara, M.B., Zhou, Z., Dakyo, B.: Energy management in hybrid electric boat based on frequency distribution approach—using diesel, lithium-battery and supercapacitors. In: IEEE Vehicle Power and Propulsion Conference, pp. 1–6 (2015)
7. Yufang, Z., Peng, Q., Hong, Y.: Fuel free ship, design for next generation. In: 2013 Eighth International Conference and Exhibition on Ecological Vehicles and Renewable Energies (EVER), pp. 1–5 (2013)

8. Sauze, C., Neal, M.: Long term power management in sailing robots. In: Oceans 2011, pp. 1–8. IEEE, Spain (2011)
9. Tani, A., Camara, M.B., Dakyo, B.: Energy management based on frequency approach for hybrid electric vehicle applications: fuel-cell/lithium-battery and ultracapacitors. IEEE Trans. Veh. Technol. **61**(8), 3375–3386 (2012)
10. Vu, T.L., Dhupia, J.S., Ayu, A.A., Kennedy, L., Adnanes, A.K.: Optimal power management for electric tugboats with unknown load demand. In: American Control Conference, pp. 1578–1583 (2014)
11. Tang, X.-J., Wang, T., Zhi, C., Huang, Y.-M.: The design of power management system for solar ship. In: International Conference on Transportation Information and Safety, pp. 548–553 (2015)
12. http://navaltboats.com/navalt-products-solar-electric-ferry/
13. Lutful Kabir, S.M., Alam, I., Rezwan Khan, M., Hossain, M.S., Rahman, K.S., Amin, N.: Solar powered ferry boat for the rural area of Bangladesh. In: International Conference on Advances in Electrical, Electronic and Systems Engineering, pp. 38–42 (2016)
14. Postiglione, C.S., Collier, D.A. F., Dupczak, B.S., Heldwein, M.L., Perin, A.J.: Propulsion system for an all electric passenger boat employing permanent magnet synchronous motors and modern power electronics. In: 2012 Electrical Systems for Aircraft, Railway and Ship Propulsion, pp. 1–6 (2012)
15. Freire, T., Sousa, D.M., Costa Branco, P.J.: Aspects of modeling an electric boat propulsion system. In: IEEE Region 8 SIBIRCON, Irkutsk Listvyanka, Russia, 11–15 July 2010
16. Singh, G.K.: Multi-phase induction machine drive research—a survey. Electr. Power Syst. Res. **61**, 139–147 (2002)
17. Levi, E., Bojoi, R., Profumo, F., Toliyat, H.A., Williamson, S.: Multi-phase induction motor drives—a technology status review. IET Electr. Power Appl. **1**(4), 489–516 (2007)
18. Levi, E.: Multi-phase machines for variable speed applications. IEEE Trans. Ind. Elect. **55**(5), 1893–1909 (2008)
19. Shaw, P., Sahu, P.K., Maitl, S., Kurnar, P.: Modeling and control of a battery connected standalone photovoltaic system. In: 1st IEEE International Conference on Power Electronics. Intelligent Control and Energy Systems (2016)
20. Beriber, D., Talha, A.: MPPT techniques for PV systems. In: 4th International Conference on Power Engineering, Energy and Electrical Drives Istanbul, Turkey, 13–17 May 2013, pp. 1437–1442
21. Abu-Rub, H., Iqbal, A., Guzinski, J.: High Performance Control of AC Drives with MATLAB/Simulink Models, 1st edn. Wiley, New York (2012)
22. Tsai, H.-L., Tu, C.-S., Su, Y.-J.: Development of generalized photovoltaic model using MATLAB/SIMULINK. In: Proceedings of the World Congress on Engineering and Computer Science 2008 WCECS 2008, 22–24 October 2008
23. Saxena, N., Hussain, I., Singh, B., Vyas, A.L.: Single phase multifunctional VSC interfaced with solar PV and bidirectional battery charger. In: 7th India International Conference on Power Electronics (IICPE), pp. 1–6 (2016)
24. Iqbal, A., Ahmed, S.M., Khan, M.A., Khan, M.R., Abu-Rub, H.: Modeling: simulation and implementation of a five-phase induction motor drive system. In: Joint International Conference on Power Electronics, Drives and Energy Systems & 2010 Power India, pp. 1–6 (2010)
25. Liu, Y.-H., Huang, S.-C., Huang, J.-W., Liang , W.-C.: A particle swarm optimization-based maximum power point tracking algorithm for PV systems operating under partially shaded conditions. IEEE Trans. Energy Convers. **27**(4), 1027–1035
26. Hanafiah, S., Ayad, A., Hehn, A., Kennel, R.: A hybrid MPPT for quasi-Z-source inverters in PV applications under partial shading condition. In: 2017 11th IEEE International Conference on Compatibility, Power Electronics and Power Engineering, pp. 418–423 (2017)
27. Tang, W.J., Wu, Q.H., Saunders, J.R.: A bacterial swarming algorithm for global optimization. IEEE Congress on Evolutionary Computation Year **2007**, 1207–1212 (2007)
28. Chen, Z., Luo, Y., Cai, Y.: Optimization for PID control parameters on hydraulic servo control system based on bacterial foraging oriented by particle swarm optimization. In: 2009 International Conference on Information Engineering and Computer Science, pp. 1–4 (2009)

29. XiaoLong, L., RongJun, L., YangPing: A bacterial foraging global optimization algorithm based on the particle swarm optimization. In: 2010 IEEE International Conference on Intelligent Computing and Intelligent Systems, vol. 2, pp. 22–27 (2010)
30. Korani, W.M., Dorrah, H.T., Emara, H.M.: Bacterial foraging oriented by particle swarm optimization strategy for PID tuning. In: 2009 IEEE International Symposium on Computational Intelligence in Robotics and Automation, pp. 445–450 (2009)

Intelligent Refrigerator

Ishank Agarwal

Abstract In this research, models have been made to predict "Cuisines from Ingredients." Many applications like prediction of top 100 cuisines based on mood and keeping in mind the health of the user, predicting similar dishes to the query dish, and comparison between optimized models on the basis of money loss that they incur if the food intake is not right are covered in this research. Formation of different patterns and combinations resulted when various machine learning models were applied to the given Yummly dataset of a wide variety of cuisines. These models provided a range of accuracies and the best one was used for each purpose according to the motto "survival of the fittest."

Keywords Multinomial naïve Bayes · Logistic regression · Random forest classifier · K-nearest neighbors classifier · K-fold cross-validation · Sentiment classifier · TF-IDF · Linear regression · Multiple regression · Gradient descent · Polynomial regression · Ridge regression · L2 penalty

1 Introduction

Food is an integral part of one's life besides clothing and shelter, no matter whether the person is young or old but the intakes or the right amount of servings largely varies from person to person depending on his or her health condition, biological clock, and choices. Many times, we waste time in deciding what to cook so that it fits our appetite and fulfills our taste buds. This choice making can take a lot of time and moreover, it depends on the ingredients we have at the moment. So, to reduce the time and overhead of choice making, various machine learning algorithms come into play.

To test the efficiency and usefulness of these algorithms, Yummly dataset was used throughout the research work [1]. Yummly is a Web site and mobile app as well,

I. Agarwal (✉)
Jaypee Institute of Information Technology (JIIT), Noida, India
e-mail: ishankagarwal62@gmail.com

S. M. Thampi et al. (eds.), *Intelligent Systems, Technologies and Applications*, Advances in Intelligent Systems and Computing 910,
https://doi.org/10.1007/978-981-13-6095-4_18

providing recipe recommendations with individual's taste personalization, recipe search, digital recipe box, shopping list, and one-hour grocery delivery. It contained a huge dataset of almost 8339 recipes with each recipe mentioned against its list of ingredients. Throughout the research work, this dataset proved immensely important in calculations while using different machine learning models. The dataset contained the most popular and widely used cuisines in the American world. The dataset has been played around with and exploited to its extremities to measure the accuracies relative to each major purpose that was tried to accomplish through the research, for instance, being able to predict the most suitable recipe for a person keeping in mind of his dietary levels such as calorie count.

2 Datasets Related Work

There are several companies with proprietary APIs and databases were acquired through scraping these recipe companies' Web sites. These companies then sell licenses to developers. These APIs vary in their quality, database size, and fees [1, 2]:

BigOven: 350,000 recipes; REST JSON and XML output.
Yummly: 1 million recipes, ratings, ingredients; REST JSON output; Academic plan available with 30,000 lifetime calls [3].
Food2Fork: 250,000 recipes; JSON output; Free plan with 500 calls per day.
Spoonacular: 350,000 recipes; UNIREST JSON output; Free plan for hackathons and academic purposes.
Edamam: 1.5 million recipes; Free Trial with up to 1000 lifetime calls.

Initially, the two APIs were selected as the choices for the predictive analysis, namely Spoonacular API and Yummly API but while Spoonacular's API had more bells and whistles, and advanced features, Yummly's database of recipes was generally of higher quality than Spoonacular's [1]. It also had access to over 1 million recipes compared to Spoonacular's and had better-formatted ingredients for each recipe. So, finally, Yummly API was chosen to carry out the task of prediction. Yummly's API is accessed by querying an API end point and returns JSON-encoded recipe data. JSON data was translated to Pandas dataframe structures using the Requests package and Pandas. Initially, cuisine data was written into 25 separate CSVs before assembling them into cuisines_data.csv. For each recipe, the recipe name, rating, cooking time, course, cuisine, and ingredients were stored against it. In total, there are 25 cuisines for 8339 recipes. Cuisine dataset basic stats [1]:

- 8037 recipes.
- 25 cuisines: Asian was the best represented, with 1400 recipes. English the least, with just 32 recipes.
- Ratings: overwhelmingly 4 star (62.1% of the 8339), some 3 star (18.9%) and 5 star (17.2%), very few 0, 1, or 2 stars (1.8%).

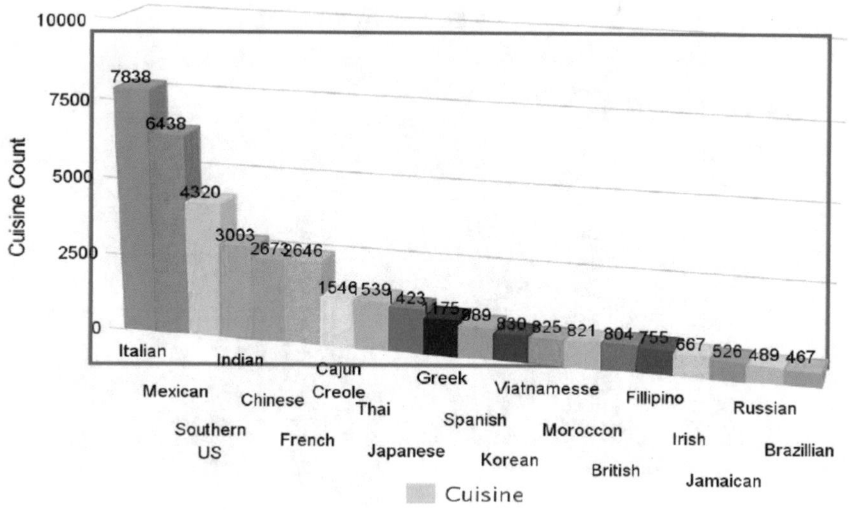

Fig. 1 Counts of all cuisines in training dataset

- Cooking time: mean of 65 min, max of 1970 min (33 h), min of 1 min, median of 40 min.
- Ingredients per recipe: mean of 10.1, max of 35 (Thai chicken tacos), min of 1, median of 9.
- 3958 unique ingredients:
  - Most common ingredients: Salt (3325 occurrences, 0.41 frequency), garlic (0.22), onions (0.20), olive oil (0.18).
  - Least common ingredients: 1401 unique ingredients are only used in one recipe throughout the dataset. Examples: chocolate candy, melon seeds, canned tuna, vegan yogurt, tarragon vinegar, unsalted almonds [1] (Figs. 1, 2 and 3).

Unique Ingredient Analysis was done by calculating the ingredient frequency obtained by dividing the instances by the number of recipes in the dataset [1].

Next, recipes within a given dataset were identified which were the most "unique" in terms of their ingredients on the basis of relative scoring using two recipe uniqueness formulas. The first method of scoring was to take the mean of each recipe's ingredient frequencies (e.g., (0.3 + 0.1 + 0.2)/3). For each recipe's ingredients, the sum of the ingredients' unique ingredient frequency score was calculated as above and then divided by the number of ingredients in the recipe so as to not bias the scoring toward recipes with dozens of ingredients. The second method of scoring was to take the product of each recipe's ingredient frequencies (e.g., 0.3 × 0.1 × 0.2). The product tends to "reward" recipes with extremely rare ingredients whereas the mean method rewards these recipes to far lower extent [1].

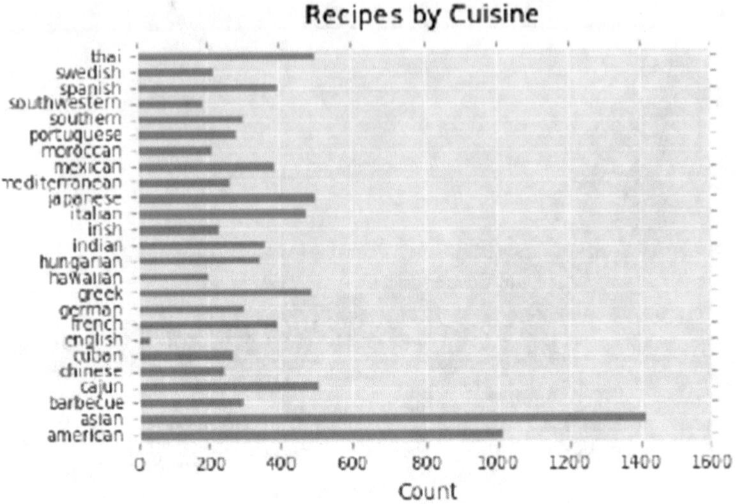

Fig. 2 Distribution of no. of recipes versus cuisines

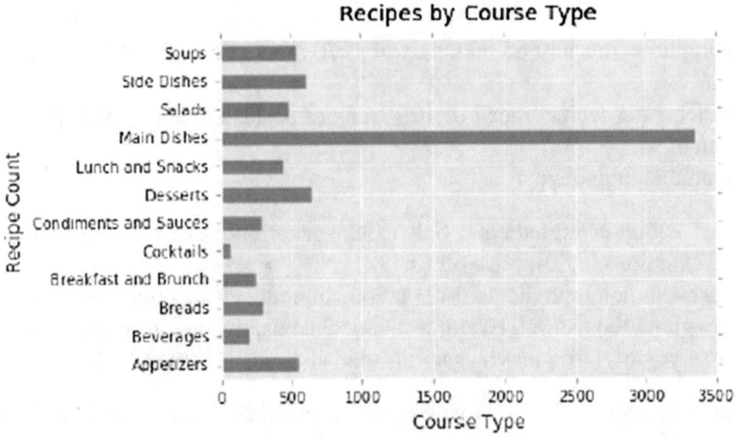

Fig. 3 Distribution of no. of recipes versus course type

2.1 Data Pre-Processing

Several processing steps were conducted on the master cuisine data CSV file.

(1) Ingredient Parsing:
Each recipe's ingredients were encoded as a single, unseparated string, including brackets: "[ingredient 1, ingredient 2, ingredient 3]." So, in order to examine the ingredient separately, brackets from each ingredient string are dropped while traversing through [1:−1]. Next, the list of the ingredient string was returned

separated by commas. This research treats essentially similar ingredients as unique, for instance, "pasta," "Giorgino pasta," and "fresh pasta" were considered as separate ingredients.

(2) Ingredient Counts:
After parsing the ingredients from a single string to a list of strings, the number of ingredients in each recipe was computed by creating a new column named "ingredCount" and setting each value to len (ingredients).

(3) Time Conversion:
Each recipe's cooking time was converted from seconds to minutes by dividing it by 60 and populating it in a column named "timeMins."

(4) Munging:
To make sure that data was sufficiently clean for analysis, some measures were taken.

A. The rows of cuisines data that had an empty "cuisine" value are dropped. Yummly's "Search by Cuisine" API call returns recipes that have "Chinese" in the recipe in some shape or form—even in the ingredients! So a sandwich recipe that employs "French bread" or an "English muffin" might have a cuisine value of "French" or "English." To avoid these ambiguities, the empty recipes without an explicit cuisine were dropped. This greatly reduced the number of recipes for some cuisines like English.

B. Empty course values were filled with "Unknown" rather than dropping these recipes.

C. Finally, duplicate recipes from the dataset were dropped.

3 Methodology

(1) **Logistic regression**
The proposed algorithm was used for binary classification to estimate the probability of binary response based on predictors (independent variables) like features such as ingredient count and cooking time.

This algorithm makes use of predicting the logit transformation function based on characteristic of interest as,

$$\text{logit}(p) = b_0 + b_1 X_1 + b_2 X_2 + b_3 X_3 + \cdots + b_k X_k$$

where p denotes the probability that characteristic of interest is present or not.

$$\text{odds} = \frac{p}{1-p} = \frac{\text{probability of presence of characteristic}}{\text{probability of absence of characteristic}}$$

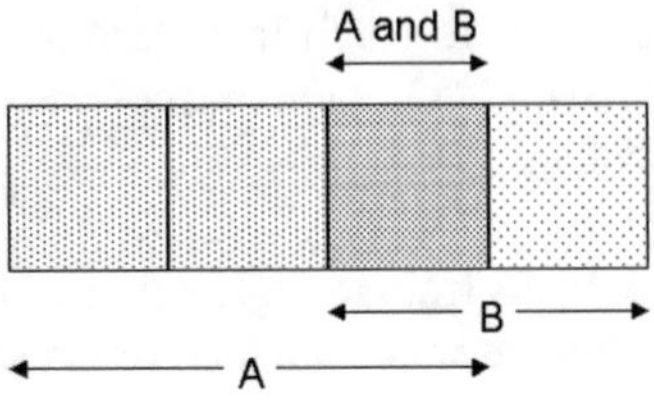

P(A) = 3/4

P(B) = 2/4

P(A and B) = P(AB) = 1/4

P(A|B) = P(AB) / P(B) = (1/4) / (2/4) = 1/2

P(B|A) = P(AB) / P(A) = (1/4) / (3/4) = 1/3

Fig. 4 Conditional probabilities in Bayes' theorem

(2) **The Naive Bayes' Theorem**
It works on conditional probability to find the probability of an event given that the another event has already occurred. It is a powerful algorithm which even works on millions of records with some attributes, e.g., calories and ingredient count [4] (Fig. 4).

(3) **Random forests or random decision forests**
Random forest is ensemble learning method which constructs a multitude of trees to get more accurate and stable prediction. It can be used for classification, regression, and other tasks in which it outputs the mode of the classes (classification) and mean prediction (regression) of the individual trees. It was also useful in dealing with the problems of over-fitting of the decision trees to their training set [5].

(4) ***K*-fold cross-validation**
It is a type of cross-validation and is a machine learning algorithm. It works as follows. First, we do partitioning of the dataset into k subsets; each one is called a fold. Out of the k folds, 1 fold is the validation set and the other $k - 1$ folds are the cross-validation sets. We train our model on the cross-validation set and the predicted results are used to calculate the accuracy of the model with the validation set. Then the average of all the k accuracies of the model is the accuracy of the machine learning model (Fig. 5).

(5) ***K*-Nearest Neighbors (KNN)**
K-nearest neighbors algorithm was used for classification and regression type of problems. In this algorithm, a majority vote of the k neighbors classifies a new case from the stored cases. The case is chosen by the class that appears the most in the k neighbors using a distance function. There are many distance functions: Manhattan, Euclidean, Minkowski, and Hamming distance. The first three functions are used for continuous variables and the Hamming distance measure is used for categorical variables. Value of K chosen is the topic of high importance which can be a challenge for better performance of the algorithm.

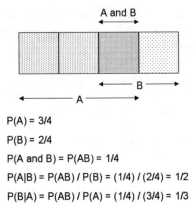

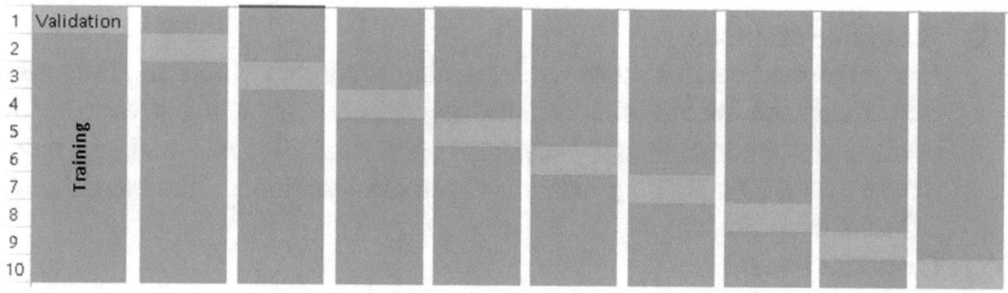

Fig. 5 Showing how training subsets of dataset are trained by model

(6) **Linear Regression**
Linear regression was used to find relationship between continuous variables, i.e., predictor or independent variable an response or dependent variable. We can get the idea that did the predictor correctly predict the outcome variable and which variables are significant predictors that impact the outcome variable.

(7) **Polynomial Regression**
Polynomial regression function fits its model to the powers of predictor like features by linear least squares method. We can fit a k order/degree polynomial to data where Y is the predicted value of the model with regression coefficients $b1$ to k for each degree with intercept b. It is made up of linear regression models with k predictors raised to the power of i where i is equal to 1 to k.

(8) **Ridge Regression**
This algorithm was used in the problems which do not have a unique number of solutions, and provide additional information to the problem and thus, predicting the best solution. It was used to deal with multicollinearity (existence of near-linear relationship among independent variables) for multiple regression data. In this condition, variances are so large so as to be far from the true value. By adding a degree bias to regression estimates, it reduces the errors.

(9) **Gradient Descent**
This algorithm was used to find the minimum of a function which is a first-order optimization algorithm which involves iterations. It finds the parameters (coefficients) for a function that minimizes the cost function. Steps proportional to the negative of the gradient are taken in order to get the local minimum at the current point. To get the local maximum, one has to take step proportional to the positive gradient. This process is called gradient descent.

(10) **Decision Trees**
A decision tree is a tool for classification and prediction whose each node does a test on an attribute, whereas each branch corresponds to the result of the test, and leaf nodes have a class label. It involves recursive partitioning in which the tree is split up into subsets undergoing recursion. Usually, decision tree classifier has good accuracy.

(11) **Ensemble Methods**
Ensemble method is a supervised learning algorithm where multiple algorithms are used instead of constituent algorithms alone which gives us better performance and accuracy. They have high flexibility like a single model would over-fit the training data but ensemble techniques (like bagging) can reduce such problems.

(12) **Gradient Boosting**
Gradient boosting is a supervised learning model which produces a prediction model which consists of an ensemble of weak prediction models like decision trees. It has a loss function (MSE) which should be minimum that means the sum of the residual values should be close to zero which means that the predicted values are almost the correct actual values. There are many other popular boosting algorithms like XGBoost, AdaBoost, and GentleBoost.

4 Features and Results

1. The proposed algorithms by Conway Yao included predefined ML models, namely logistic regression and multinomial naïve Bayes with accuracies of 44 and 54%, respectively.
2. To improve the accuracy, random forest classifier was chosen with predefined ensemble scikit-learn library and we achieved accuracy of 52%.
3. Then, *K*-nearest neighbors classifier was applied with number of neighbors ranging from 1 to 13 and maximum accuracy of 44% was achieved [6].
4. It was observed that the recipe names were too diverse from each other and training on some part of the data and leaving the other for testing will prevent the model from learning the whole data. Henceforth, *K*-fold cross-validation was applied which trains and tests on the whole data with the number of folds equal to 100 and achieved an improved accuracy of 56%.
5. A model was built to find recipes that we can make given a set of supplies in the pantry. To implement this, the first 100 most frequent ingredients used in most recipes worldwide were chosen and got 114 recipes as the output of the model [7].
6. Another fact observed was that there was no relationship between the recommended cuisines we obtained from the above model, so, a sentiment classifier was built based on the ratings and probability of the presence of particular ingredient. The number of recipes obtained was 100.
7. The next aim was to predict similar dishes for a particular dish like pizza. So, Tf-Idf algorithm was applied to increase the weights of unique ingredients of dishes and thus, by applying nearest neighbor model, 9 similar dishes closest to "grilled buffalo chicken pizza" (Fig. 6) [8].
8. The distribution of recipes by rating, cuisine versus number of recipes, cooking time and rating versus ingredient count, rating versus cooking time were visualized (Figs. 7, 8 and 9).

query_label	reference_label	distance	rank
0	Grilled Buffalo Chicken Pizza ...	0.0	1
0	Pizza Pinwheels	0.7	2
0	Weeknight Skillet Lasagna	0.730769230769	3
0	Classic Hawaiian Pizza	0.791666666667	4
0	Italian Alfredo Broccoli Strata ...	0.8	5
0	Chicken Parmesan Alfredo	0.8	6
0	Lasagna With A Twist	0.8	7
0	Easy Beef Lasagna (featuring Ragu 2 Lb. 13 ...	0.806451612903	8
0	Italian Baked Pasta	0.826086956522	9
0	Chicken Parmigiana	0.827586206897	10

[10 rows x 4 columns]

Fig. 6 9 nearest similar dishes to "Grilled Buffalo Chicken Pizza"

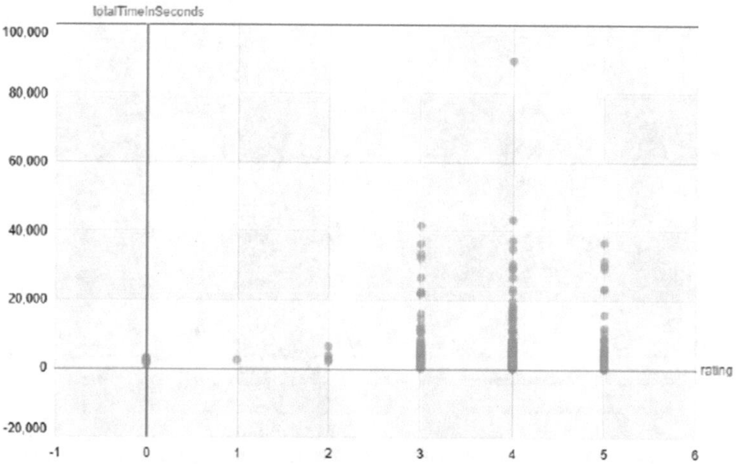

Fig. 7 Relationship of totalTimeInSeconds (cooking time) versus rating

9. Using a generic simple linear regression function, cuisines were predicted for random ingredient lists (e.g., 127). This model was evaluated by the model using RSS (Residual Sum of Squares).
10. Initially, the cuisines were being predicted directly from ingredients in the pantry using "issubset" function. But since, there were lots of new cuisines emerging in the modern world today, the aim was to predict those cuisines using these

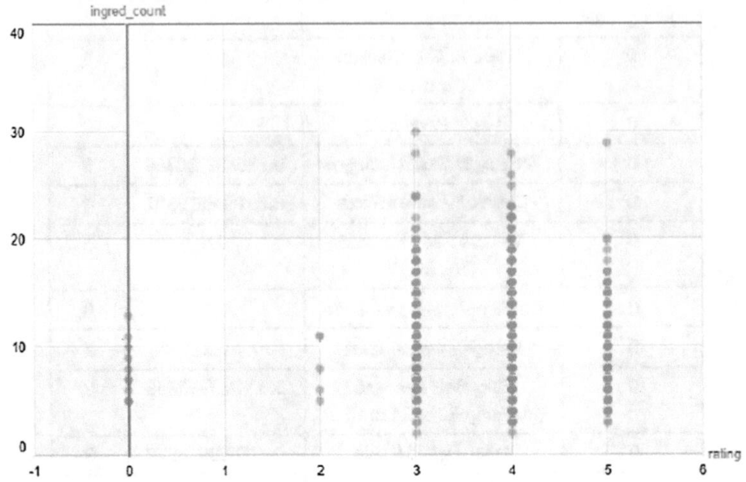

Fig. 8 Relationship of ingred_count (ingredient count) versus rating

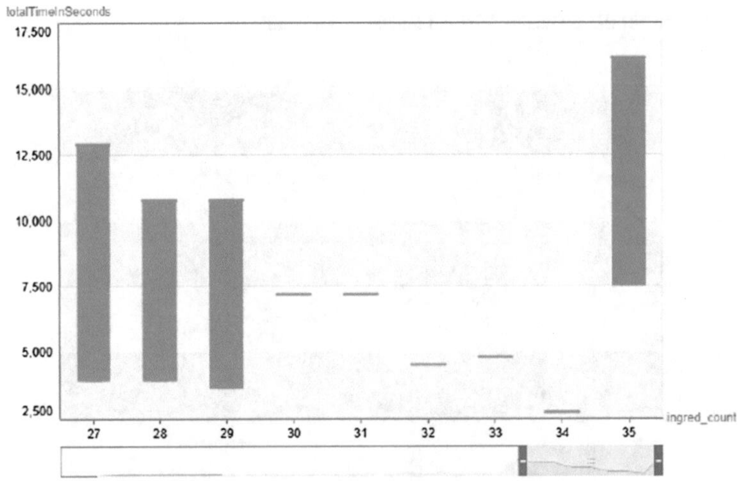

Fig. 9 Relationship of totalTimeInSeconds (cooking time) versus ingred_count (ingredient count)

ingredients by making regression model more efficient fitting it using different degree polynomials.

11. Compared different regression models in order to assess which model fits best and using polynomial regression as a means to examine the same. To implement this, a polynomial SFrame function was made to create an SFrame consisting of powers of an SArray up to a specific degree.

12. Different degree polynomials were created ranging from 1 to 15 to predict cuisines, then plotted the predicted plot, and found out the difference between their respective standard errors.
13. To make the plots more clean to visualize, the data was divided into 4 subsets of roughly equal size and estimated a 15th degree polynomial model on all 4 subsets and plotted the resulting fit.
14. At this stage, the degree polynomial which fits the best was not clear. So, a well-known way to select parameters was applied like degree by splitting the dataset into training, testing, and validation set. Using these to compute the RSS on validation data for degree polynomials ranging from 1 to 15. Thus, the degree polynomial which gives minimum RSS among all 15, and that is degree 2 polynomial was obtained.
15. Further, ridge regression (Ridge regression aims to balance the bias-variance tradeoff was run on the dataset.

    – Large tuning parameter (lambda) corresponds high bias, low variance.
    – Small tuning parameter (lambda) corresponds low bias, high variance.

    Thus, tuning parameter controls model complexity).
    Multiple times with different L2 penalties to see which one produces best fit. With so many features and so few data points, the solution can become highly numerically unstable which can sometimes lead to strange unpredictable results. Thus, rather than using no regularization, a tiny amount of regularization was introduced, (L2 penalty = 1e−5) to make the solution numerically stable.
16. Recalling that the polynomial fit of degree 15 changed wildly whenever the data changed. So, the data was split into 4 subsets and was fit with the model of degree 15 and the result came out to be very different for each subset. The model had high variance. To reduce this variance, now, ridge regression with small test L2 penalty upon the 4 datasets was used and visualized their plots.
17. Since variance still came to be high, ridge regression aims to address this issue by penalizing large weights and choosing L2 penalty = 1e5 on all 15th order polynomial. On visualizing the plots of 4 different datasets, variance significantly decreased. So, these curves vary a lot less, now that we have applied a high degree of regularization (regularization aims to minimize total cost to fit function which is equal to measure of fit + measure of magnitude of coefficients) (Fig. 10).
18. Just K-fold cross-validation was implemented before; it was again implemented by creating a K-Cross Function via polynomial regression using L2 penalty as parameter because leaving any observations for training data was not giving optimal solution.
19. A function to compute the average validation error for the above model was made which finds the model that minimizes the average validation error. The loop does the following:

    – Fit a 15th order polynomial model.
    – For L2 penalty in [10^1, $10^{1.5}$, 10^2, $10^{2.5}$,…, 10^7].

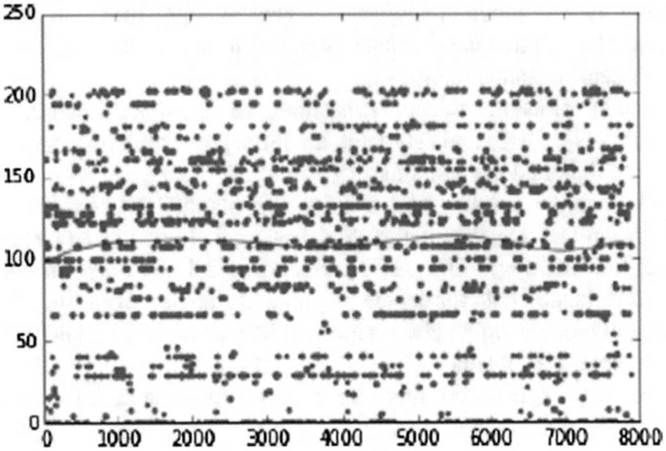

Fig. 10 Graph showing 15th order polynomial with L2 penalty = 1e5

– Run tenfold cross-validation with L2 penalty.
– Report which L2 penalty produces lowest average validation error.

So, L2 penalty = 10^1 produces minimum RSS value.

20. To compute the derivate of regression cost function (cost function is the sum over the data points of the squared difference between an observed output and a predicted output + L2 penalty term):
$\text{cost}(w) = \text{SUM}\left[(\text{prediction} - \text{output})^2\right]$
Derivative = $2 * \text{SUM}[\text{error}[\text{feature\_}i]]$
21. A function that performs gradient descent algorithm was needed in our ridge regression model.
22. So, the effect of L2 penalty was visualized and how large weights get penalized was seen. To do this, first a model was made considering no regularization and calculating the weights, then, a model was made considering high regularization to get weights. The two learned models were plotted and compared both using the weights obtained before and observed that working with high regularization produced less RSS value as compared to the model with no regularization (RSS is equal to summation of square of differences between real value and predicted value) (Fig. 11).
23. For predicting similar cuisines for a particular cuisine of interest, in computing distances, it is crucial to normalize features otherwise, e.g., the cooking time feature (typically on the order of 1000s) would exert a much larger influence on distance than the ingredients feature (typically on the order of 1s). To calculate this, Euclidean distance was defined which is used in the nearest neighbor regression model to compute the distance between the query cuisine and all cuisines in our training set.

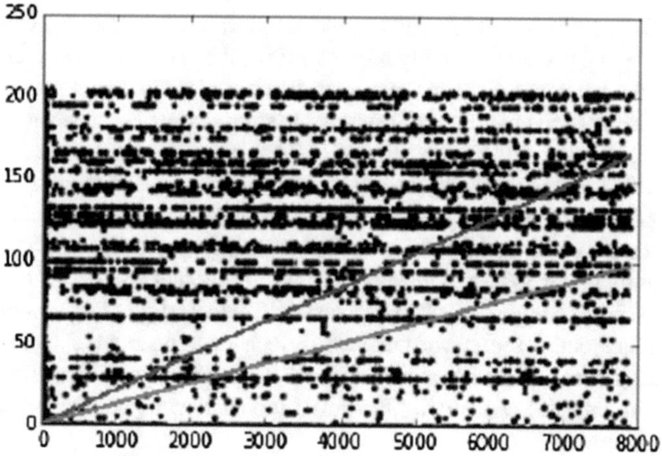

Fig. 11 Graph showing RSS comparison (red: high regularization, blue: low regularization)

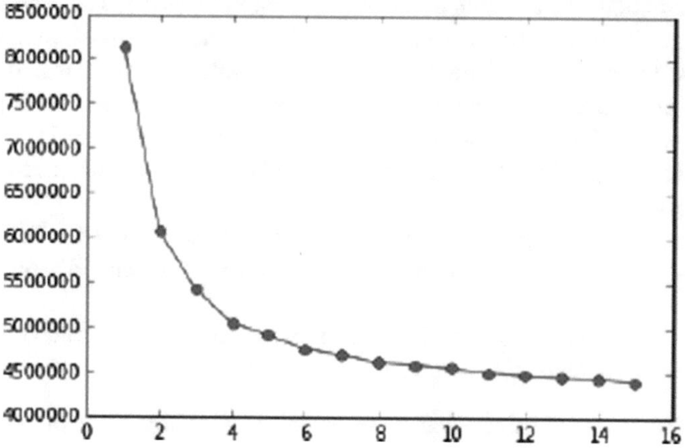

Fig. 12 Graph showing variation of RSS with different values of k

24. A function to find K-nearest neighbors and another function to take the average of the cuisines of the K-nearest neighbors in the training set was made to return a predicted value of the query cuisine.
25. In order to make multiple predictions, a function was made to predict the most similar dish corresponding to each and every dish in the dataset.
26. Choosing the best value of k using validation set, values of k were taken ranging from 1 to 15 and the RSS was computed for predictions of these values and reported which k value produces the lowest RSS on validation set. $k = 14$ gave minimum RSS value on validation set. The performance of the above model was visualized as a function of k and plotted RSS on validation set for each considered k value (Fig. 12).

27. Calorie count was crawled and appended to our dataset. Usually, young adults who are in early 20s normally require 1500 cal to be taken every day in order to have a balanced diet (this much amount of calorie value is generally taken from a renowned health Web site www.livestrong.com). Since we normally take 3 meals per day, this implies almost 500 cal per meal. So, taking 500 cal as a threshold in calorie ranging from 50 to 999, a target column (healthy) was made in our dataset in which 1 signified healthy meal and −1 signified unhealthy. The number of cuisines which came out to be healthy was 4423, i.e., 53.046% and unhealthy being 3915 which was 46.954% of the total number of cuisines.
28. A classification algorithm was proposed with features ingredients, cooking time, ingredient_count, rating, and cuisines. A decision tree model of depth 6 and also a small model of decision tree with depth 2 were made and visualized both the trees, finding that visualizing the small model is comparatively easy to gain some intuition.
29. Some predictions were made taking two +ve and two −ve examples from the validation set and saw what the decision tree model predicts. The percentage of predictions the decision tree model got correct was 0.5. The probability of a cuisine being classified as healthy was 52% for the above 4 examples.
30. In the visualization of the decision tree, the values at the leaf nodes are not class predictions but scores of being healthy. If score > 0, class +1 is predicted. Otherwise, if score < 0, we predicted class −1 (Figs. 13 and 14).
31. The accuracy of both the small (depth 2) and the big (depth 6) was 54 and 57%, respectively.
32. The accuracy of complex decision tree model was evaluated with max depth equal to 10 which came out to be 64% on trained data and 53% on validation data.
33. Every mistake the model makes costs money. So, the cost of each mistake was quantified made by the model according to:

    - False –ve's: Cuisines that were actually unhealthy were predicted to be unhealthy.
    - False +ve's: Cuisines that were actually unhealthy but predicted to be healthy.

    A ROC curve of the model was plotted to visualize the variation of True +ve versus False +ve rate (Fig. 15).
34. The number of False +ve came out to be 663 whereas number of False −ve's was 111.
35. Assuming each mistake costs money, cost of $10,000 per False −ve, cost of $20,000 per False +ve; the total cost obtained by decision tree model turned out to be 14.37 million. Ensemble methods: To further explore the use of boosting, pre-implemented gradient boosting trees were used. To do this, subsample of the dataset was used to make sure the classes become balanced [9].
36. Models were now trained to predict healthy cuisines with training and ensemble of 5 trees. With the same technique as before by using two +ve and two −ve examples from the validation data, the accuracy achieved was 52%.

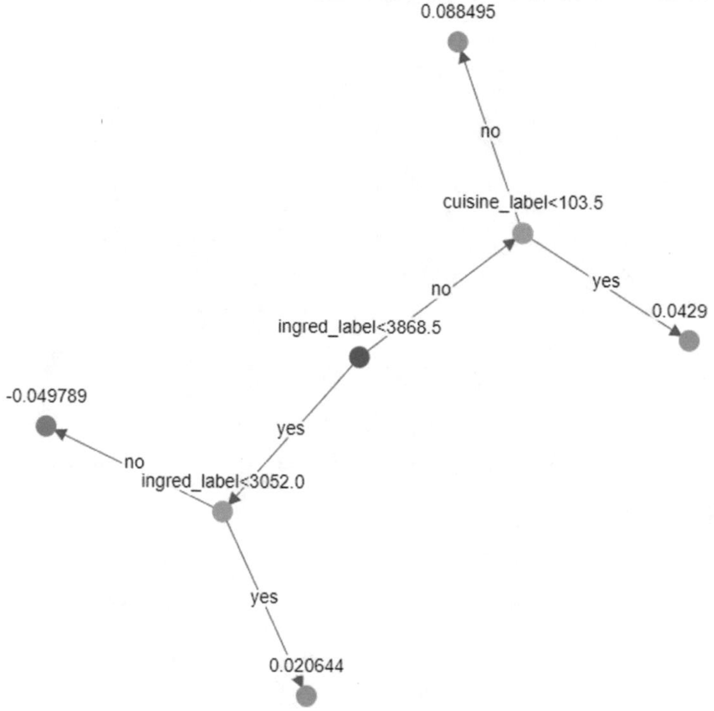

Fig. 13 Decision tree capped to depth = 2 for better visualization

37. Confusion Matrix:

    True +ve(1,1) = 415
    True −ve(−1, −1) = 404
    False −ve(1,−1) = 392
    False +ve(−1,1) = 387

38. The benefit of extra 3% increase in accuracy in decision tree from cap 2 to cap 6.
39. The cost of mistakes of the above ensemble model using false −ve and false +ve comes out to be 11.66 million which is 3.1 million less than the model without boosting.
40. To find the cuisines which were most likely to be predicted healthy an ensemble model with 5 trees was used and probability predictions were made for all the cuisines and the data was sorted in decreasing order by probability predictions. "Southern cornbread dressing" was found as a result along with the top 5 cuisines with lowest probability of being predicted as a healthy cuisine.

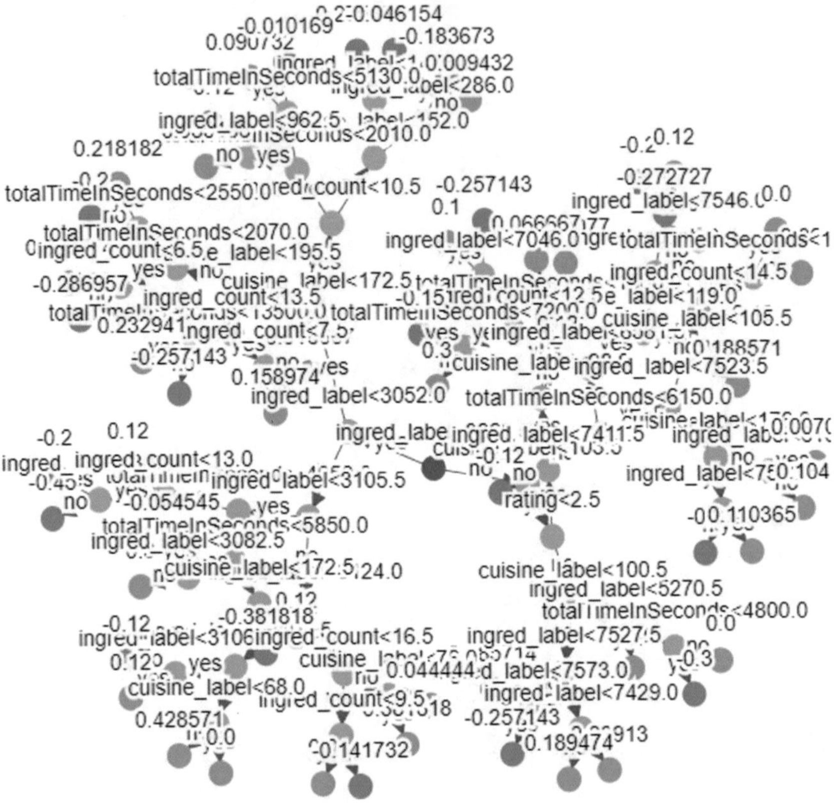

Fig. 14 Decision tree visualization (depth = 10)

41. The effect of adding more trees was seen in the form of gradient boosted trees with 10, 50, 100, 200, 500 trees and calculated their respective accuracies in which the model with 100 trees came out to be the best.
42. Finally, the training and validation errors were plotted versus the number of trees to see the effect of adding more trees to the model (Fig. 16).

5 Conclusion

To conclude my research, I assert that these important features and research objectives highlighted by me added something very important to the modern-day food industry and also will help other researches in their research work as well. I aimed to build models that were efficient keeping in mind the various aspects related to humans like health by predicting healthy cuisines, mood by predicting similar dishes to a particular type of dish. Also assessing the popularity, cooking times and ingredients

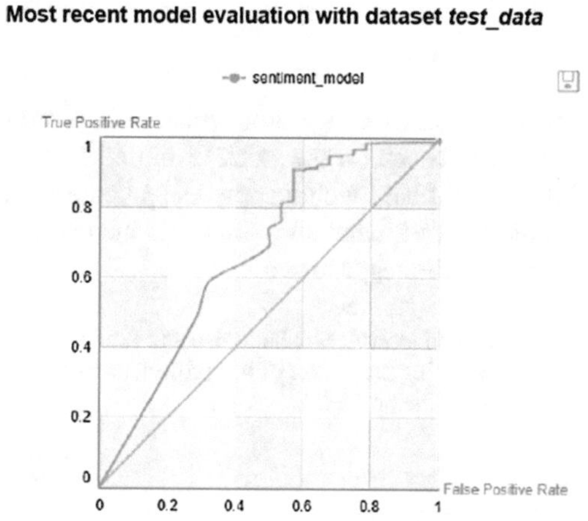

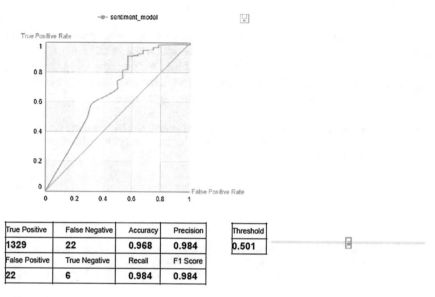

True Positive	False Negative	Accuracy	Precision
1329	22	0.968	0.984
False Positive	True Negative	Recall	F1 Score
22	6	0.984	0.984

Threshold
0.501

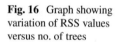

Fig. 15 False positives and negatives for quantification of mistakes

Fig. 16 Graph showing variation of RSS values versus no. of trees

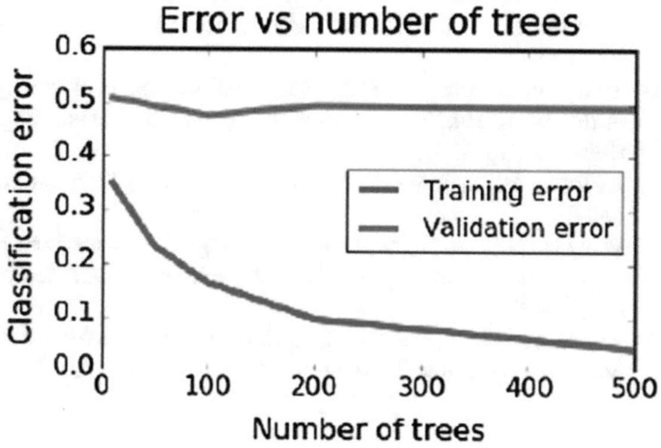

counts were kept in mind while developing the models which seems to be the biggest achievement in my work. I also found relationships between many attributes and their dependence on each other like cooking time and rating of dishes increased with increase in number of ingredients, also dishes with higher ratings had greater cooking times.

6 Future Scope

In the future, I aim to deploy an app Intelligent Refrigerator which will use our predictive models to predict healthy cuisines based on the stock of ingredients the user has, and the mood of the user at that time. Also I would apply Deep Learning to capture the images of ingredients to directly sense what all ingredients the user has in stock, rather than typing them manually in the respective app.

Acknowledgements I would like to show our gratitude to Prof. Shikha Mehta and Prof. Prashant Kaushik for sharing their pearls of wisdom with me during the course of this project, and I hope to receive their continuous insights till the deployment of the project.

References

1. McCallum, A., Nigam, K.: A comparison of event models for naive Bayes text classification. In: AAAI-98 Workshop on Learning for Text Categorization, vol. 752 (1998)
2. Kaggle.com: Kaggle: Your home for data science [online]. Available at https://www.kaggle.com/ (2017). Accessed 28 Nov 2017
3. Yummly.com: Yummly: Personalized recipe recommendations and search [online]. Available at https://www.yummly.com/ (2017). Accessed 28 Nov 2017
4. Safreno, D., Deng, Y.: The recipe learner. Tech. N.p. (2013)
5. Breiman, L.: Random forests. Mac. Learn. **45**(1), 5–32. https://github.com/conwayyao/Recipe-Analysis (2001)
6. Trudgian, D., Yang, Z.: Spam classification using nearest neighbour techniques. In: Proceedings of the 5th International Conference on Intelligent Data Engineering and Automated Learning (Revised), pp. 578–585
7. Agarwal, R., Jachowski, D., Shengxi, D. RoboChef: automatic recipe generation. Tech. N.p. (2009)
8. Wang, L., Li, Q., Li, N., Dong, G., Yang, Y.: Substructure similarity measurement in Chinese recipes. In: Proceedings of the 17th International Conference on World Wide Web, ACM, 2, pp. 979–988
9. Kumar, R.M.R.V., Kumar, M.A., Soman, K.: Cuisine prediction based on ingredients using tree boosting algorithms. Indian J. Sci. Technol. 9(45) (2016)

Online Knowledge-Based System for CAD Modeling and Manufacturing: An Approach

Jayakiran Reddy Esanakula, J. Venkatesu Naik, D. Rajendra and V. Pandu Rangadu

Abstract To hold and have their share in the present unpredictable market, the companies are heading toward the alternative techniques for designing and manufacturing. Web-based technologies have the capability to blend into the context of engineering design and manufacturing. This paper illustrates a preliminary approach for developing an online knowledge-based system for CAD modeling and manufacturing. The purpose of developing this system is to cut down the CAD modeling and manufacturing time and thereby reducing the production cost. The parametric modeling technique has been incorporated in the proposed system along with the macro codes for generation of the CAD model. The generated CAD model is the intermediate output of the proposed system and is the input to the CNC code generator to carry out the manufacturing processes. An inference engine and relevant Web UI were developed for assisting the design engineers.

Keywords KBE · KBS · SolidWorks · Online · CAD · Modeling

1 Introduction

In the present volatile market, CAD/CAM is accepted as a considerable tool to achieve better design and manufacturing. The present-day manufacturers are aiming to produce more customized products of better quality, in small batches, with little lead time and at reduced costs. In the knowledge-based CAD modeling system, decision logic, empirical formulas, algorithms, and geometrical data of the component directs the generation procedure of the CAD model. The generation of the automatic CAD model is carried out with the help of the data in the database and predefined framework. In concise, the system is defined as a structured method which automatically generates

J. R. Esanakula (✉) · J. V. Naik · D. Rajendra
Sri Padmavati Mahila Visvavidyalayam, Tirupati, Andhra Pradesh, India
e-mail: ejkiran@gmail.com

V. P. Rangadu
Jawaharlal Nehru Technological University Anantapur, Anantapur, Andhra Pradesh, India

S. M. Thampi et al. (eds.), *Intelligent Systems, Technologies and Applications*,
Advances in Intelligent Systems and Computing 910,
https://doi.org/10.1007/978-981-13-6095-4_19

the CAD model of the component. Currently, for manufacturing globalization the requirement of collaboration between geographically scattered business branches and international customer association has increased. Hence, an online CAD modeling approach provides a significant benefit over the desktop applications. Recently, the online applications established themselves as the beneficial in various business areas and could extend the benefits to the area of CAD too. So, in view of economic benefits of Web-based technologies, online CAD modeling and manufacturing have been proposed as a new paradigm in the modern way of modeling and manufacturing.

CAD is an application of computer technology which is used to develop product concept for engineering. SolidWorks is CAD software which supports parametric modeling, additionally, which also supports secondary development tool called Visual Basic for Application (VBA). Parametric modeling technique is efficient enough to recreate the CAD model based on the already explained tasks in decreased time when compared with the humans [1]. SolidWorks Application Programming Interface (API) offers the modeling automation by making use of specific programming codes and functions. Macro is a specific code for storing the operations which are done on screen of CAD software and will be used to regenerate the same tasks later.

The developing of the online knowledge-based system for CAD modeling and manufacturing is aimed at finding an efficient way to generate the CAD model and CNC code in a lesser time and cost when compared with the conventional method.

2 Literature Review

A variety of frameworks for developing online CAD modeling and manufacturing systems has been proposed. Cyber View System is a tool to collaborate various product development partners existing in different remote places. It uses a Web browser to develop the same product for different designers at different locations. Later, it was enhanced using the technologies like HTML, ASP, and JAVA to become a robust platform for collaborative design and it was called Cyber-Eye system [2]. Anumba and Duke [2] introduced the concept of People and Information Finder (PIF) for acquiring the information regarding projects and people using Web browser so as to continue the communication between team members of the project. In the year 1998, Reed and Afjeh [3] developed an Internet-based tool for simulating the engineering process and which was developed as interactive tool. Similarly, the researchers like Jones and Iuliano [4], Hardwick et al. [5], Kenny et al. [6], Yu and Chen [7] attempted and shown the progress in developing the online engineering and manufacturing applications. The development of WebCAD, a Java-based CAD software, revolutionized the Internet-based accessing the CAD/CAM tool. The Web-based micromachining service provides the free-form tool path generation to the remote designer [8]. Kim et al. [8] proposed an online tool for designing and manufacturing purpose to work on SolidWorks. For semantic Web services, Liu et al. [9] introduced a fuzzy match making method to maintain extra automated service in the environments of collaborative manufacturing. Web-based CAD/CAM systems use special environments for design visualization like Java technologies, Flash, X3D, or VRML. But, this method has certain restrictions that the user needs to bound to certain software to enable full functionality [10]. Sypek et al. [11] developed Web-based virtual grinding machine tool system using these technologies. But, the additional tools like these are not required for the browser-based CAD/CAM systems. The application developed in this scenario is highly compatible [12–17]. Few of them are WebCAD, PunchCAD, SketchUP, Tinkercad, Tripodmaker, Mydeco 3D Planner, Autodesk Homestyler, Aftercad online, etc.

Though many Web-based design and manufacturing systems are available, a user-friendly knowledge-based system (KBS) is still underdeveloped. Few of the already existing similar systems are generating the unalterable CAD models in the formats such as *.step, *.iges, and *.stl, and the remaining systems are able to generate the 2D model or sketch or animated Graphics Interchange Format (GIF) file, but not the real CAD models. So, the authors believed in developing a system that can generate the fully featured accessible CAD model. Therefore, with the intention of achieving a robust system, KBE approach is crucial and hence it has been incorporated to develop the system.

3 Tools and Brief Development Process

With the day-to-day enhancement of the Internet and the client–server model, distributing and sharing of decentralized CAD became possible. Currently, the majority of commercial CAD software supports automatic CAD model generation with a computer program written in some specific programming languages like VBA and C++. The commercial CAD software SolidWorks uses the VBA as the programming language and is equipped with macro code to support the automation. So, it is believed that SolidWorks is the better option to execute the proposed system. Moreover, to store the standard data regarding the components to be modeled a database is required. Hence, in order to retrieve the data from the database it is required to store in a logical manner. Here in the proposed system, Microsoft Access is chosen as the database management system.

For executing the entire process in a simple and easy way, a knowledge-based system (KBS) was created with the help of the domain experts. The created KBS is given in Fig. 1. This KBS comprises of 5 sub-systems: Decider (DR), Intelligent Finder (IF), CAD modeler (CM), macro maker (MM), and CNC code generator (CG).The knowledge to the KBS has been obtained from the domain experts and from the recognized standard organizations like ISO, AGMA, and ASTM. The field professional or domain expert is the person who is from academic institution and/or from the industry. The means of knowledge acquisition from the field professionals is by having the face-to-face meetings with appropriate questions and/or by observing them. The significance of macro & CNC codes' database in the proposed system is to store the generated respective codes.

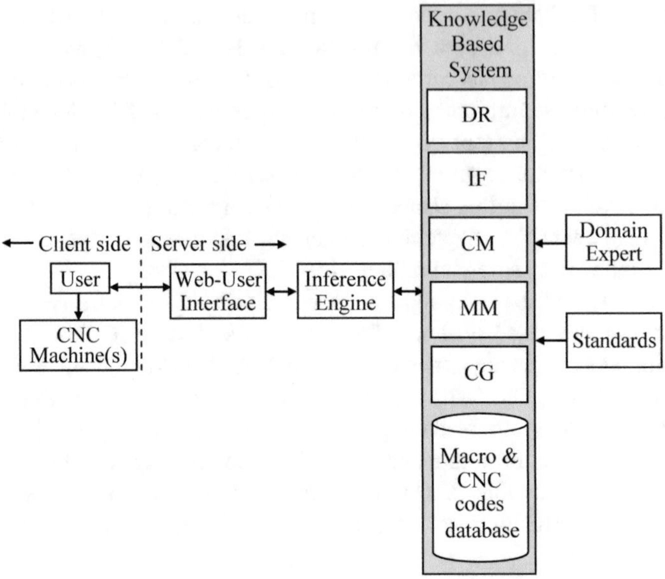

Fig. 1 Framework of the online KBS for CAD modeling and manufacturing

Through Web User Interface (WUI), the user queries the system to reduce the production time. Then, the WUI will send the query to the inference engine without any data loss. Inference engine (IE) is the module which works on the artificial intelligence procedures to use the knowledge-base for achieving the result for the user posed query. For generating the CAD model and CNC Code, IE utilizes the user query and searches the knowledge in the knowledge-base of the proposed system. If the user entered data for posing the query is inadequate or irrelevant, then the DR sub-system can appropriately guide the user for further action. On obtaining the adequate data from the user, the IF sub-system will formulate the procedure for computing the design calculations by using the in-built knowledge in the knowledge-base. Soon after, based on the in-built knowledge of standard organizations data, the proposed system computes the design calculations. Later, the proposed system generates the procedure for CAD modeling by using the CG sub-system. Afterward, based on this generated procedure, the MM sub-system generates the macro code. On the other hand, at this stage of the process, the proposed system blends the parametric modeling technique with the MM sub-system to achieve the effective modeling. Afterward, the user desired CAD model can be generated by using the MM sub-system generated macro code. Afterward, the proposed system will send the CAD model to the CG for generation of the CNC code. For optimal CNC code generation, the critical constrains, such as geometry and material of cutting tool and work piece, machine tool, wet or dry machining, clamping device, and machining operation, will be taken into account. Finally, the generated CNC code will be sent to the network connected CNC machine(s) by the user for manufacturing the component. Therefore,

Table 1 Sample input data for developing the CAD model of industrial battery stacks

Parameter	Input data
Backup capacity	48 V; 600 Ah
Maximum allowable weight (kg)	485
Type of the channel	C
Height of the channel (mm)	50
Maximum roof level (m)	60
Available floor area (m^2)	3

the proposed system is able to achieve model, design, and manufacture processes in reduced time than the traditional methods. Finally, for future use, the generated CAD model and CNC code are stored in the macro & CNC codes database. The workflow in the proposed system is shown in Fig. 2.

For the proposed system, three-tier architecture was adopted, having client browser, Web server, and database components. Figure 3 is the architecture of the proposed system. It consists of two major modules called expert module and user module. The expert module is to create and manage the knowledge-base, whereas the user module is to accept the needs of the user though WUI.

4 Case Study and Validation

In the process of the verifying the usefulness of the proposed system, many case studies are executed on it. One in them was executed at battery manufacturing company in India. The aim of the task was to generate the CAD model and manufacture the industrial battery stack (IBS). The available input data for the task is the power backup requirement of the customer or user and the accessible space at the installation location.

In IBS, the stack is the assembly of many similar modules as shown in Fig. 4a whereas the module is the assembly of the same type of cells. In order to deliver the user required power, these cells need to be connected in series and/or parallel. It is obvious that, the need of the power backup vary from one customer to another. Hence, it is not economical to produce the standard size stack in the scenario where optimal production is insisted. Therefore, customized stack is the best alternative to meet the needs of the industry as well as the customer. Figure 4b shows various components of IBS assembly.

As per the user requirements, such as battery power, accessible installation space, and kind and size of the channel, the DR sub-system verifies the validity of the user requirements to carry out the design calculation. Table 1 shows the sample user requirements and Fig. 5 shows the WUI to feed the user requirements to the proposed system. After obtaining the valid data, the IF sub-system determines the requirement of number of cells to be chosen.

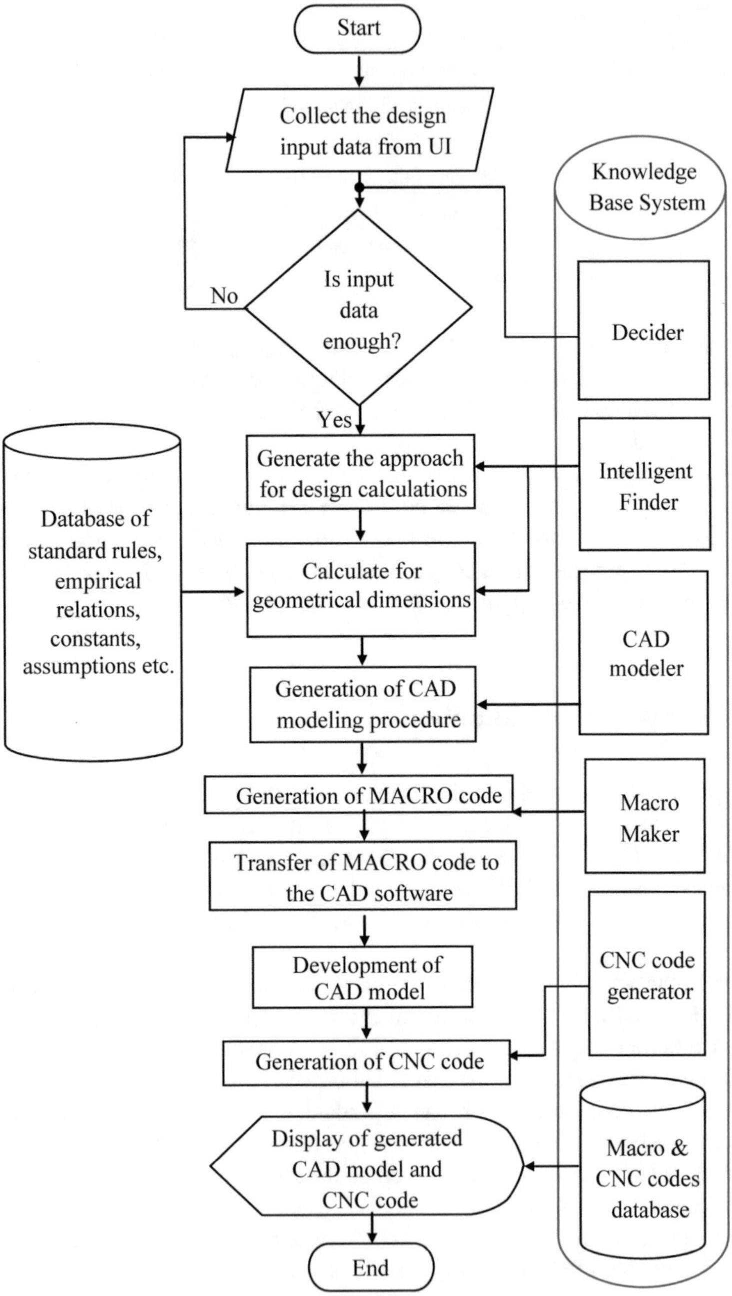

Fig. 2 Workflow of the online KBS for CAD modeling and manufacturing

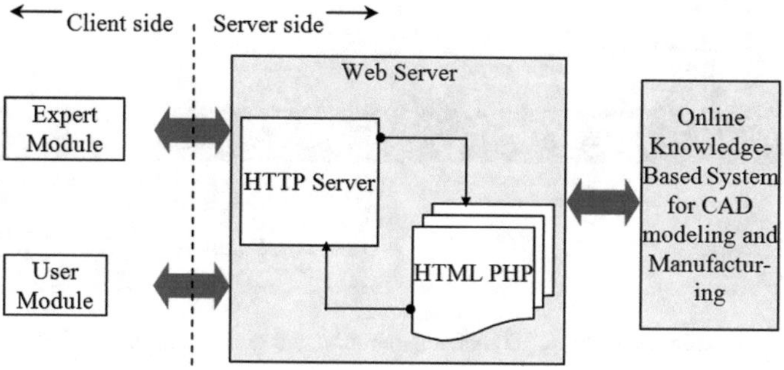

Fig. 3 Architecture of the online KBS for CAD modeling and manufacturing

Fig. 4 **a** Various arrangements of IBS. **b** Different parts of IBS

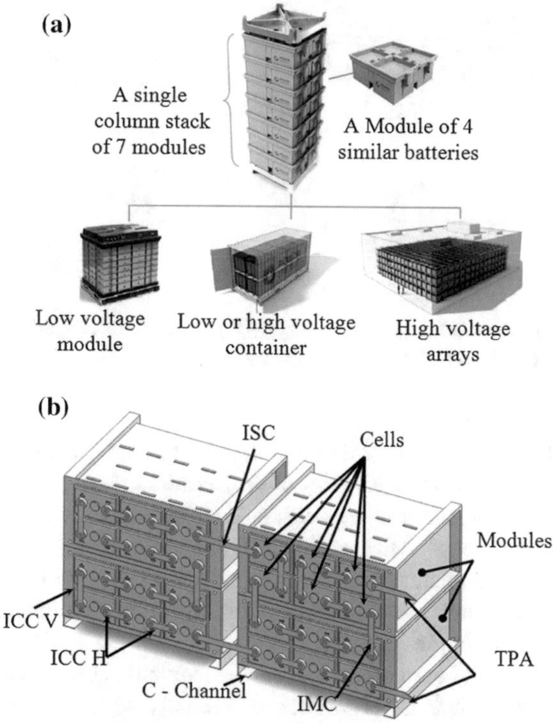

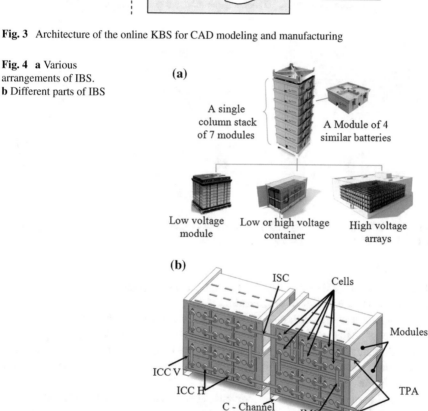

Soon after, as per the recommendation of the IF sub-system, all the chosen cells are connected in a strategic way with the different connectors namely intercell connector horizontal (ICC H), inter stock connector (ISC), intercell connector vertical (ICC V), inter-module connector (IMC), for achieving the required power output. The proposed system determines the geometry and total quantity of connectors needed for the purpose. The function of the connector is the central factor for deciding the length of the connector, but the width and the thickness are always constant.

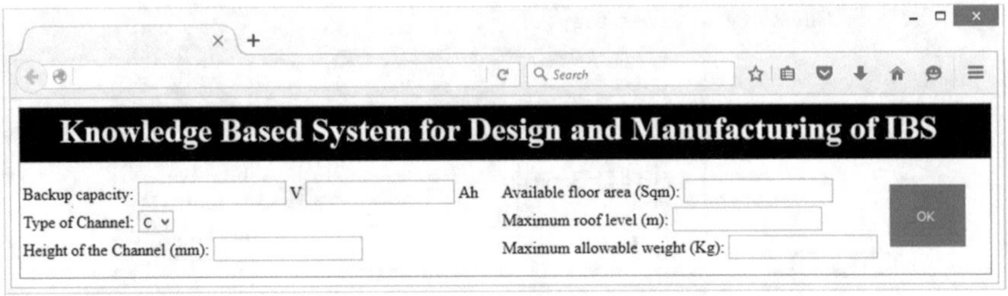

Fig. 5 WUI for online KBS for CAD model generation and manufacturing of IBS

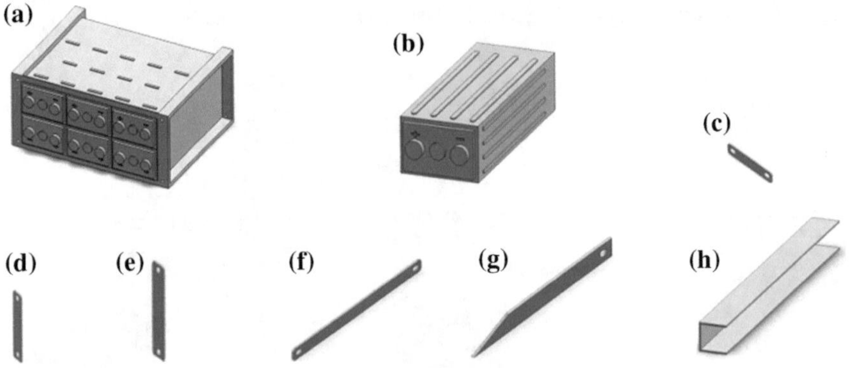

Fig. 6 CAD models of the industrial battery stack assembly components: **a** module; **b** single cell; **c** ICC H; **d** ICC V; **e** IMC; **f** ISC; **g** TP A; **h** C-channel

For manufacturing the module, sheet metal process is preferred as per the industry standards. The geometry of the Terminal plate assembly (TPA) is predefined by the industry, and it is the constant for every configuration or combination of the power requirement. Afterward, the IF sub-system computes and determines the geometry of the C-channel by considering the total weight of the single stack. For all the above calculation, industrials standards and related empirical formulas are used which are present in the database.

Later, based on the respective geometries of the IBS components, CG sub-system creates their CAD models individually which are shown in Fig. 6. Soon after the completion of the individual CAD model creation, the assembly of IBS has been done by the CG sub-system. Figure 7 is the assembly of IBS. Based on the industry records, the total time required for generating the CAD model and CNC code for the given data was 573–727 man-hours if it is done by the humans. But, it takes only 51 s to accomplish the task if it uses the proposed system. Hence, it is proven that the proposed system is better than the industry approach.

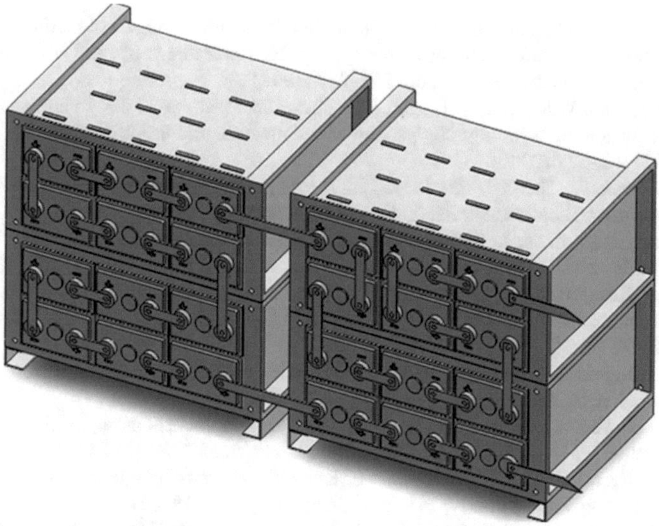

Fig. 7 CAD model of the industrial battery stack

5 Conclusion and Future Scope

With the intension to reduce the production cost, an approach is presented for developing CAD model and to manufacture the components by using the Internet. The working procedure of the propose system has been demonstrated and evaluated with the case study. Based on the results of the case study, it was concluded that the propose system is faster and better than the existing system in the industry.

References

1. Reddy, E.J., Sridhar, C.N.V., Rangadu, V.P.: Development of KBS for CAD modeling of a two wheeler IC engine connecting rod: an approach. Adv. Intell. Syst. Comput. **530**, 597–606 (2016)
2. Anumba, C.J., Duke, A.K.: Telepresence in concurrent lifecycle design and construction. Artif. Intell. Eng. **14**, 221–232 (2000)
3. Reed, J.A., Afjeh, A.A.: Developing interactive educational engineering software for the World Wide Web with Java. Comput. Educ. **30**, 183–194 (1998)
4. Jones, A., Iuliano, M.: Controlling activities in a virtual manufacturing cell. In: Proceedings of the 1996 Winter Simulation Conference (WSC'96), NJ, USA, pp. 1062–1067 (1996)
5. Hardwick, M., Spooner, D.L., Rando, T., Morris, K.C.: Sharing manufacturing information in virtual enterprises. Commun. ACM **39**, 46–54 (1996)
6. Kenny, K.B., Erkes, J.W., Lewis, J.W., Sarachan, B.D., Sobolewski, M.W., Sum, R.N.: Implementing shared manufacturing services on the world-wide web. Commun. ACM **39**, 34–45 (1996)

7. Yu, S.S., Chen, W.C.: Java based multitier architecture for enterprise computing: a case study from a university academic information system. In: Proceedings of the 17th Conference on Consumer Electronics, NJ, USA, pp. 252–253 (1998)
8. Kim, H.J., Chu, W.S., Ahn, S.H., Kim, D.S., Jun, C.S.: Web-based design and manufacturing systems for micromachining: comparison of architecture and usability. Comput. Appl. Eng. Educ. **14**, 169–177 (2006)
9. Liu, M., Shen, W., Hao, Q., Yan, J., Bai, L.: A fuzzy matchmaking approach for semantic web services with application to collaborative material selection. Comput. Ind. **63**, 193–209 (2012)
10. Malahova, A., Butans, J., Tiwari, A.: A web-based cad system for gear shaper cutters. In: Industrial Informatics, INDIN 2009, 7th IEEE International Conference, pp. 37–42 (2009)
11. Sypek, P., Kozakowski, P., Mrozowski, M.: Internet as a new platform for delivery of microwave CAD services. In: Tech. University of Gdansk, Department of Electronics, Gdansk, vol. 2, pp. 574–577 (2004)
12. Reddy, E.J., Sridhar, C.N.V., Rangadu, V.P.: Knowledge based engineering: notion, approaches and future trends. Am. J. Intell. Syst. **5**, 1–17 (2015)
13. Reddy, E.J., Venkatachalapathi, N., Rangadu, V.P.: Development of an approach for knowledge-based system for CAD modelling. Mater. Today Proc. **5**, 13375–13382 (2018)
14. Reddy, E.J., Sridhar, C.N.V., Rangadu, V.P.: Knowledge—based parametric modeling for bolts, nuts and bearings using SolidWorks. Int. J. Appl. Eng. Res. **10**, 16111–16120 (2015)
15. Reddy, E.J., Sridhar, C.N.V., Rangadu, V.P.: Development of KBS for CAD modeling of industrial battery stack and its configuration: an approach. Adv. Intell. Syst. Comput. **530**, 607–618 (2016)
16. Reddy, E.J., Sridhar, C.N.V., Rangadu, V.P.: Research and development of knowledge based intelligent design system for bearings library construction using SolidWorks API. Adv. Intell. Syst. Comput. **385**, 311–319 (2016)
17. Reddy, E.J., Rangadu, V.P.: Development of knowledge based parametric CAD modeling system for spur gear: An approach. Alexandria Eng. J. **57**, 3139-3149 (2018)

Hepatoprotective Activity of the Biherbal Extract in Carbon Tetrachloride (CCl_4) Induced Hepatoxicity—A Study of Histopathological Image Analysis

K. Sujatha, V. Karthikeyan, R. S. Ponmagal, N. P. G. Bhavani, V. Srividhya, Rajeswari Hari and C. Kamatchi

Abstract Hepatotoxicity implies chemical-driven liver damage. Liver disease is still a worldwide health problem. Traditional drugs for the treatment of liver diseases are at times inactive and can have adverse effects. So there is a worldwide trend to go back to traditional medicinal plants. The present investigation was aimed for a comparative study of conventional histopathological examination with modern electronic image processing studies with respect to evaluating the hepatoprotective action of ethanolic extract of Melia azedarach (MAE) and Piper longum (PLE) with their combination biherbal extract (BHE) against carbon tetrachloride (CCl4) induced hepatic damage in rats. The three ethanolic extracts at a dose level of 50 mg/kg body weight each were administered to three different groups of rats orally once daily for 14 days. The hepatoprotective effect of the biherbal extract (BHE) was assessed by histopathological photograph of the cells.

K. Sujatha (✉) · V. Karthikeyan
Department of EEE, Dr. MGR Educational and Research Institute, Chennai, India
e-mail: drksujatha23@gmail.com

V. Karthikeyan
e-mail: karthikeyan.ei@drmgrdu.ac.in

R. S. Ponmagal
Department of CSE, Dr. MGR Educational and Research Institute, Chennai, India
e-mail: rsponmagal@gmail.com

N. P. G. Bhavani · V. Srividhya
Department of EEE, Meenakshi College of Engineering, Chennai, India
e-mail: sbreddy9999@gmail.com

V. Srividhya
e-mail: sripranav2007@gmail.com

R. Hari · C. Kamatchi
Department of IBT, The Oxford College of Science, Bengaluru, India
e-mail: rajihari@gmail.com

C. Kamatchi
e-mail: ckamatchi@gmail.com

S. M. Thampi et al. (eds.), *Intelligent Systems, Technologies and Applications*, Advances in Intelligent Systems and Computing 910,
https://doi.org/10.1007/978-981-13-6095-4_20

Keywords Hepatoprotective activity · Biherbal extract · Carbon tetrachloride · Gray-level co-occurence matrix · Back propagation algorithm · Radial basis function network · Artificial neural networks

1 Introduction

The liver is the largest gland and also the largest internal organ which has an important role in metabolism. A number of functions in the body, including glycogen storage, decomposition of red blood cells, plasma protein synthesis, and detoxification are controlled by liver apart from which it regulates a wide variety of biochemical reactions. Liver diseases are a broad term describing any number of diseases affecting the liver. Almost all the liver diseases are accompanied by jaundice leading to increased bilirubin. Important among the liver diseases are classified as Fatty liver, Hepatitis, and Cirrhosis.

Meleppat et al. [1] state that the Optical Coherence Microscopy (OCM) plays an important role in monitoring the growth of biofilm. A swept source-based Fourier domain OCM system was used to analyse the representative biofilm, Klebsiella pneumonia (KP-1). The dynamic behavior and the time response of the KP-1 biofilms were investigated using the enface visualization of microcolonies and their spatial localization by subjecting it to disturbances and measured the colony forming units using standard procedures.

Ratheesh et al. [2] proposed Fourier Domain Optical Coherence Tomography (FD-OCT) for high resolution, precise, and complex images. This system was found to improve accuracy and speed as compared to the existing automatic schemes. The automatic calibration in the course of scanning operation does not need any auxiliary interferometer for generation and acquisition of calibration signal in additional channel.

1.1 Fatty Liver, Cirrhosis, and Hepatitis

In case of Fatty liver, a reversible condition where large vacuoles of triglyceride accumulate in liver cells known as fatty liver disease (FLD), steatorrhoeic hepatosis, or steatosis hepatitis. Normal liver may contain as much as 5% of its weight as fat. Lipiotic liver may contain as much as 50% of its weight as fat, most of being triglycerides. Steatohepatitis is a situation where there is liver inflammation. The first and foremost factor leading to cirrhosis is due to large fat content.

Cirrhosis can be defined as a chronic disease condition presenting morphological alteration of the lobular structure characterized by destruction and regeneration of the parenchyma cells and increased connective tissue. It leads to the formation of fibrous tissue replacing the dead liver cells. Toxic chemicals cause death of liver cells due to alcoholism.

Some poisons, autoimmunity or hereditary conditions induced by viruses causes liver inflammation. Cells undergo inflammation due to hepatitis. Autoimmune process induces toxins and other infections.

1.2 *Hepatotoxicity*

Almost all drugs are identified as foreign substances by liver (i.e., *xenobiotics*) and subject them to various chemical processes (i.e., metabolism) to make them suitable for elimination. Reduction in fat solubility and change in biological activity causes chemical transformations induced by drugs, identified as foreign substances. Liver acts as a metabolic clearinghouse for all the tissues in the body and has some ability to metabolize chemicals. The liver is prone to drug-induced injury during the elimination of chemicals.

The liver produces large amounts of oxygen free radicals in the course of detoxifying xenobiotic and toxic substances. Reactive oxygen species (ROS) has been shown to be linked to liver diseases, such as hepatitis, cirrhosis, portal hypertension, viral infections and other liver pathological conditions [3]. They play an important role in the inflammation process after intoxication by ethanol, carbon tetrachloride, or carrageenan. These radicals react with the cell membrane of the liver cells and induce lipid peroxidation which leads to drastic results. Cytotoxicity is due to ROS induced alterations and loss of structural/functional architecture in the cell and/or indirectly leading to genotoxicity, with various unknown and dangerous diseases [4]. Hepatic injury caused by chemicals, drugs, and virus is a well-known toxicological problem to be taken care of by various therapeutic measures.

1.3 *CCl_4 Induced Liver Damage*

Carbon tetrachloride (CCl_4), a non-inflammable liquid with a characteristic odor is a potent and well-established substance. CCl_4 is used for detecting liver injuries which is xenobiotic-induced hepatotoxicity. By its administration, chronic liver injury occurs along with fibrotic changes in the anatomy of the liver, degeneration, and necrosis of liver cells, fibrous tissue proliferation, and fatty liver production which constricts the blood flow in the liver sinusoids. There is also accumulation of interstitial fluids in the liver cells. The toxicity of CCl_4 on liver relies on the amount and period to which it is exposed [5]. The free radicle ($^{\bullet}CCl_3$ and/or $^{\bullet}CCl_3OO$) generation causes hepatic injury which is a result of biotransformation when treated with CCl_4 [6]. The cytochrome P-450 system is encased in phospholipids membrane rich in polyenoic fatty acid. Hence, these polyenoic fatty acids are the most likely immediate target for the initial lipid peroxidative attack to occur. The organic fatty acid radical rearranges, yielding organic peroxy and hydroxyl peroxy radicals. The radical destroys the cytochrome P-450 homoprotein, thus compromising the mixed-function oxygenase activity. The rapid decomposition of the endoplasmic reticulum and its function as a direct result of this lipid peroxidative process was observed in CCl_4 damage [7].

The initial step of the chain reaction is membrane lipid peroxidation which will occur when trichloromethyl free radicals react with sulfhydryl groups, such as glutathione (GSH) and protein thiols leading to the formation of covalent bonds and finally to cell necrosis [8]. The ethanol-inducible isoform (Koop 1992) has greater importance because it can metabolize CCl_4. Moreover, alterations in the activity of P-450 2E1 can affect the susceptibility to hepatic injury by CCl_4 [9]. P-450 2E1 is active in the metabolism of small organic molecules including acetaminophen, aliphatic organic alcohols, nitrosamines, benzene, phenol, 4-nitrophenol, and pyrazole [10].

Covalent bonding of macromolecules are possible because of the formation of reactive intermediates formed during the metabolism of therapeutic agents, toxins, and carcinogens by this enzyme causing liver damage [11] (Eaton et al. 1995). Reactive metabolites are reduced due to the suppression of P-450 with less tissue injury. P-450 2E1 activity reduces the hepatotoxicity induced by CCl_4 [12].

1.4 CCl_4 Toxicity and Lipid Peroxidation

Several mechanisms have been proposed for CCl_4 induced fatty liver and necrosis. Important mechanisms include damage to endoplasmic reticulum, mitochondrial lysosomes, disturbances in hepatocellular calcium homeostasis, and lipid peroxidation. All are mediated by free radicals. Lipid peroxidation may be looked upon as occurring in two steps. Some toxic event initiates lipid peroxidation and organic free radical generated by the initiation process serve to propagate the reaction. Liver cirrhosis is caused by CCl_4-induced liver dysfunction in rats which is also a similar case with the human being (Pérez-Tamayo 1983) [13].

1.5 Mechanism of Lipid Peroxidation

The steps involved in lipid peroxidation are described below and shown schematically [14].

Initiation:

$$H_2O \rightarrow HO^\bullet, H^\bullet, e^-{}_{aq}, O_2{}^{\bullet -}, H_2O_2$$

$$LH + {}^\bullet OH \rightarrow L^\bullet + H_2O$$

Propagation:

$$L^{\bullet} + O_2 \rightarrow LOO^{\bullet}$$

$$LOO^{\bullet} + LH \rightarrow LOOH + L^{\bullet}$$

Termination:

$$L^{\bullet} + L^{\bullet} \rightarrow L{-}L$$

$$LOO^{\bullet} + LOO \rightarrow LOOH + O_2$$

$$LOO^{\bullet} + L^{\bullet} \rightarrow LOOL$$

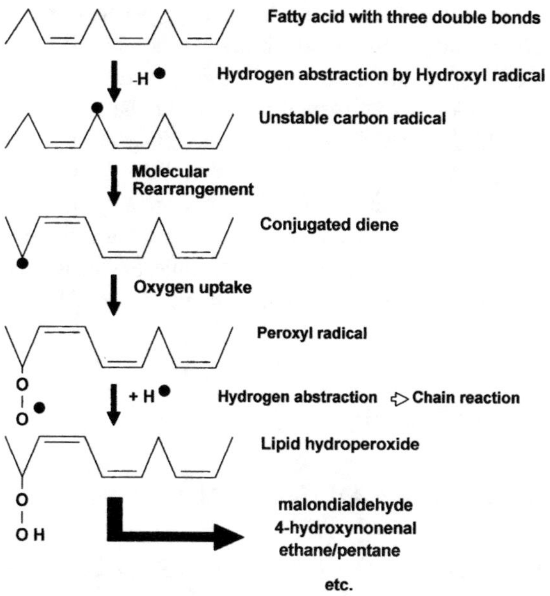

The major reactive aldehyde (Malondialdehyde) resulting from the peroxidation of polyunsaturated fatty acid (PUFA) helps to reduce hepatotoxicity following acute exposure to CCl_4 which is manifested as necrosis. This is also accompanied by inflammation in central lobular areas of the rodent's liver [15]. The lipoprotein secretion leads to steatosis, whereas the reaction with oxygen leads to the formation of $^{\bullet}CCl_3{-}OO$, initiating lipid peroxidation. The latter process results in loss of calcium homeostasis and, ultimately, apoptosis and cell death (Boll et al. 2001). The reactive species-mediated hepatotoxicity can be effectively managed upon administration of such agents possessing anti-oxidants, free radical scavengers, and anti-lipid peroxidants (Sadanobu et al. 1999; Attri et al. 2000; Lim et al. 2000).

1.6 Importance of Medicinal Plants

Since long time the liver diseases are treated using herbal products [16]. These herbal plants are to be systematically evaluated for if it is used for treatment of liver diseases. Many plants such as *Silybum marianum* (milk thistle), *Picrorhiza kurroa* (kutkin), *Curcuma longa* (turmeric), *Camellia sinensis* (green tea), and *Glycyrrhiza glabra* (licorice) have been clinically proved and patented. *Melia azedarach*, a member of the family *Meliaceae* commonly called as 'Malai vembu' is widely grown as an ornamental tree, being used against intestinal worms in skin diseases, stomach ache, intestinal disorders, uterine illnesses, cystitis, diuretic, and febrifuge.

It has antiviral, antimalarial, anthelmintic, and cytotoxic activities and is also used for the treatment of hepatitis and cirrhosis. *Piper longum* an important medicinal plant belonging to the family of Piperaceae is known as 'Thippali' being used in traditional medicine by many people in Asia and Pacific islands, especially in Indian medicine.

Both *M. azedarach* and *P. longum* have immense therapeutic properties especially for the treatment of liver-related disorders. But the practice is only at the traditional level because of the lack of experimental proof to standardize the optimum dosage, efficacy, and toxic effects. The aim of the present study is to provide experimental proof for the hepatoprotective efficacy of the ethanolic extracts of *M. azedarach* and *P. longum* and the combined biherbal formulation made up of equal concentrations of *M. azedarach* and *P. longum*. Generally, polyherbal formulations are considered more effective than the single drug, and hence, the biherbal formulation has been used in the study and compared with the individual plants *M. azedarach* and *P. longum*.

2 Materials and Methods

2.1 Preparation of Biherbal Extract (BHE)

The leaves of *Melia azedarach* (1 kg) and seeds of *Piper longum* (1 kg) were dried in shade and grounded into coarse powder. Both the substances are to taken in equal measures. This is then passed through 40-mesh sieve and extracted with the addition of ethanol in soxhlet apparatus at a temperature of 60 °C. The evaporation of the extract was carried out under reduced pressure using rota flash evaporator till all the solvent has evaporated. This extract was lyophilized and stored in refrigerator for phytochemical and pharmacological studies. The lyophilized material was administered to the animals by dissolving each time with 2% v/v aqueous Tween 80.

We have taken all necessary approvals for the use of animals in our study (Ethical clearance no. 1/243/2007).

2.2 *Hepatoprotective Efficacy Studies—an Experimental Protocol*

The rats were grouped into seven categories of six animals which were given a dose as scheduled below.

Group I served as control which received 0.5 ml corn oil p.o. daily for 28 days. Group II served as negative control and received single dose of 40 μg/kg b.w. of TCDD in corn oil p.o. on 7th day. Group III: Animals were pretreated with EETT (200 mg/Kg b.w.) daily in corn oil for one week prior to the single dosing of TCDD (40 μg/kg b.w.) on 7th day and throughout the experimental period of 28 days. Group IV: Animals were pretreated with SFTT (100 mg/Kg b.w.) daily in corn oil for one week prior to the dosing of TCDD(40 μg/kg b.w.) on 7th day and throughout the experimental period 28 days. Group V: Animals were pretreated with standard drug silymarin (50 mg/Kg b.w.) daily in corn oil for one week prior to the dosing of TCDD and throughout the experiment. This group served as positive control.

Group 1: served as a control received 0.5 ml 2% v/v aqueous Tween 80 p.o. daily. Duration is 14 days. This group served as normal control.
Group 2: served as negative control received 2 ml/kg, p.o. CCl_4 in 2% v/v aqueous Tween-80 a single dose daily for a duration of 7 days which served as negative control.
Group 3: Animals were pretreated with 50 mg/kg, p.o. of BHE in 2% v/v aqueous Tween-80 for 14 days intoxicated with CCl_4 on days 7–14.
Group 4: Animals were pretreated with 50 mg/kg, p.o. of MAE in 2% v/v aqueous Tween-80 for 14 days intoxicated with CCl_4 on days 7–14.
Group 5: Animals were pretreated with 50 mg/kg, p.o. of PLE in 2% v/v aqueous Tween-80 for 14 days intoxicated with CCl_4 on days 7–14.
Group 6: Animals received 50 mg/kg, p.o. Silymarin in 2% v/v aqueous Tween-80 daily for 14 days and administered with CCl_4 on days 7–14. This group served as positive control.

After 15 days of experimentation, the animals were dissected after 12 h fasting under mild pentobarbitone anesthesia. Liver was excised from the animals, washed in ice-cold saline, and dried gently on the filter paper. The liver sections from different experimental groups were then treated in 10% formalin-saline and finally embedded in paraffin. Serial portions were taken out at 5 cm and stained with hematoxylin and eosin. The liver sections were examined under light microscope for the morphological changes and photographs were taken.

3 Results and Discussion

3.1 Laboratory Results

Image processing in the field of biotechnology is playing an important role in diagnosing and treatment of various types of diseases in early stages. As a traditional approach for identifying bone features is microscopic images were used, these images are acquired from the culture test done in the laboratory, where it is needed to repeat, time consuming and labor intensive process. This method is not sufficient to handle low resolution and noisy images. Hence, there is a need for automated and reliable techniques to carry out the image processing analysis. The histology of liver from the control and treated groups (the 10× and 40× Magnifications) are shown in Fig. 1.

Figure 1a: Section of liver from control rats showing normal architecture.
Figure 1b: Section of liver from CCl_4 induced rats showing reactive hyperplasic hepatocytes with fatty changes and necrosis in the centrilobular position.
Figure 1c: Section of liver from BHE treated and CCl_4 induced rats showing near normal architecture with less fatty changes and necrosis.
Figure 1d: Section of liver from MAE treated and CCl_4 induced rats showing less necrosis and regenerating cells.
Figure 1e: Section of liver from PLE treated and CCl_4 induced rats showing less fatty infiltration and Necrotic cells.
Figure 1f: Section of liver from silymarin treated CCl_4 induced rats showing near normal architecture.

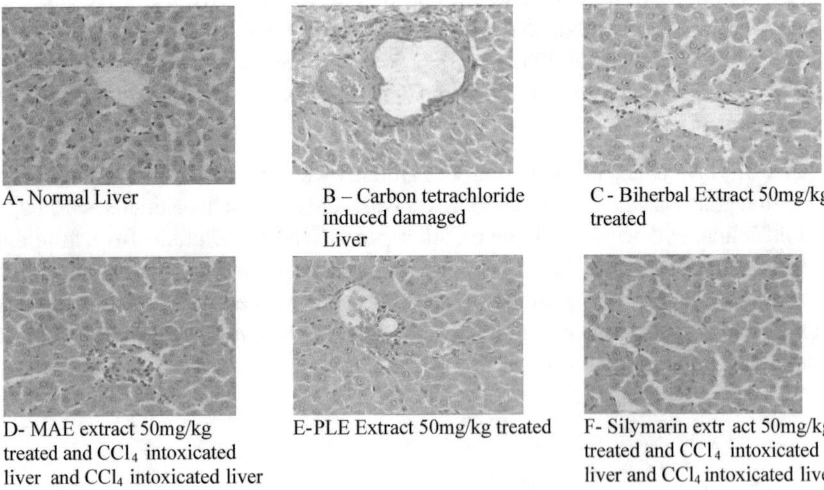

A- Normal Liver

B – Carbon tetrachloride induced damaged Liver

C - Biherbal Extract 50mg/kg treated

D- MAE extract 50mg/kg treated and CCl_4 intoxicated liver and CCl_4 intoxicated liver

E-PLE Extract 50mg/kg treated

F- Silymarin extr act 50mg/kg treated and CCl_4 intoxicated liver and CCl_4 intoxicated liver

Fig. 1 Section of liver from rats

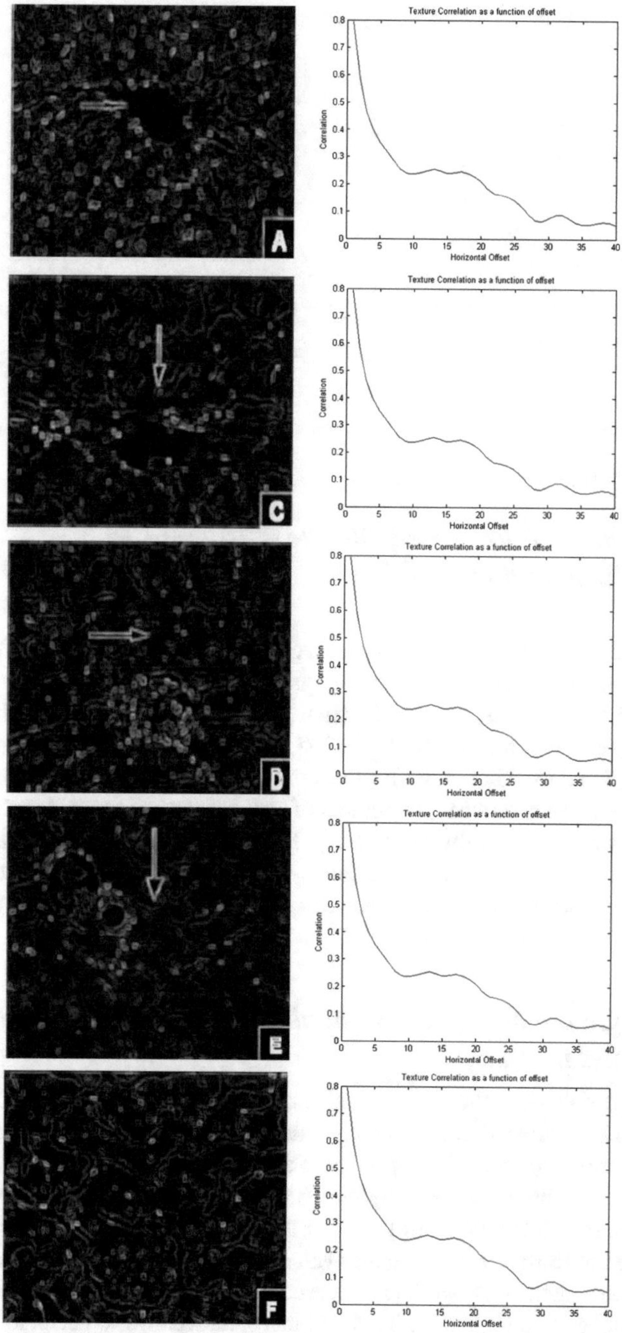

Fig. 2 Outputs for GLCM

```
To get started, select MATLAB Help or Demos from the Help menu.
>> glcms

glcms(:,:,1) =

   591     45      6      8      4      0      1
    52     35     39     43     11      2      0
    22     59    260    337    115     14      0
    13     30    367   3080   2336    118      0
     1      7    111   2279  18903   2401     33
     1      4     19     83   2251   7759    126
     1      0      0      0      7    101     30
     4      2      0      2      2      3      5


glcms(:,:,2) =

   555     30      8     10     16      5      1
    32     14     12     41     49     29      1
    24     21     81    227    337    110      2
    26     33    285   1896   3129    540     12
    12     63    314   3187  16523   3425     46
    25     19     96    411   3387   6157    118
     2      0      3      8     22     94      9
     9      2      2      2      3      2      6
```

Fig. 3 GLCM values as feature input

3.2 Texture Feature Extraction Based on GLCM for Histopathological Study

This method yields a co-occurrence matrix from image 'I' which is called as gray-level co-occurrence matrix (GLCM) or spatial dependence matrix. For each value of the pixel '*i*,' a corresponding '*j*' exists. The offset parameters will be used to represent spatial parameters. Each pixel (*i*, *j*) in the GLCM denotes that the pixel with value '*i*' occurred horizontally adjacent to a pixel with value '*j*.' This method of finding the adjacency reduces the number of redundant pixels considered for training the feed-forward neural network (FFNN). The correlation with the horizontal offset serves as the texture feature. Figure 2 shows correlation values for the texture feature.

Figure 2 denotes the graphical representation of GLCM for extracting the nominal values. The screenshot to record the GLCM feature inputs are shown in Fig. 3.

3.3 Simple Decorrelation Stretching for Histopathological Studies in Rats

This method enhances the color which is separated in three color planes (RGB plane). The embellished colors increase the elucidation and make it accurate. In Linear Contrast Stretching, it is optional to add a linear contrast stretch to the decorrelation stretch. There are normally three color bands and NBANDS in the image, respectively. It is independent of the color bands. The true RGB values are mapped to a fresh set of RGB values of greater value. The color intensities of each pixel are converted into Eigen values in Eigen space corresponding to NBANDS by covariance matrix, stretched to equalize the band variances, then converted back to the original color bands. This method is dependent on decorrelation stretching using principal component analysis and weighted summation. The decorrelation values are indicated in Table 1 (Fig. 4).

Table 1 Correlation values of RGB plane

185	191	189	185	184	190	200	206	215	218	210	198	191	196	205	210	211	212	215
177	180	176	170	167	174	184	196	211	215	200	174	162	172	188	203	207	213	216
170	173	168	158	154	158	168	185	203	209	187	155	139	152	173	196	204	212	217
153	156	167	160	142	139	154	172	179	179	178	166	141	135	157	179	202	210	207
151	156	167	160	138	124	129	143	151	157	169	172	152	142	155	176	190	193	193
154	156	163	155	132	113	112	120	127	135	158	176	168	158	165	170	173	170	171
165	160	157	144	125	113	113	120	122	128	148	170	170	164	171	167	163	152	152
170	169	160	143	126	118	116	123	127	133	147	162	157	147	151	170	167	156	148
172	179	174	159	141	125	112	118	128	138	151	159	145	127	125	172	179	171	153
179	188	188	180	164	137	112	111	120	132	148	163	153	135	131	167	188	185	158
192	196	197	195	183	151	119	111	117	125	145	168	168	157	154	159	190	194	160
208	212	208	205	196	161	124	126	132	139	149	161	169	164	154	156	169	176	173
219	222	213	206	192	159	126	138	146	151	151	150	151	148	143	143	152	158	167
226	227	217	207	195	167	138	152	158	158	146	132	124	124	126	123	128	134	157
226	229	223	215	207	187	168	167	168	160	140	117	106	107	113	108	113	120	149

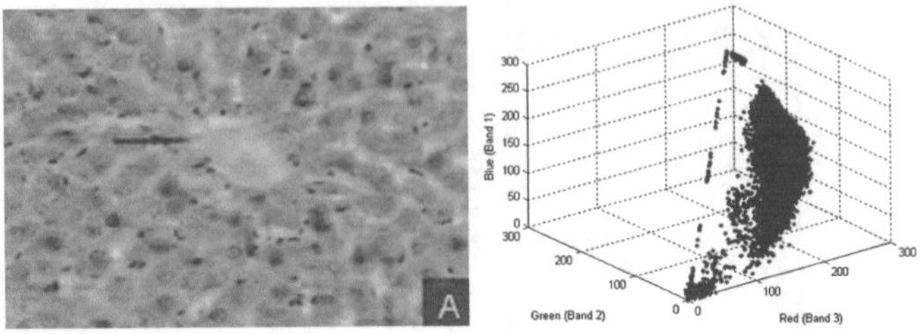

Fig. 4 Simple decorrelation stretching

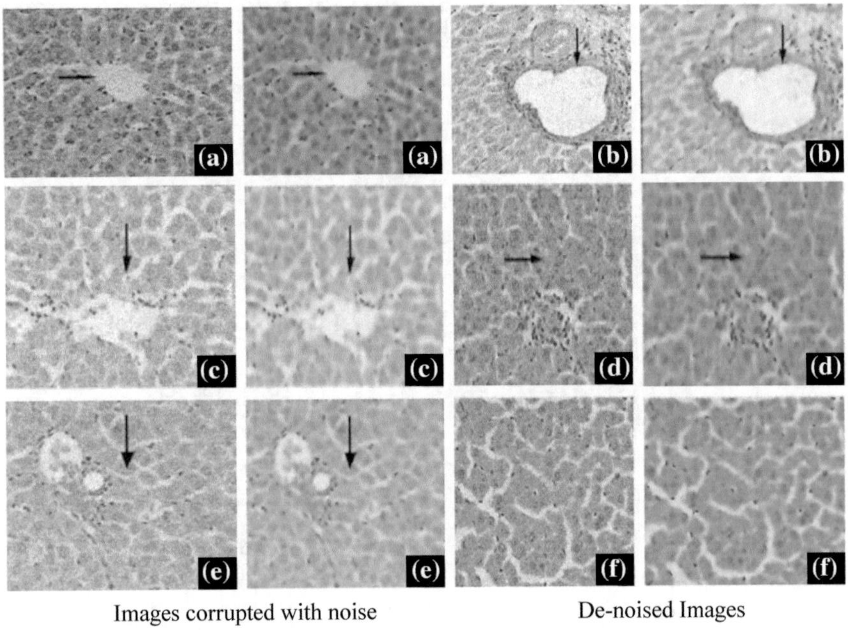

Fig. 5 Output for adaptive filtering

3.4 *Noise Removal Using Adaptive Filtering in Histopathological Studies*

An adaptive filter is a linear filter with the transfer function controlled by changeable parameters and an algorithm to optimize the set of parameters. The closed loop form has a feedback which helps to refine the error signal. The optimal performance of the adaptive filter is dependent on the filter coefficients which minimizes the cost function on iterative basis. The images corrupted with noise and the corresponding de-noised image is shown in Fig. 5.

Table 2 Features extracted from the samples images

Area	Mean	StdDev	Mode	X	Y	XM	YM	Feret's diameter	Median	Kurtosis	Entropy
42,168	159.414	31.123	168	125.5	84	124.444	83.143	302.035	163	7.566	7.5737
48,280	196.768	38.814	242	142	85	141.434	84.199	330.992	198	7.13	7.1522
42,168	196.469	31.901	190	125.5	84	124.165	83.746	302.035	196	14.754	7.0333
47,544	168.321	29.066	168	141.5	84	140.528	83.187	329.109	169	11.419	7.3774
54,467	185.927	29.103	186	125.5	108.5	123.621	107.605	331.798	188	20.381	7.0553
62,480	175.641	29.828	170	142	110	140.871	108.653	359.244	173	10.931	7.2871

3.5 Feature Extraction

The features are the basic patterns present in the image which is present in all directions in a image. The extraction of features includes GLCM, RGB values from simple decorrelation, area, mean standard deviation mode, centroid X, centroid Y, orientation X, orientation Y, and Feret's as in Table 2. A technique of edge detection and line finding for linear feature extraction is described. Edge detection is by convolution with small edge-like masks. The resulting output is thinned and linked by using edge positions and orientations and approximated by piecewise linear segments. Some linear features, e.g., roads and airport runways, are suitably described by 'antiparallel' pairs of linear segments. Experimental results on aerial images are presented.

3.6 Estimation of Intoxication by FFNN

Imaging techniques have been widely used in the field of biotechnology. Neural network (NN) plays an important role in this respect [17], especially in the application of hepatoprotective activity. The developments of these algorithms enhance the detection techniques with respect to specificity and sensitivity. The back propagation algorithm (BPA), perceptron learning rule, and radial basis function (RBF) are used for levels of intoxication with biherbal extract in Carbon tetrachloride (CCl_4).

Figure 6 shows the usage of ANN in MATLAB toolbox. The network is created for a FFNN as in Fig. 7 (Figs. 8 and 10).

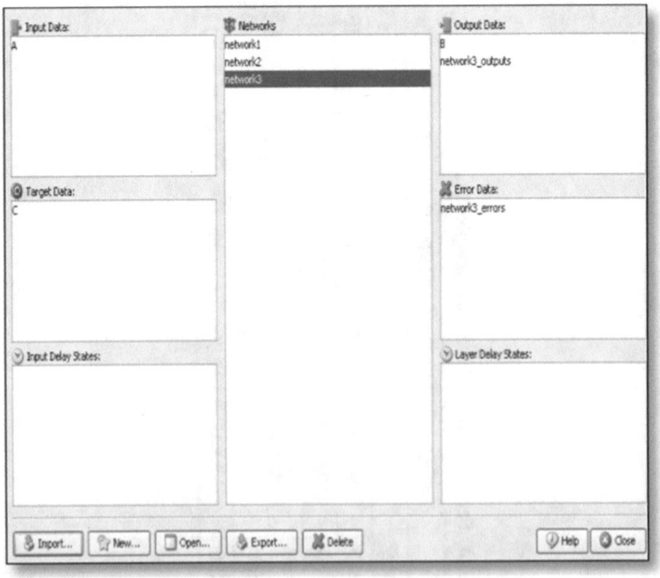

Fig. 6 MATLAB screenshot for using ANN toolbox

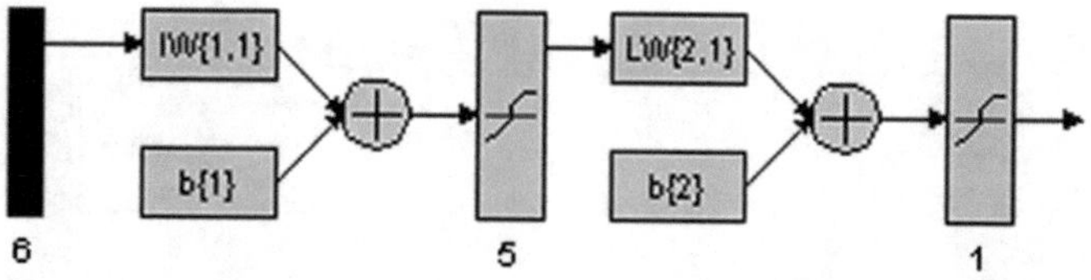

Fig. 7 Formulation of FFNN trained with BPA

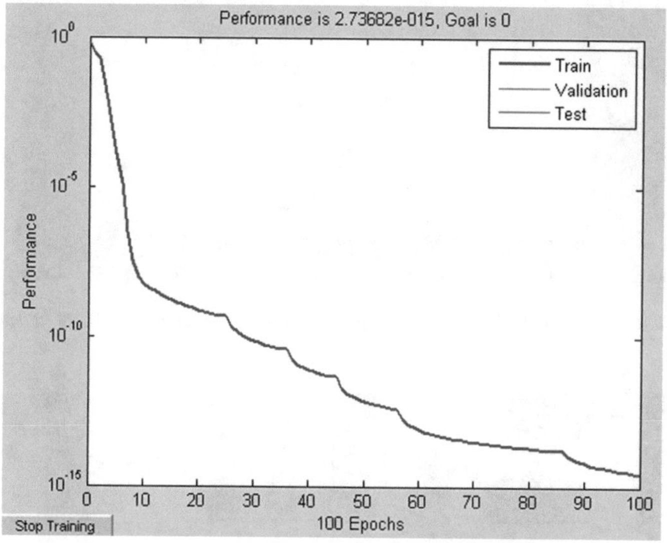

Fig. 8 Training result for FFNN using BPA for estimation of hepatoprotective activity

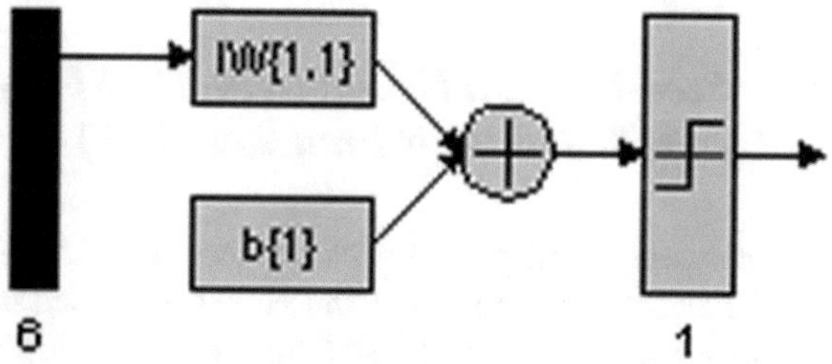

Fig. 9 Formulation of ANN with perceptron learning rule

The next algorithm is perceptron learning rule for estimation of hepatoprotective activity. The network architecture is shown in Fig. 9.

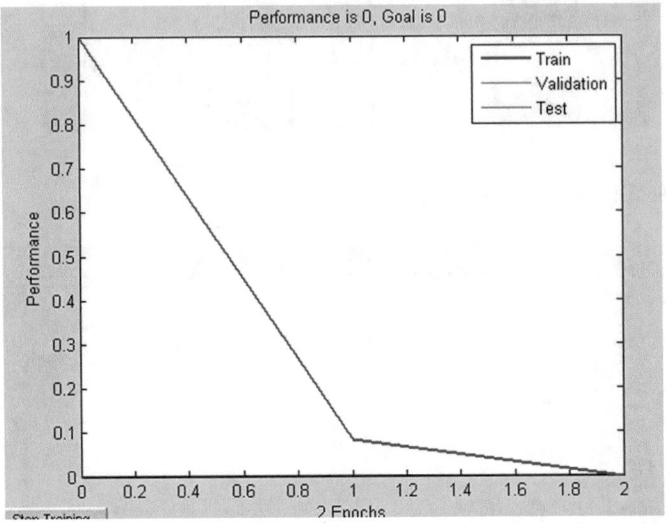

Fig. 10 Training result for ANN using perceptron learning rule for estimation of hepatoprotective activity

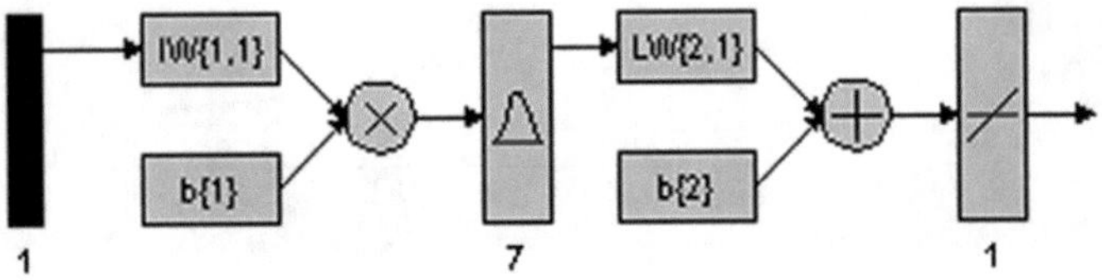

Fig. 11 Formulation of ANN with RBF

3.7 *Feed-Forward Neural Network (FFNN) Trained with Radial Basis Function (RBF)*

The network architecture for estimation of hepatoprotective activity is shown in Fig. 11 and results for training the same is indicated in Fig. 12.

The classification results state that the precision and recall values for various types of algorithms almost matches the laboratory scale results. The value of '1' for precision and recall indicates that the animals treated with BHE are less prone to r hepatocellular necrosis, mononuclear infiltration, and loss of cell architecture in comparison with liver section from control animals.

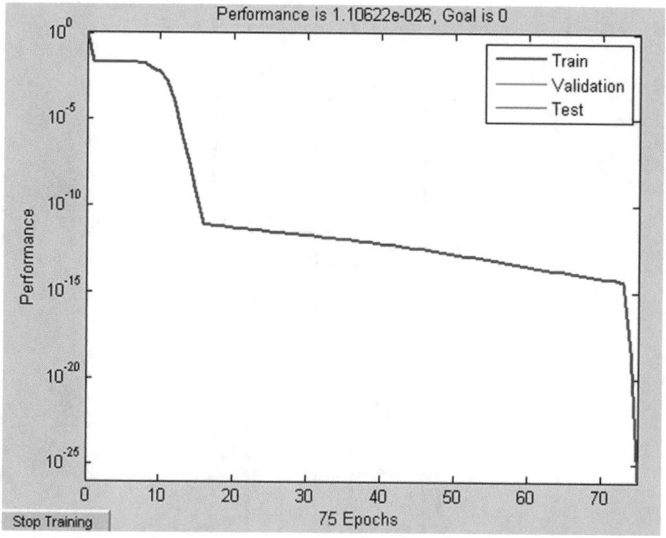

Fig. 12 Training result for ANN using RBF for estimation of hepatoprotective activity

4 Conclusion

The various liver diseases induced by CCl_4 can be successfully treated with BHE, to prevent hepatocellular necrosis, mononuclear infiltration, and loss of cell architecture in comparison with liver section from control animals. The BHE was able to treat the hepatotoxicity which was identified from histopathological observations as compared to the treatment with synthetic drug, silymarin to protect against liver dysfunction. Table 3 supports these findings which were also cross validated by the proposed image processing algorithms and ANN-based classifiers.

Table 3 Performance measures for classification by various types of ANN

Type of algorithm	Precision					Recall				
	Class B	Class C	Class D	Class E	Class F	Class B	Class C	Class D	Class E	Class F
Perceptron learning rule	0.90	0.894	0.85	0.8	0.5	1	1	1	1	1
BPA	1	0.9	0.94	0.95	0.96	1	1	1	1	1
RBF	1	1	1	1	1	1	1	1	1	1
Laboratory results	1	1	1	1	1	1	1	1	1	1

References

1. Meleppat, R.K., Shearwood, C., Seah, L.K., Matham, M.V.: Quantitative optical coherence microscopy for the in situ investigation of the biofilm. J. Biomed. Opt. **21**, 9 (2016). (SPIE)
2. Ratheesh, K.M., Seah, L.K., Murukeshan, V.M .: Spectral phase-based automatic calibration scheme for swept source-based optical coherence tomography systems. Phys. Med. Biol **61**(21), 7652–7663 (2016)
3. Mehendale, H.M., Roth, R.A., Gandolfi, A.J., Klaunig, J.E., LemastersJJ, Curtis L.R.: Novel mechanisms in chemically induced hepatotoxicity. FASEB J. **8**, 1285 (1994)
4. Sies, H.: Oxidative stress: introductory remarks. In: Sies, H. (ed.) Oxidative Stress, p. 1. Academic Press, Orlando (1985)
5. Junnila, M., Rahko, T., Sukra, A., Linderberg, L.A.: Reduction of carbon tetrachloride induced hepatotoxic effects by oral administration of betaine in male Hans–Wister rats: a morpho metric histological study. Vet. Pathol. **37**, 231 (2000)
6. Brent, J.A., Rumack, B.H.: Role of free radicals in toxic hepatic injury. Clin. Toxicol. **31**, 173 (1993)
7. Zangar, R.C., Benson, J.M., Burnett, V.L., Springer, D.L.: Cytochrome P450 2E1 is the primary enzyme responsible for low-dose carbon tetrachloride metabolism in human liver microsomes. Chem. Biol. Interact. **125**, 233 (2000)
8. Recknagel, R.O., Glende Jr., E.A., Britton, R.S.: Free radical damage and lipid peroxidation. In: Meeks, R.G. (ed.) Hepatotoxicology, p. 401. CRC Press, Boca Raton, FL (1991)
9. Jeong, H.G.: Inhibition of cytochrome P450 2E1 expression by oleanolic acid: Hepatoprotective effects against carbon tetrachloride-induced hepatic injury. Toxicol. Lett. **105**, 215 (1999)
10. Lee, S.S.T., Buters, J.T.M., Pineau, T., Femandez-Salguero, P., Gonzalez, F.J.: Role of CYP2E1 in the hepatotoxicity of acetaminophen. J. Biol. Chem. **271**, 12063 (1996)
11. Guengerich, F.P., Kim, D.H., Iwasaki, M.: Role of human cytochrome P-450 IIE 1 in the oxidation of many low molecular weight cancer suspects. Chem. Res. Toxicol. **4**, 168 (1991)
12. Allis, J.W., Brown, B.L., Simmons, J.E., Hatch, G.E., McDonald, A., House, D.E.: Methanol potentiation of carbon tetrachloride hepatotoxicity: the central role of cytochrome P450. Toxicol **112**, 131 (1996)
13. Wensing, G., Sabra, R., Branch, R.A.: Renal and systemic hemodynamics in experimental cirrhosis in rats: relation to hepatic function. Hepatology **12**, 13 (1990)
14. Anjali, K., Kale, R.K.: Radiation induced peroxidative damage: mechanism and significance Ind. J. Exp. Biol. **39**, 309 (2001)
15. Germano, M.P., D'angelo, V., Sanogo, R.: Hepatoprotective activity of *Trichilia roka* on carbon tetrachloride induced liver damage in rats. J Pharm Pharmacol **53**, 1569 (2001)
16. Mitra, S.K., Seshadri, S.J., Venkataranganna, M.V., Gopumadhavan, S., Udupa, U.V., et al.: Effect of HD-03-a herbal formulation in galactosamine -induced hepatopathy in rats. Ind. J. Physiol. Pharmacol. **4**, 82 (2000)
17. Sujatha, K., Pappa, N.: Combustion quality monitoring in PS boilers using discriminant RBF. ISA Trans. **2**(7), 2623–2631 (2011)

Author Index

A
Agarwal, Ishank, 241
Amritha, P. P., 117
Anand, Deepa, 53
Anand Kumar, M., 1, 39
Anjana, K., 117

B
Begam, S. Sabeena, 219
Bhavani, N. P. G., 269

C
Chandra, Pravin, 141
Chandra Sekhar, K., 227

D
Dhage, Sudhir N., 65
Dixit, Veer Sain, 77

E
Esanakula, Jayakiran Reddy, 259

G
Gandhiraj, R., 181
Geetha, P., 129
Gupta, Shalini, 77

H
Hari, Rajeswari, 269

J
John, Jisha, 91
Jude, Vivek M., 91

K
Kamatchi, C., 269
Karthikeyan, V., 269
Karthi, R., 129
Keerthana, K., 155
Khan, Khaleel Ur Rahman, 101
Kouser, K., 17

L
Lavanya, P. G., 17

M
Malave, Nitin, 65
Menon, Vijay Krishna, 39
Mittal, Apeksha, 141

N
Naik, J. Venkatesu, 259
Nayana, A., 91

P
Pandipriya, AR., 219
Peng, Xindong, 219
Pillai, Reshma M., 91
Ponmagal, R. S., 269
Poornachandran, Prabaharan, 1
Praveen, K., 117
Priyanga, G., 197

R
Rajendra, D., 259
Rajendran, S., 39
Rangadu, V. Pandu, 259
Rithin Paul Reddy, K., 129

S. M. Thampi et al. (eds.), *Intelligent Systems, Technologies and Applications*,
Advances in Intelligent Systems and Computing 910,
https://doi.org/10.1007/978-981-13-6095-4

S
Sachin Kumar, S., 1
Sanjanasri, J. P., 39
Sethumadhavan, M., 117
Shanmugha Sundaram, G. A., 155, 181, 197
Singh, Amit Prakash, 141, 173
Singh, Shyamli, 173
Soman, K. P., 1, 39
Srija, Suda Sai, 129
Srivastava, Chavi, 173
Srividhya, V., 269
Sujatha, K., 269
Suresha, Mallappa, 17
Suri Babu, Y., 227

T
Taranum, Fahmina, 101

V
Vignesh, R., 181
Vimala, J., 219

W
Wagh, Rupali Sunil, 53

Printed in the United States
By Bookmasters